ABOUT THE EDITORS

Richard J. Field is Professor and Chairman in the Department of Chemistry at the University of Montana. His research interests are in the area of kinetics and mechanisms of complex, especially oscillating, chemical reactions. His major efforts over the last 20 years have been directed toward understanding the oscillatory Belousov-Zhabotinsky reaction. This work has involved both classical methods of experimental chemical kinetics as well as methods of qualitative and quantitative analysis of the sets of ordinary and partial differential equations resulting from chemical mechanisms. He is coeditor with Maria Burger of *Oscillations and Traveling Waves in Chemical Systems* (Wiley, 1985). He has been visiting Professor at the University of Notre Dame and Universität Würzburg and is a member of the Editorial Advisory Board of *The Journal of Physical Chemistry*.

László Györgyi is Assistant Professor in the Department of Inorganic and Analytical Chemistry at Eötvös Loránd University, Budapest, Hungary. His main research interest is the modeling and analysis, by numerical techniques, of complex chemical systems, in particular, bromate-ion-driven, oscillating chemical reactions. Most recently, with R. J. Field, he has developed a series of simple models of the Belousov-Zhabotinsky reaction that both reproduce very well and demonstrate the dynamical source of the observed chaotic behavior in this reaction. His other major interest is in applying various methods from computer science, e.g., expert systems and neural networks, to solving problems in chemistry and medicine. He has been a visiting researcher at the University of Montana and at Brandeis University.

CHAOS IN CHEMISTRY AND BIOCHEMISTRY

Editors

Richard J Field
University of Montana
USA

László Györgyi
Eötvös Loránd University
Hungary

World Scientific
Singapore • New Jersey • London • Hong Kong

Published by

World Scientific Publishing Co. Pte. Ltd
5 Toh Tuck Link, Singapore 596224
USA offic 27 Warren Street, Suite 401-402, Hackensack, NJ 07601
UK offic 57 Shelton Street, Covent Garden, London WC2H 9HE

British Library Cataloguing-in-Publication Data
A catalogue record for this book is available from the British Library.

CHAOS IN CHEMISTRY AND BIOCHEMISTRY

ISBN-13 978-981-02-1024-3
ISBN-10 981-02-1024-8

PREFACE

Chaos is the word we apply to oscillatory but aperiodic, apparently random behavior appearing in a system not subject to stochastic perturbation but entirely governed by a deterministic dynamic law. It is intimately related to periodic oscillation, and periodicity may decompose to chaos when some varying parameter constraining an oscillatory system crosses a critical value. The resulting complex, chaotic waveform may be regarded as periodic but with a repetition time approaching infinity. Indeed, one of the major routes from periodicity to chaos involves a repetitive doubling of the period as a parameter is monotonically varied.

The amplitudes and/or periods of the individual cycles of a chaotic waveform look to be random and are unpredictable and irreproducible over an extended period of time. A chaotic system will remain apparently *noisy* regardless of how well experimental conditions are controlled. However, a chaotic waveform results from a quite ordinary deterministic dynamic law and has considerable order related to the presence of a so-called *strange attractor* that attracts trajectories in the same way as do simpler attractors such as a steady state or limit cycle. Order is present because a chaotic waveform stays within a finite region in phase space in the close neighborhood of the *strange attractor*. Various maps, *e.g.*, return maps constructed from Poincaré sections and next-amplitude maps, may be constructed and used to show the underlying orderliness of chaos, which often has a fractal nature.

The unpredictability of a chaotic waveform results because trajectories starting from arbitrarily close initial conditions diverge. The measure of this divergence is the Lyapounov exponent. The fundamental mathematical definition of a chaotic system is one with a positive Lyapounov exponent. This definition and similar measures can be calculated for an experimental waveform, but practical problems often arise that cloud their interpretation. Chaos is thus usually identified in an experimental system or simulation by construction of the sorts of maps mentioned above and by investigation of the route from periodicity to aperiodicity.

A class of chemical reactions exists during which the concentration of various chemical species may oscillate in time and space. These are referred to as oscillating chemical reactions and provide some of the best experimental examples of chaos and other phenomena related to the presence of nonlinear dynamic laws. Oscillatory chemical reactions are easily controlled experimentally, even though their behavior is very complex. They are a particularly fruitful arena in which to investigate complex dynamical behaviors such as chaos because it is often possible to determine their governing dynamic law as described below.

An elementary chemical reaction is one that occurs via a single collision involving a single transition state. However, very few observed chemical changes actually result from a single elementary reaction. The overall conversion of reactants to products more often occurs via a sequence of elementary reactions (called a mechanism) involving reactants and products as well as a number of intermediate and sometimes catalyst species. Methods of determining the mechanism of a chemical reaction are highly developed. The rate of the observed overall chemical reaction is composed of the rates of the elementary reactions constituting its mechanism through a dynamic law that is easily written down from the mechanism as a set of coupled differential equations. If these equations are nonlinear, *i.e.*, if they contain terms that are not simply first order in a single chemical species, and if suitable feedback loops exist in the chemical mechanism, then the space-time behavior of intermediate and catalyst species may become very complex. Nonlinear dynamic laws are the rule rather than the exception in both chemical and biochemical systems.

A chemical system also must be beyond a certain critical distance from equilibrium for the driving force to be sufficient for complexity to develop. For example, the concentrations of some intermediate or catalyst species in a batch reactor may oscillate many hundreds of times as the overall reaction rushes toward equilibrium. However the final approach to equilibrium must be monotonic. Continuous-flow, stirred tank reactor (CSTR) technology has been introduced to maintain chemical systems far enough from equilibrium for complex behavior to appear. Much of the work described here involves CSTRs. The dynamics of the CSTR also interact with the chemical dynamics and dramatically increases the range of complex behaviors observed. The use of CSTRs allows the observation of not only simple oscillations but of other complex behaviors, *e.g.*, multiple steady states, hysteresis, excitability, quasiperiodicity and chaos. Furthermore, the continuous addition of reactants and removal of products allows states exhibiting complex behavior to be stabilized and studied in detail. Equally as important is that the transitions (bifurcations) between two states may be studied carefully as some parameter of the system, *e.g.*, the flow rate, is varied. The best evidence for chaos in chemical systems is indeed that the observed transitions from periodic to chaotic states fall into a few categories, *e.g.*, quasiperiodicity, intermittency and the period-doubling cascade, that are mathematically well understood. Some basic mathematics of chaos is developed here by Derrick and by Gontar.

The best known and understood oscillatory chemical system is the Belousov-Zhabotinsky (BZ) reaction, and it is here that chaos in a homogeneous (single phase) chemical system is perhaps best investigated and understood in terms of the chemistry occurring. Roux reviews the experimental characterization of BZ-CSTR chaos and its interpretation on the basis of dynamical systems theory. Györgyi and Field treat BZ-CSTR chaos within the framework of the chemical mechanism of the BZ reaction. BZ-CSTR chaos may be thought of as resulting from the interaction of two cycles that are relatively independent because the period of one is mainly controlled by the homogeneous BZ chemistry while the period of the other is strongly affected by the flow rate. The introduction of chaos by the coupling and forcing of chemical oscillators is discussed by Marek and Schreiber.

New types of nonlinearities are introduced into dynamic laws when adsorption of species onto surfaces is an intimate part of the overall chemical reaction. The appearance of chaos in heterogeneous electrochemical systems and systems involving a catalytic surface is considered by Hudson and by Eiswirth, respectively. The dynamical laws describing biochemical systems are closely related to those of chemical origin, and they also offer a rich sampling of complex experimental behavior. Chaos in the horseradish peroxidase system is discussed by Larter, *et al.* and in other biochemical systems by Goldbeter.

These topics taken together give the reader a good overview of chaos itself as well as its appearance in chemical and biochemical systems. We thank the authors as well as those people too numerous to mention who helped introduce us to chemical chaos.

April, 1992

Richard J. Field
Missoula, MT, USA

László Györgyi
Budapest, Hungary

CONTENTS

CHAPTER 1: MATHEMATICS OF CHAOS IN PERTURBED OSCILLATORS

WILLIAM R. DERRICK
Department of Mathematical Sciences
University of Montana, Missoula, MT 59812, U.S.A.

ABSTRACT

This chapter provides a quick introduction to the mathematics of chemical oscillators and the mechanisms by which chemical chaos is produced. Phase-plane analysis is used to explain the large-amplitude oscillations of a 2- and 3-dimensional version of the Field-Noyes model of the Belousov-Zhabotinsky (BZ) reaction. We then describe the type of forcing that is required to change these limit cycles into chaotic oscillations. We also examine a mathematical model that has been proposed for the BZ reaction in a continuous-flow, stirred tank reactor. Here a Hopf-hysteresis bifurcation produces toroidal flows, that, when wrinkled, lead to chaotic oscillations. Finally, we explain the strengths and weaknesses of seven tests that are commonly used to detect chaos in simulations and experiments.

1.1 Introduction

The purpose of this chapter is to provide a quick introduction to the mathematics of chemical oscillators and to describe the mechanisms by which forcing or coupling two oscillators, or controlling flow, produces chaotic oscillations. An oscillator must be perturbed in some way in order to exhibit complicated dynamical behavior. Coupling, forcing, and controlling flow are the physical mechanisms by which chemical chaos has been produced.

The prototype oscillating chemical reaction was discovered by B. P. Belousov[1] in 1951, but his paper was contemptously rejected because most chemists believed at that time that, even far from equilibrium, concentrations of intermediate species had to behave monotonically. Eventually he was able to publish a brief note in the proceedings of an obscure radiation medicine conference. His work attracted the attention of S. E. Schnoll, who suggested the need for further investigation of this reaction to his student A. M. Zhabotinsky. An interesting description of the early work on oscillating chemical reactions has been given by Zhabotinsky.[2] Belousov shared the Lenin prize posthumously with Zhabotinsky in 1980.

Although many chemical oscillators are now known,[3,4] the Belousov-Zhabotinsky (BZ) reaction[5,6] is the prototypical and best understood chemical oscillator. The principal five steps in a reduction of the complex chemical mechanism of the BZ reaction, called the Field-Noyes[7] (FN) model, are given by the following chemical reactions:

$$A + Y \xrightarrow{k_1} X + P; \qquad X + Y \xrightarrow{k_2} 2\,P$$

$$A + X \xrightarrow{k_3} 2\,X + \rho Z; \qquad 2\,X \xrightarrow{k_4} A + P,$$

$$Z \xrightarrow{k_5} f\,Y,$$

where ρ is a constant and $X \equiv HBrO_2$, $Y \equiv Br^-$, $Z \equiv Ce(IV)$, $A \equiv BrO_3^-$, $P \equiv HOBr$. The values of the rate constants k_1, k_2, k_3, and k_4 are known. The values of k_5 and the stoichiometric factor, f, are treated as expendable parameters but usually with k_5 in the range 0.01 to 1 and $f \approx 1$ for $\rho = 1$ or $f \approx 0.5$ for $\rho = 2$. The concentration of BrO_3^-, A, is taken to be constant, and the concentration HOBr, P, does not appear in the governing kinetic equations. The Law of Mass Action (See Chapter 3) yields the following third-order system of kinetic equations, where concentrations are denoted by lower case letters:

$$\begin{aligned} dx/dt &= k_1 a y - k_2 x y + k_3 a x - 2 k_4 x^2, \\ dy/dt &= -k_1 a y - k_2 x y + f k_5 z, \\ dz/dt &= \rho k_3 a x - k_5 z. \end{aligned} \tag{1.1}$$

System 1.1 often is called the "Oregonator" because it oscillates and was invented by Field and Noyes at the University of Oregon.

We discuss in Section 1.2 the large-amplitude oscillations of System 1.1 and of a 2-dimensional system derived from it. Section 1.3 deals with forced oscillations of these systems and gives an indication of the conditions that lead to chaotic oscillations in the 2-dimensional system. We consider in Section 1.4 a mathematical model proposed for the BZ reaction in a continuous-flow, stirred tank reactor (CSTR). Here a combination of a Hopf and a hysteresis bifurcation produce toroidal flows, that, when wrinkled, lead to chaotic oscillations. Section 1.5 describes several tests that are commonly used to detect chaos in complicated simulations and experiments. Many of the ideas discussed here are treated in more detail in several of the other chapters in this book.

1.2 Oscillating Reactions

It is useful to non-dimensionalize the equations in order to analyze system 1.1. Several ways of doing this have been proposed;[8,9,10] We shall follow that proposed by Tyson.[10] Let

$$x = x_0 \bar{x}, \qquad y = y_0 \bar{y}, \qquad z = z_0 \bar{z}, \qquad t = t_0 \bar{t}$$

where $\bar{x}$, $\bar{y}$, $\bar{z}$ and $\bar{t}$ are dimensionless variables, and

$$x_0 = \frac{k_3 a}{2\,k_4}, \quad y_0 = \frac{k_3 a}{k_2}, \quad z_0 = \frac{\rho (k_3 a)^2}{2\,k_4 k_5}, \quad t_0 = \frac{1}{k_5}$$

$$\epsilon = \frac{k_5}{k_3 a}, \quad \eta = \frac{2\,k_4 k_5}{k_2 k_3 a}, \quad q = \frac{2\,k_1 k_4}{k_2 k_3} \sim 3.15\times 10^{-4}.$$

Then system 1.1 becomes (omitting the overbars and denoting d/dt by ":)

$$\epsilon x'' = qy - xy + x(1 - x)$$

$$\eta y'' = -qy - xy + \rho f z$$

$$z'' = x - z \tag{1.2}$$

It is useful to approximate system 1.2 by a 2-dimensional model, where nullcline analysis may be used to understand the behavior of the system. Assume that $0 < \eta << \epsilon << 1$, so that the left-hand side of the second equation in system 1.2 is very small. Setting $\rho f = 1$ and $\eta y'' = 0$, we get $z = y(x + q)$, and use this expression to eliminate y from system 1.2 obtaining,

$$\epsilon x' = x(1 - x) + z\frac{q - x}{q + x}$$

$$z' = x - z \tag{1.3}$$

The nullclines for system 1.3 are the line $z = x$ and the curve

$$z = \frac{x(1 - x)(x + q)}{(x - q)} . \tag{1.4}$$

The curve in Eq. 1.4 is asymptotic to the vertical line $x = q$, and, for $x >> q$, the term $(x + q)/(x - q) \approx 1$, so that Eq. 1.4 resembles the parabola $z = x(1 - x)$. Thus, Eq. 1.4 has an "S" shape (See Fig. 1.1). The nullclines intersect to the right of $x = q$ at the positive root of

$$x^2 + qx - 2q = 0, \tag{1.5}$$

that is at $z_{ss} = x_{ss} \approx \sqrt{2q} = 0.025$.

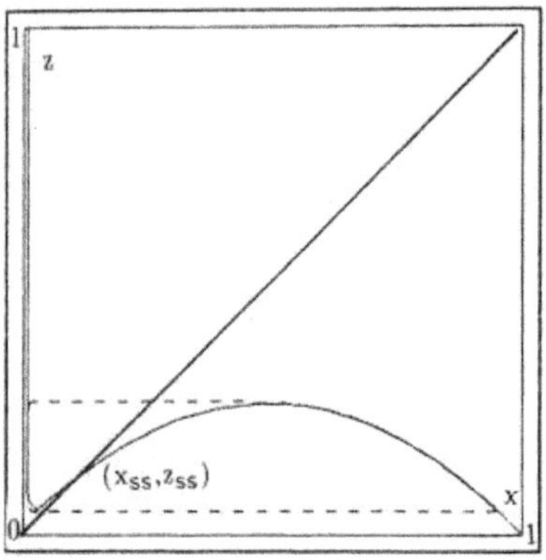

Figure 1.1. Nullclines and direction field for System 1.3. The dashed line represents the attracting limit cycle.

The steady state (x_{ss}, z_{ss}) occurs when the slope of Eq. 1.4 is positive. Calculating the Jacobian matrix, **J**, at this point, and using Eq. 1.5, we get the trace

$$\mathrm{tr}\,\mathbf{J} = \epsilon^{-1}(1 - 3x_{ss} + (1/2)x_{ss}^2) - 1,$$

and the determinant

$$\det J = (1/2)\epsilon^{-1} x_{ss}(4 - x_{ss}),$$

and both are positive, so that the steady state is unstable. Since ϵ is small, the first variable in System 1.3 is fast and the second is slow. Below the line $z = x$, the direction field points upward, while above this line it points downward. Below Eq. 1.4 it points to the right, while above it points left. Thus, any trajectory in the region $\{(x, z): x > q, z > 0\}$ quickly approaches the limit cycle shown in Fig. 1.1, which hugs the $x' = 0$ isocline when the slope is negative, and dashes across horizontally as soon as the trajectory clears the local maximum or minimum of Eq. 1.4. It follows from this analysis that the unforced 2-dimensional BZ reaction exhibits only periodic oscillations.

Hastings and Murray[11] proved that System 1.2 also has a large-amplitude periodic solution. Their technique can be generalized, requiring only that the following conditions be satisfied:

(i) The basin of attraction of all trajectories in $\Re^3_+$ is a convex, compact region **B** (typically a cube).

(ii) **B** contains a single steady state that has one negative eigenvalue and two complex eigenvalues with positive real part.

(iii) Exactly two trajectories enter **B** and go to this steady state as $t \to \infty$.

THEOREM. A periodic orbit will exist whenever conditions (i), (ii), (iii) hold for a 3-dimensional system of differential equations.

Proof. For $\delta > 0$ sufficiently small, construct a δ-tube around the 1-dimensional stable manifold $\mathbf{W}^s$ guaranteed by (iii). All orbits but those on $\mathbf{W}^s$ leave the δ-tube as $t \to \infty$. Homotopically contract **B** along $\mathbf{W}^s$, obtaining a pinched torus, **T**, that is positively invariant by (i). Consider any cross-section of **T** through the steady state. Since $\mathbf{W}^u$ is unstable, and all remaining trajectories in the deformed δ-tube leave it as $t \to \infty$, all trajectories move away from the steady state in **T**. By (ii), the steady state is an unstable focus on $\mathbf{W}^u$. Thus, trajectories exist that begin in the cross-section (with the deformed δ-tube deleted) **G** and return to **G**. This is the *return* or *Poincaré map* **P** of the beginning point. By continuity, we must have a surface of section, and a result of Fuller[12] then yields that **P(G)** $\subset$ **G** because **T** is positively invariant. Since **G** is homotopic to a disk, the Brouwer fixed-point theorem guarantees the existence of a fixed point x of **P** in **G**. But then the trajectory beginning at x must be a periodic orbit for the system.

Observe that the cube $\mathbf{B} = [0, X_0] \times [0, Y_0] \times [0, Z_0]$, where $1 < X_0 < Z_0 < qY_0$ is invariant and contains a unique steady state (x_{ss}, y_{ss}, z_{ss}). Here $x_{ss} = z_{ss}$ and $y_{ss} = x_{ss}/(x_{ss} + q)$. Following the same analysis as in Murray[13] [p. 184], the characteristic equation for system 1.3 at the steady state is

$$\lambda^3 + A\lambda^2 + B\lambda + C = 0,$$

where

$$A = 1 + \eta^{-1}(x_{ss} + q) + \epsilon^{-1} D, \qquad D = (2x_{ss} + y_{ss} - 1),$$

$$B = \eta^{-1}(x_{ss} + q) + \epsilon^{-1} D\,[1 + \eta^{-1}(x_{ss} + q)] + \epsilon^{-1}\eta^{-1} y_{ss}(q - x_{ss}),$$

$$C = \epsilon^{-1}\eta^{-1}[D\,(x_{ss} + q) + (y_{ss} - 1)\,(q - x_{ss})].$$

Using Eq. 1.5, it follows that $D = 3qy_{ss}/x_{ss}$, and $\epsilon\eta C = x_{ss}^2 + 2q$, so A, C, D > 0. By Descartes' rule of signs there is at least one negative eigenvalue. When AB < C the characteristic equation can have two complex eigenvalues with positive real part, so conditions (i)-(iii) can be verified for certain choices of the parameters with some additional work.

1.3. Forced Oscillations

We now consider the case where the 2-dimensional BZ reaction is subject to periodic forcing. For simplicity, we look at the case where the forcing affects the concentration of Ce(IV), that is, the z variable, and we rewrite system 1.3 in the form

$$\epsilon x' = x(1 - x) + z\frac{q - x}{q + x} = f(x, z)$$

$$z' = x - z + \beta \cos \omega t = g(x, z) \tag{1.6}$$

The forcing term simply raises or lowers the $z'' = 0$ isocline. If β is sufficiently large, this isocline may intersect Eq. 1.4 with a **moving** steady state (x_m, z_m) at places where its slope is negative. From Figs. 1.2a and 1.2c we see that $f_x, f_z < 0$ at (x_m, z_m), since we pass from a region where $f > 0$ to one where $f < 0$ as we move horizontally or vertically through (x_m, z_m). Similarly, $g_x > 0 > g_z$, so that tr $J <$ $0 <$ det J, and the moving steady state (x_m, z_m) is stable. Trajectories which happen to be near (x_m, z_m) at that time, will chase this stable moving steady state: if it moves slowly they may wind around it if it is a focus. It is not necessary that both Figs. 1.2a and 1.2c occur: if either occurs, we will have very complicated dynamics.

For simplicity, assume Fig. 1.2a occurs and that (x_m, z_m) is on the stable branch of Eq. 1.4 whenever $\cos \omega t > \alpha$. The two periodic functions involved in this process, $\cos \omega t$ and the stable limit cycle derived in Section 1.2, need not be synchronized. If $\omega << 1$, a solution to System 1.6 may go around the limit cycle several times while the moving steady state is on the unstable branch of Eq. 1.4 before being forced to follow the steady state while it is on the stable branch. Once the steady state reaches the minimum of Eq. 1.4 (near $2q$), the trajectory again follows the limit cycle, until the steady state again reaches that minimum. A time series for $x(t)$ would show several oscillations, followed by a long quiescent interval of low levels of x, followed again by oscillations. This phenomenon is called *bursting* or *intermittency*. If $\omega >> 1$, a trajectory might be trapped by the moving steady state every time, or every so many times, it comes around to the left stable branch of Eq. 1.4. In the former case it would force the system to oscillate with the same frequency as $\cos \omega t$, a phenomenon known as *entrainment*. In the latter case the oscillations might be *r*:*s phase-locked*. It is convenient to visualize this process on a torus T with t the axial parameter going around T, while the limit cycle describes the motion

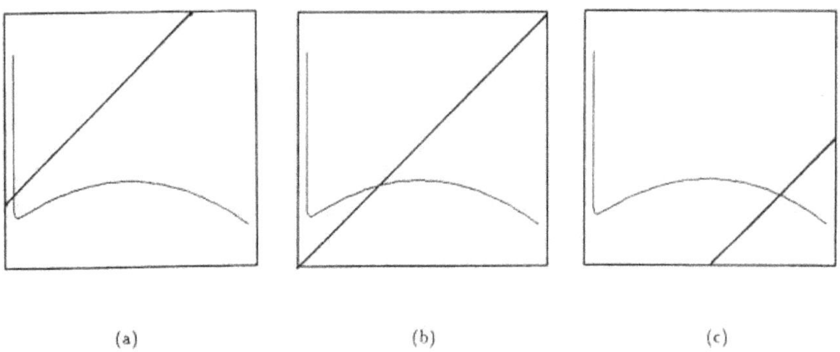

(a) (b) (c)

Figure 1.2. Moving steady states. (a) Steady state when $\cos \omega t > \alpha$. (b) Steady state when $\beta = 0$ or $|\cos \omega t| << \alpha$. (c) Steady state when $\cos \omega t < -\alpha$.

around a cross-section of T. A trajectory corresponds to an orbit on the surface of T: if it winds around once as it goes once around T, then it is entrained (or 1: 1 phaselocked), but if it winds around r times as it goes s times around T, then it is r: s phase locked. If it covers T, then it is *quasiperiodic*.

We may view a collection of periodic solutions of the unforced BZ system as a moving "strip" following the limit cycle described in Section 1.2. Suppose that the moving steady state is a focus, so that when we force the system, a portion of this strip winds around the stable focus until it is released. The loop so formed then continues around the limit cycle. We visualize the strip as having many consecutive loops, since motion along the stable arms of Eq. 1.4 is slow (See Fig. 1.3a). Assume each loop returns to the stable branch in time to once again wind around the stable focus. Since motion in the x-direction is fast, the loop will have been stretched lengthwise both times it travelled horizontally, and compressed laterally as it travelled upward and downward on the stable branches of Eq. 1.4. The first half and the last half of two strip segments might come back as the thinner S-shaped substrip inside the larger segment shown in Fig. 1.3b. We have labelled some substrips in the larger segment as V_0, V_1, V_2, V_3, and the corresponding returning parts of the two segments as W_0, W_1, W_2, W_3. Thus, $F(V_i) = W_i$, where $F = \mathrm{col}(f, g)$. We now construct the directed graph shown in Fig. 1.3c, called "Mickey," by drawing an arrow from node i to node j whenever substrip W_i intersects substrip V_j.

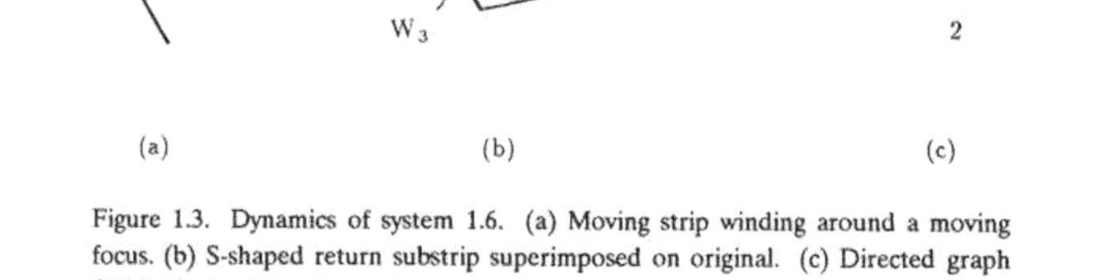

(a) (b) (c)

Figure 1.3. Dynamics of system 1.6. (a) Moving strip winding around a moving focus. (b) S-shaped return substrip superimposed on original. (c) Directed graph (Mickey) showing effects of returning maps.

Any loop in Fig. 1.3c (except the repetition of a single loop) reveals the existence of a periodic orbit. For example, the loop from node 1 to itself (Mickey's left ear) implies that there is a fixed point P in $V_1 \cap W_1$. Thus, for some trajectory through P, F(P) = P, and this trajectory has period 1. The loop 12011 indicates the existence of a point P in $V_1 \cap W_1$ such that $F^4(P) = P$, that is, there is a period-4 trajectory. Note there are loops of **all** periods: for example period 2 is the loop 131, period 5 is the loop 131111, *etc.*

The importance of having loops of all periods for this mapping comes from the study of one-hump, 1-dimensional mappings. The typical example is the logistic mapping $x_{n+1} = r x_n(1 - x_n)$, with $0 \le r \le 4$. This mapping has zero as an attracting steady state for $0 \le r \le 1$, the point $(r - 1)/r$ as an attracting steady state for $1 \le r \le 3$, and then for $r > 3$ undergoes a pitchfork bifurcation where the steady state becomes repelling but casts off an attracting 2-cycle, which later becomes unstable and casts off an attracting 4-cycle, leading to a *period-doubling* cascade. The resulting Sharkovskii sequence: 1, 2, 4, 8, 16,..., 2^n,... , $2^k \cdot 5$, $2^k \cdot 3$,..., 7, 5, 3 is well-known[14] [p. 311] and is interpreted in the following way: if an attracting p-cycle exists, then repelling q-cycles exist for all q to the left of p in the Sharkovskii sequence. In particular, an attracting 3-cycle implies the existence of repelling q-cycles of all orders $\neq 3$. Li & Yorke[15] in a well-known paper proved that (repelling) period 3 implies chaos (*aperiodicity*): therefore, when Mickey occurs, the forced BZ system will exhibit chaotic oscillations.

The S-shaped segment in Fig. 1.3b is a Smale "horseshoe" (See Guckenheimer & Holmes,[14] p. 106-116). A Smale horseshoe always produces an invariant hyperbolic set Λ on which the mapping **F** is structurally stable, with Λ containing:

(i) a countable set of periodic orbits of all periods which is dense in Λ,

(ii) an uncountable set of aperiodic (chaotic) trajectories, and

(iii) a dense orbit.

While chaos is present in such a system (in (ii) and (iii)), the trajectory produced by the initial conditions may not pass near Λ, and may undergo only regular oscillations. Trajectories passing close to Λ exhibit *transient chaos* (possibly for very long time intervals). In order to guarantee the observation of chaos, it is necessary that Λ be an attractor, that is, a set containing a dense orbit that attracts all trajectories in its basin of attraction. Mickey or the Smale horseshoe do not guarantee that Λ is an attractor.

The forced 2-dimensional BZ reaction in System 1.6 closely resembles the forced van der Pol system which has been studied in detail by Cartwright &

Littlewood[16] and Levi.[17] A quick introduction to this work can be found in Guckenheimer & Holmes[14] [p.74-82 and p. 253-255], where it is shown that the forced van der Pol system has an attracting invariant hyperbolic set Λ, confirming the existence of chaotic oscillations. Because of the close similarity between the forced van der Pol and the BZ reaction, it has been assumed that the invariant hyperbolic set of the latter is attracting.

Evidence of this winding of strips for the forced BZ reaction has been given by Crowley & Field.[18] Numerous experiments exploring the existence of chaotic oscillations in both coupled and forced BZ reactions have been conducted,[19,20] and described in Chapter 4 here by Marek & Schreiber. While forcing is required to produce chaotic oscillations in a 2-dimensional autonomous system, unforced autonomous 3- and higher-dimensional systems exhibiting chaotic oscillations are known (*e.g.*, the Lorenz equations). Chaos has not been observed in unforced, closed BZ experiments.

1.4 BZ-CSTR Chaos

Complicated dynamical behaviors which include multistability, hysteresis, bursting, complex periodicity, and chaos, have been observed experimentally for the BZ reaction in continuous-flow, stirred tank reactors (CSTR). The flow through the reactor is used both to maintain the BZ system far from equilibrium and to couple with the chemical dynamics. Aperiodic oscillations were first reported in BZ-CSTR experiments by Schmitz *et al.*,[21] but the most reliable and best documented experiments have been performed by various researchers at the University of Texas in Austin,[22] and are called "Texas chaos."

Several models have been proposed to explain this behavior. While the models reproduce all or most of the above phenomena, they suffer from various defects: those based on the chemistry generally involve so many variables that a clear understanding of the dynamics is difficult and speculative, while those based on mathematics give calculated results that do not agree well with the experiments. In this section we will briefly describe a mathematical model of the latter type proposed by Richetti *et al.*;[23] reference to others may be found in several of the Chapters in this book.

To describe the model we need to consider both the Hopf and the hysteresis bifurcations. A *Hopf bifurcation* occurs when the steady state has a pair of complex conjugate eigenvalues $\lambda, \bar{\lambda}$ with Re $\lambda = 0$. Consider the system

$$x' = a\,x - y - x\,(x^2 + y^2),$$

$$y' = x + a\,y - y\,(x^2 + y^2), \tag{1.7}$$

with a steady state at the origin and having the Jacobian matrix at the origin

$$J = \begin{bmatrix} a & -1 \\ 1 & a \end{bmatrix}$$

with characteristic equation $\lambda^2 - 2\,a\lambda + (a^2 + 1) = 0$. Here $\lambda = a \pm$ i, so a Hopf bifurcation occurs at the origin when $a = 0$. When a Hopf bifurcation happens in 2-dimensional space, the steady state changes from a stable to an unstable focus or *vice versa*. For example, if we were increasing the parameter a in System 1.7 from negative to positive values, the origin changes from a stable focus to an unstable focus. However, the global behavior does not change, so if the steady state was globally attracting (as the origin is for $a < 0$ in the example above), but is now repelling, it casts off an attracting limit cycle. This is often called a *supercritical bifurcation*, while the reverse situation (a globally repelling steady state that becomes attracting and casts off a repelling limit cycle) is called a *subcritical bifurcation*. In both cases oscillations occur arbitrarily close to the Hopf bifurcation. To see what is happening in System 1.7, it is convenient to rewrite it in polar coordinates

$$r' = r\,(a - r), \quad \theta' = 1.$$

For $a < 0$, the term in parenthesis in the r′ equation is negative, so all radial motion is towards the origin, but for $a > 0$ radial motion in a neighborhood of the origin is outward towards the attracting limit cycle at $r = \sqrt{a}$.

In spaces of dimension greater than 2, the limit cycle itself can be viewed as a fixed point of the Poincaré map, so it too may undergo a Hopf bifurcation (this is called a *secondary Hopf bifurcation*) and cast off a limit cycle for the Poincaré map, which would be a torus in phase space. Thus, a stable focus could cast off a stable limit cycle, which in turn could cast off a stable torus. An example of this is provided by the system in polar coordinates

$$z' = f(z) - r^2$$

$$r' = r\,(z - a)$$

$$\theta' = 1, \tag{1.8}$$

with $f(z) > 0$ on the interval $z_1^* < z < z_2^*$, with $f(z_j^*) = 0$ for $j = 1, 2$. Since the rotation is constant, we can ignore the last equation in System 1.8 and consider the dynamics only in the rz-plane (where the return map functions as a Poincaré map). The nullclines for $r' = 0$ are the lines $r = 0$ and $z = a$, and for $z' = 0$ the segments $r = \sqrt{f(z)}$. Steady states exist at $(0, z_1^*)$ and $(0, z_2^*)$ and any intersections of $z = a$ with $r = \sqrt{f(z)}$. The Jacobian shows $(0, z_1^*)$ has the eigenvalues $f'(z_1^*)$ (> 0) and (rotating about the z axis) $(z_1^* - a) \pm i$, while $(0, z_2^*)$ has the eigenvalues $f'(z_2^*)$ (< 0) and $(z_2^* - a) \pm i$. If $a < z_1^*$, both pairs of complex conjugate eigenvalues have real part positive. If $z_1^* < a < z_2^*$, the steady state at $(0, z_1^*)$ now has two complex conjugate eigenvalues with negative real part. Thus, $(0, z_1^*)$ undergoes a Hopf bifurcation as a passes through z_1^*, and casts off a limit cycle obtained by revolving the point $(\sqrt{f(a)}, a)$ around the z axis (See Fig. 1.4a). We now check the stability of the point $(\sqrt{f(a)}, a)$: the 3-dimensional Jacobian matrix yields the characteristic equation

$$\lambda^3 - f'(a)\lambda^2 + (1 + 2f(a))\lambda - f'(a) = 0,$$

which undergoes a secondary Hopf bifurcation when $f'(a) = 0$, for $z_1^* < a = z_{crit} < z_2^*$. (Note that the bifurcation is degenerate since one eigenvalue is zero; we will take care of this problem later.) For a range $z_{crit} < a$, System 1.8 has an invariant torus, a limit cycle, and two steady states (See the bifurcation diagram in Fig. 1.4b).

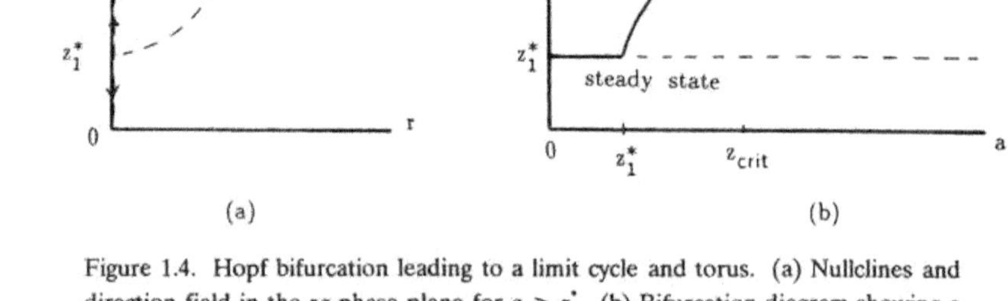

Figure 1.4. Hopf bifurcation leading to a limit cycle and torus. (a) Nullclines and direction field in the rz-phase plane for $a > z_1^*$. (b) Bifurcation diagram showing a transition from a steady state to a limit cycle and then to a torus.

The oscillations on the torus progress through various phase-locking regimes and quasiperiodic flows. The torus can be destroyed[24] if the separatrix leaving $(0, z_2^*)$ goes to $(0, z_1^*)$ forming a *heteroclinic connection*.

An easy example of the *hysteresis bifurcation* is given by the equation $z' = f(z)$, where $f(z) = b - (z - z_0)(z - z_1)(z - z_2)$, with $z_0 < z_1 < z_2$. When $b = 0$, the steady states z_0 and z_2 are stable, while z_1 is unstable, since $z' > 0$, for $z < z_0$ and $z_1 < z < z_2$, and $z' < 0$, for $z_0 < z < z_1$ and $z > z_2$. As b is increased to a value b_1, the steady states z_0 and z_1 coalesce and vanish for $b > b_1$. A similar result happens with z_1 and z_2 as b is decreased to the value b_0. Observe that if we begin with $b = 0$ at $z = z_0$ and slowly increase b, the solution follows the steady state z_0 until $b = b_1$, and then moves rapidly to z_2 as b increases beyond b_1. If we now decrease b, the solution follows z_2 until b decreases below b_0, when it moves back toward z_0. Thus, the motion from z_0 to z_2 occurs at b_1, while the motion from z_2 to z_0 occurs at b_0. We say that the equation undergoes a *hysteresis bifurcation* at the values b_0 and b_1. In particular, there is a range in which *bistability* occurs.

The mathematical model proposed by Richetti *et al.*[23] requires that both bifurcations occur simultaneously: we call this a hysteresis-Hopf bifurcation. If we substitute the cubic above in the hysteresis bifurcation into System 1.8 and let $b = 0$, $z_0 = z_1 = -a$, and $z_2 = 2a$, then $f(z) = -z^3 + 3a^2z + 2a^3$ and $f'(a) = 0$. Thus a Hopf bifurcation is occurring at $(\sqrt{f(a)}, a)$ and a hysteresis bifurcation at $(0, z_0)$ simultaneously. Now let $a \to 0$, so that a hysteresis bifurcation also occurs at $(0, z_2)$. Because both Hopf and hysteresis bifurcations occur, we know there are oscillations, invariant tori, and multiple steady states near the origin. Richetti *et al.*[23] then add two additional terms to the equation for z' in System 1.8 to take care of the degeneracy of the secondary Hopf bifurcation that occurs when $f'(a) = 0$, and to destroy the axisymmetry. The latter, as we approach the heteroclinic connection mentioned above, leads to fractalization of the torus and the appearance of chaotic dynamics because of the creation of Smale horseshoes. Essentially, we "wrinkle" the torus, and if it is attracting, then chaotic oscillations will occur. Richetti *et al.*[23] show that this system exhibits quasiperiodicity and a trajectory ergodically covers the torus surface, and that phase-locking occurs in narrow bands when the level of nonaxisymmetry is small. As the latter is increased, chaotic oscillations are observed, as well as periodic-chaotic sequences. They speculate that if the torus wrinkles before other attracting steady states can influence the dynamics in a set containing the torus, then both chaotic oscillations and stable steady states may coexist, but if the torus wrinkles after this set falls into the basin of attraction of the steady state, then only transient chaos occurs. Richetti *et al.*[23] define a seven-variable BZ-CSTR Oregonator type model which reproduces many of the steps to chaos that have been found in experiments[25] as well as in the system described above.

Barkley *et al.*[24] found quasiperiodicity as well as periodic-chaotic sequences using a six-reaction, modified-Oregonator model. They have shown that while this model has a bifurcation diagram resembling that in Fig. 1.4b, it cannot arise through the hysteresis-Hopf bifurcation mechanism of Richetti *et al.*[23] because the unstable limit cycle (from which the torus was cast off) terminates in a Hopf bifurcation for the hysteresis-Hopf mechanism, while for their model it terminates in a saddle-loop bifurcation. It would be very interesting to find a mathematical mechanism (other than hysteresis-Hopf) that could be used to interpret their model's dynamics.

The two models above are chaotic only for small flow-rate ranges, or for unrealistically high concentrations of one of the variables. They use outdated rate constants, making it very difficult to compare their results to experiments. Györgyi *et al.*[26] devised an 11-variable model which combines both the traditional inorganic chemistry of the Oregonator (simplified in reactions k_1 to k_4), and the organic chemistry of the reactions of malonic acid and its derivatives represented by the single stoichiometric factor f in reaction k_5. This model reproduces virtually all the dynamical behavior in the Texas experiments at comparable low flow rates. Györgyi and Field [27,28] reduce this model to two 4-variable models, and to a single 3-variable model, respectively, in non-polynomial form, that reproduce the major features of the experimentally observed chaos at low flow rates, including period-doubling cascades of maxima similar in appearance to the deterministic chaos of a one-hump, 1-dimensional map, as well as many of the observed experimental features at high flow rates.

1.5 Detecting Chaos

The last two sections have detailed some of the theoretical concepts giving rise to chaotic oscillations in BZ-CSTR and forced or coupled BZ oscillators. These included period-doubling cascades, Smale's horseshoe, Hopf bifurcations to quasiperiodicity and wrinkled tori, heteroclinic connections leading to Smale horseshoes, and transient chaos (possibly due to intermittency), which constitute virtually all currently known avenues to chaos in a dissipative system. These mathematical concepts are frequently difficult to discover in a known multivariable nonlinear system of differential equations, and can only be speculated about from experimental data. So how can a researcher determine that the system being studied, be it a numerical simulation or a chemical experiment, is exhibiting chaotic oscillations? The answer is by subjecting the data to several tests, each of which determines some characteristic of chaos. If the data survives all these tests, the researcher can claim with **some** confidence that chaos exists in the system.

The main tests for determining the presence of chaos in a dynamical system or experiment involve the route by which periodicity decomposes to aperiodicity as well as an analysis or calculation of: 1) trajectories in phase space, 2) the Poincaré map, 3) the maximum-return map, 4) the fractal dimension of the attractor, 5) the Lyapounov exponents, 6) the Fourier spectrum, and 7) the autocorrelation function. In this section we shall briefly describe each of these tests, and the characteristics chaos exhibits for each test. References are provided in a few subsections to more extensive discussions; other information appears in several other chapters of this book and in a review by Scott.[29]

1.5.1 Trajectories in Phase Space

Since chaotic oscillations are ergodic on some strange attractor, we would expect a trajectory in a simulation to "fill" at least some region in the phase space. If we are working with experimental data, we can use the method of Takens[30] to construct a *pseudo-phase space* (or *embedding space*) by using an n-tuple of time-delayed measurements from the time series of one of the physical variables: $(x(t), x(t + T), x(t + 2T), ..., x(t + (n - 1)T))$. This n-tuple, viewed as a moving time series, provides topologically equivalent trajectories to those that exist in an n-variable reaction. The time delay, T, is often taken to be a tenth of the mean orbital period, and n is usually taken to be less than 5 because the BZ reaction is known to be of "low dimensionality" (we will say more about this when we discuss Poincaré mappings). However, the main idea here is that if the strange attractor is of low dimension, it serves no purpose to take n large, as all the motion is limited to a few dimensions, and we might as well project our trajectories into a low dimensional space.

As the trajectories are plotted, it is useful to look for horseshoes in the orbit, particularly if they return transversally to previous motion, as these can be indications of chaos. Any "foldings" should be carefully analyzed, as this flow mixing is the principle that leads to chaos in the Lorenz equations. Any jumps may provide evidence of wrinkling on some manifold that may lead to chaos. Note that even if a trajectory appears to fill a torus ergodically, it may still not be chaotic: quasiperiodicity, or even a periodic oscillation with a large period yield similar results. As a parameter is varied, one should look for period-doubling cascades of the orbits, or transitions from steady state to limit cycle, to toroidal motion, as precursors of chaotic regimes. Observation of one of the standard routes from periodic to aperiodic behavior is strong evidence that the aperiodicity is indeed chaos. It is also useful to plot trajectories starting at close-by initial conditions comparing the resulting dynamics: chaos is probably there if the dynamics differ substantially.

1.5.2 Poincaré Mappings

A Poincaré map is constructed by selecting a (hyper) plane in the phase space that the trajectory crosses repeatedly, and plotting all points (after transients are eliminated) where the trajectory intersects the (hyper) plane. This method is particularly useful for 3-dimensional problems, as the Poincaré map is now planar. If the Poincaré map consists of just a few points, this indicates the orbit is periodic with period equal to that number of points. When a Poincaré map describes a locus, such as a closed curve or a line, it gives evidence of quasiperiodicity and possibly of chaos. For example, if the trajectory lies on a wrinkled torus, the Poincaré map may appear as a scalloped circle. When a Poincaré map fills a region containing a disk there is good evidence of chaos.

If the Poincaré map lies on a straight line, as frequently happens with the BZ reaction, the dynamics occur in a low-dimensional space. Then, it is useful to use the "return" map: plot the pairs of points (x_n, x_{n+1}), where x_n is the n th point in the Poincaré map. Often this return map will exhibit a one-humped appearance, from which chaos can sometimes be inferred using the Sharkovskii sequence.

1.5.3 Maximum-Return mappings

This method is similar to the return map mentioned in Section 1.5.2. Here the map (x_n, x_{n+1}) maps two consecutive maximum values of the time series for some variable. The return map is used in the same way as that for Poincaré maps. Stroboscopic maps can also be used for similar analyses.

1.5.4 Fractal Dimension

This method assumes that the trajectory is so close to a strange attractor that it describes it. The procedure involves calculating the fractal dimension of the attractor. There are numerous algorithms for calculating fractal dimensions: it is worth noting that they do not give the same values. In this brief description we shall indicate one of these algorithms. The reader should refer to Moon[31] and to the literature for additional algorithms.

The *Hausdorff* or *capacity* dimension is based on the following idea. Select a minimal covering of a set A in space (say 3-space to simplify the discussion) by balls of radius ϵ. Then, for example, if we have a straight line of length 1, we should be able to cover it with approximately $1/\epsilon$ of these balls, and if we have a square of area 1, we should be able to cover it with approximately $1/\epsilon^2$ such balls. Thus, since the line is 1-dimensional, and the square is 2-dimensional, the exponent of ϵ in the number of required balls equals the dimension. Hence, a set A requiring $1/\epsilon^d$ balls to cover it should have dimension d approximately. So, for any set A we calculate the number $N(\epsilon)$ of ϵ balls required to cover it, and assume that $N(\epsilon)$ is proportional to $1/\epsilon^d$ (*e.g.*, $N(\epsilon) = c/\epsilon^d$). Then $\ln N(\epsilon) = \ln c + d \ln (1/\epsilon)$, so

$$d = \lim_{\epsilon \to 0} \frac{\ln N(\epsilon)}{\ln (1/\epsilon)} .$$

To find $N(\epsilon)$, divide the region containing the image of the trajectory in space by a cubic lattice of points $1/(2^{k-1}\sqrt{3})$ units apart. The lattice points are the centers of balls of radius $\epsilon = 1/2^k$ covering the entire region. $N(\epsilon)$ consists of the balls intersected by the trajectory. Generally, just a few values of k are needed to establish a trend, so while the computation is time consuming, it need only be repeated a few times.

Criticisms of the Hausdorff dimension include the fact that it does not consider the frequency with which the trajectory meets a ball and that it is difficult computationally. Some researchers use boxes instead of balls to simplify the computations (called the *box dimension*), and count the number of times the trajectory enters each box.

If the Hausdorff dimension is not an integer, this test provides evidence that the attractor is strange, implying that chaotic oscillations occur.

1.5.5 Lyapounov Exponents

This test measures the sensitivity of the system to changes in initial conditions. One of the important features of chaos is sensitivity to initial conditions: two trajectories that start arbitrarily close together move, on the average, exponentially away from each other over certain small intervals. Suppose two trajectories are initially a distance d_0 apart. If the oscillation is chaotic, over a time interval Δt (longer than a time step but less than a tenth of the mean orbital period), the distance may grow to

$$d_1 = d_0 2^{\lambda \Delta t}, \qquad \lambda = \frac{1}{\Delta t} \log_2 \left(\frac{d_1}{d_0}\right),$$

where λ is the (leading) Lyapunov exponent. Taking one measurement is not enough, because the exponential separation is a local phenomena, thus an average is computed:

$$\lambda = \lim_{N \to \infty} \frac{1}{N} \sum_{k=1}^{N} \frac{1}{\Delta t} \log_2 \left(\frac{d_{1k}}{d_{0k}}\right),$$

where d_{0k} is the distance between the current point on the trajectory (which may be the n-tuple $(x(t_0), x(t_0 + T), \ldots, x(t_0 + (n-1)T))$ as in Section 1.5.1), and the nearest neighboring point $(x(t_j), x(t_j + T), \ldots, x(t_j + (n-1)T))$ (not necessarily the next time step), and d_{1k} is the distance that is between the points Δt later, that is,

$$\left(\sum_{k=0}^{n-1} (x(t_0 + kT + \Delta t) - x(t_j + kT + \Delta t))^2 \right)^{1/2}$$

Wolf *et al.*[32] includes a FORTRAN program for determining the leading Lyapounov exponent from a time series using this technique.

Harding & Ross[33] provide a discussion of the differences that occur when calculating the Lyapounov exponent with noisy but periodic data and with noisy chaos. Their discussion also applies to the next two subsections.

1.5.6 Fourier Spectrum

One expects the Fourier spectrum for a chaotic oscillation to be broad; periodic oscillations (for small periods) usually produce a few sharp peaks substantially higher than the "noise" in the signal. As the signal-to-noise ratio increases, the plateau of the power spectrum rises and these peaks begin to vanish into the background, so noisy experimental data may not benefit much from this analysis. The method is not very useful for periodic orbits with large periods, as aliasing may yield a broad spectrum.

1.5.7 Autocorrelation

If chaos is present one would expect the autocorrelation function to go to zero. Periodic oscillations involve wide peak amplitudes. As noise increases in the data, the peak amplitudes are reduced, but generally stay asymptotically away from zero. Noisy chaos approaches zero asymptotically. This test is very straightforward, but can lead to poor conclusions as the noise-to-signal ratio is increased.

It is worth noting that in an experiment the true state can never be measured exactly due to limitations in the measuring equipment. Similar difficulties, due to round-off and truncation errors, also occur in simulations. Thus, some noise is always present.

The seven tests above are indicators: several (perhaps all) of them should be used and satisfied before chaos can be claimed.

Literature Cited

1. B.P. Belousov, in *Oscillations and Traveling Waves in Chemical Systems*, R.J. Field and M. Burger, Eds. (Wiley-Interscience, New York, 1965) pp. 605-613.
2. A.M. Zhabotinsky, *in Oscillations and Traveling Waves in Chemical Systems*, R.J. Field and M. Burger, Eds., (Wiley-Interscience, New York, 1985) pp. 1-6.
3. I.R. Epstein and K. Kustin, *Structure and Bonding*, **56** (1984) 1.
4. G. Rabai and I.R. Epstein, *J. Phys. Chem.* **94** (1990) 6361.
5. R.J. Field, E. Köros and R.M. Noyes, *J. Am, Chem. Soc.* **94** (1972) 8649.
6. R.J. Field, in *Oscillations and Traveling Waves in Chemical Systems*, (Wiley-Interscience, New York, 1985). pp.55-92.
7. R.J. Field and R. M. Noyes, *J. Chem. Phys.* **60** (1974) 1877.
8. J.D. Murray, *Nonlinear Differential Equations Models in Biology* (Clarendon Press, Oxford, 1977).
9. J.J. Tyson, *J. Phys. Chem.* **86** (1982) 3006.
10. J.J. Tyson, in *Oscillations and Traveling Waves in Chemical Systems*, R.J. Field and M. Burger, Eds. (Wiley-Interscience, New York, 1985), pp. 92-144.
11. S.P. Hastings & J.D. Murray, *SIAM J. Appl. Math.* **28** (1975) 678.
12. F.B. Fuller, *Amer. J. Math.* **87** (1965) 473.
13. J.D. Murray, *Mathematical Biology*, (Springer-Verlag, New York, 1989).
14. J. Guckenheimer and P. Holmes, *Nonlinear Oscillations, Dynamical Systems, and Bifurcations of*

Vector Fields (Springer-Verlag, new York, 1983).
15. T.-Y. Li and J.A. Yorke, *Am. Math. Monthly*. **82** (1975) 985.
16. M.L. Cartwright & L.E. Littlewood, *J. London Math. Soc.* **20** (1945) 180.
17. M. Levi, *Mem. Am. Math. Soc*. **214** (American Mathematical Society, Providence, 1981).
18. M.F. Crowley and R.J. Field, *J. Phys. Chem.* **90** (1986) 1907.
19. P. Lambda and J.L. Hudson, *Chem. Eng. Sci.* **42** (1987) 1.
20. M.F. Crowley R.J. Field, in *Lecture Notes in Biomathematics*, **66,** H. Othmer, Ed. (Springer-Verlag, New York, 1986). pp. 68-97.
21. R.A. Schmitz, K.R. Graziani and J.L. Hudson, *J. Chem. Phys*. **67** (1977) 3040.
22. J.-C. Roux, J.S. Turner, W.D. McCormick, H.L. Swinney, *Nonlinear Problems: Present and Future*, A.R. Bishop, D.K. Campbell, B. Nicolaenko, Eds. (North Holland, Amsterdam, 1982). pp. 409-422.
23. P. Richetti, J.-C. Roux, F. Argoul, and A. Arneodo, *J. Chem. Phys*. **86** (1987) 3339.
24. D. Barkley, J. Ringland, and J.S. Turner, *J. Chem. Phys.* **87** (1987) 3812.
25. F. Argoul, A. Arneodo, P. Richetti, and J.-C. Roux, *J. Chem. Phys*. **86** (1987) 3325.
26. L. Györgyi, S.L. Rempe, and R.J. Field, *J. Phys. Chem.* **95** (1991) 3159.
27. L. Györgyi and R.J. Field, *J. Phys. Chem.* **95** (1991) 6594.
28. L. Györgyi and R.J. Field, *Nature* **355** (1992) 808.
29. S.K. Scott, *Chemical Chaos* (Clarendon Press, Oxford, 1991).
30. F. Takens, in *Lecture Notes in Mathematics* **898**, D.A. Rand and L.S. Young, Eds. (Springer-Verlag, New York, 1981). pp.366-81.
31. F.C. Moon, *Chaotic vibrations* (Wiley-Interscience, New York, 1987).
32. A. Wolf, J.B. Swift, H.L. Swinney and J.A. Vastano, *Physica*, **16D** (1985) 285.
33. R.H. Harding and J. Ross, *J. de Chemie Phys*. **84**, (1987) 1305.

CHAPTER 2: DYNAMICAL SYSTEMS THEORY ILLUSTRATED: CHAOTIC BEHAVIOR IN THE BELOUSOV-ZHABOTINSKY REACTION

J.-C. ROUX
Centre de Recherche Paul Pascal
Université de BORDEAUX-I
Avenue A. Schweitzer F- 33600 Pessac
France

ABSTRACT

Deterministic chaos undoubtedly occurs when the Belousov-Zhabotinsky (BZ) reaction is run in a continuous-flow, stirred tank reactor (CSTR). The experimental characterization of the periodic-chaotic sequences that occur in the BZ reaction at both high and low CSTR flow rates is described. The evidence that the observed aperiodic oscillations observed are truly deterministic chaos is marshalled. The connection between the dynamic behavior of the BZ reaction and the normal form of dynamical systems theory is made.

2.1 Introduction

The idea of "deterministic chaos" began to percolate into the population of people working in the area of chemical kinetics after the original suggestion of D. Ruelle[1] in 1973 that chemical reactions might exhibit aperiodic behavior. A handful of people working on oscillating chemical reactions were the best prepared to receive these ideas because they already knew that a chemical process occurring monotonically could become unstable to yield a periodic state. Thus some of them were prepared to accept the idea that this oscillating state might, in turn, become unstable to give birth to new, even more exotic behaviors.

In this context, a lot of experimental data previously not understood but deeply sleeping in the abyss of the "lower right-hand drawer" suddenly became of great interest. For example, a few oscillatory times series (waveforms) obtained from the Belousov-Zhabotinsky (BZ) reaction in a continuous-flow, stirred tank reactor (CSTR) for some mysterious reason displayed apparently erratic behavior for certain sets of constraints, whereas for other conditions the usual periodic behavior was observed. In both situations the experimental parameters (the constraints) were equally well/badly settled. Better experiments needed to be carried out. However, the erratic times series were reproducible in that an erratic behavior always was observed for a particular set of constraints. The phenomenon became known as deterministic chaos.

The present review of experimental findings concerning chaos in the BZ reaction does not pretend to be exhaustive. It will rather emphasize the successive steps by which the characteristics of the phenomenon were elucidated. This historical approach, in turn, demonstrates the difficulties by which the existence and concept of chaos was accepted by the community of chemical kinetists.[2]

2.1.1 *Experimental Conditions*

The reactants of the BZ reaction are: malonic acid $\{CH_2(COOH)_2\}$ in the $\approx$ range 0.03 to 0.3 M, bromate (BrO_3^- as the K^+ or Na^+ salt) in the range $\approx$ 0.01 to 0.1 M, cerium-ion (Ce(IV)/Ce(III)) catalyst initially present as a SO_4^{2-} or $(NH_4^+)_xSO_4^{2-}$ salt at $\approx 10^{-4}$ M or Mn(II) in the same range. The medium is $\approx$ 1 M H_2SO_4. A redox indicator (ferrous o-orthophenanthroline) is sometimes used to make the oscillations visible and may even replace the catalyst. A constant set of constraints is provided by a continuous injection of all the above reactants into a well-stirred (?) CSTR maintained at a constant temperature. The reacting solution is maintained at a constant volume by an overflow. The evolution of the reaction is monitored by measuring one or several of the following characteristic properties (responses) of the system: electrochemical potential at a Pt-electrode, $[Br-]$ with an ion-selective electrode, changes of absorbance due to changes in [Ce(IV)]/[Ce(III)], or temperature. The behavior of the system is non-monotonic over a wide range of control-parameter values,[3] *e.g.*, feedstream concentrations, temperature and CSTR flow rate.

The use of a CSTR allows the system both to be maintained far from equilibrium and makes possible the development and observation of true steady states. The steady states are reached asymptotically at a rate mainly dependent on τ, the CSTR residence time, *i.e.*, the CSTR volume divided by the total flow rate. The dynamics in the CSTR are generally considered asymptotically stable after about 10 residence times. Whereas this assumption is easily verified visually when the asymptotic behavior is either monotonic (steady) or periodic, its generalization to non-periodic oscillation is more questionable. This is especially true when it is necessary to decide between chaos and very long periodic patterns. However, the 10-τ rule is also used here as well, mainly for practical reasons.

The evolution of the BZ dynamics from one regime (*e.g.*, periodic) to another (*e.g.*, aperiodic) is followed in virtually all studies of chaotic behavior by varying the CSTR residence time with the reactant concentrations in the feedstream kept constant. This choice of the flow rate as the "bifurcation" parameter is made not simply for the experimental convenience of controlling the system by its easily adjusted pumping speed. It also results from the reasonable assumption that for an infinite residence time the reaction approaches equilibrium, whereas at a high flow rates it is very far from equilibrium. Thus the flow rate is a "measure" of the distance from equilibrium. The branch of low flow-rate states is often called the thermodynamic or bulk branch, as opposed to the far-from-equilbrium, flow-induced branch that appears at higher flow rates.

2.2 Prehistory, The Periodic-Chaotic Sequence

The first publications[4-11] in this area reported time series showing apparently irregular oscillations, as in Fig. 2.1. Their chaotic nature was asserted only on the basis of inspection of their appearance. Thus the only conclusion that safely could be drawn from these experiments was that there exists for the BZ reaction some domains in parameter space where no clear periodic dynamics can be observed. Nothing could be said about the nature (deterministic or stochastic) of the observed aperiodicity. However, a characteristic feature of the dynamics of the BZ reaction emerged from these first studies; when the control parameter is continuously scanned one observes a succession of periodic and non-periodic states (see Fig. 2.1) referred to as a Periodic-Chaotic Sequence (PCS).

A step further was made when, following parallel studies of chaotic dynamics in hydrodynamic systems, the PCS time series were submitted to a Fourier analysis.[12,13] The visual evidence was confirmed for the chaotic states of the PCS by a typical, broad-band spectrum with no clear dominating frequencies, whereas the periodic states yielded Fourier spectra with sharp lines and a signal-to-noise ratio of about 80db.[13] This analysis, as well as ones to follow, were made possible by the digitalization and computer storage of the experimental times series.

The autocorrelation of the times series also was computed[13] in the same spirit. However, no attempts were made to quantify the chaotic states by a measure of the autocorrelation time, mainly because it was not unique.

All of these studies were performed at high flow rates. This means that they concern chaotic states reached by increasing the flow rate from values where the usual, paradigmatic, regular BZ oscillations (called P_{BZ} here) are observed. We shall

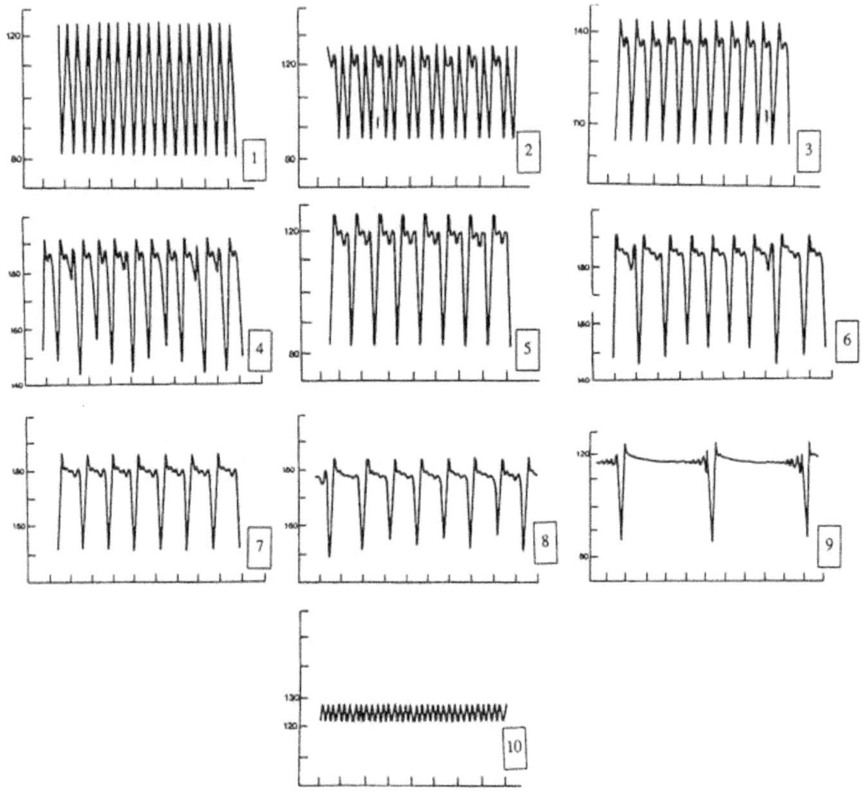

Figure 2.1. The Periodic-Chaotic Sequence (PCS) observed by Hudson *et al.* at high CSTR flow rate. The residence times (minutes) are: (1) 8.72, (2) 6.75, (3) 6.25, (4) 5.89, (5) 5.85, (6) 5.63, (7) 5.49, (8) 5.33, (9) 4.72, (10) 4.68. The ticks on the time (x)-axis represent two minutes. The y-axis is the measured potential of a Pt-electrode in millivolts. Four feedstreams were used. The initial concentrations in the CSTR before any reaction takes place are: $[CH_2(COOH)_2]_0$ = 0.3 M, $[NaBrO_3]_0$ = 0.14 M, $[Ce_2(SO_4)_3]_0$ = 0.001 M, and $[H_2SO_4]_0$ = 0.2 M. Temperature = 25 °C. Reproduced by permissiom from J.L Hudson, M. Hart and D. Marinko, *J. Chem. Phys.* **71** (1979) 1601.

now see that other chaotic states may be obtained working at flow rates below P_{BZ}, and that these states are easier to study than those appearing at higher flow rates. This distinction between high and low flow rates does not mean that all other parameters are the same in both cases. However, there are generic behaviors that are characteristic of high and low flow rates for many sets of values of the other parameters.

2.3 Deterministic Chaos

2.3.1 Attractors

We may say as a first approximation that deterministic chaos differs from stochastic aperiodicity in that it has a deeply buried order under its apparent disorder. Thus, the next goal in order to assert that the aperiodic dynamics observed in the BZ reaction is characteristic of deterministic chaos is to find this order, which reflects the existence of a so-called "strange attractor." The word attractor in this context refers to a special trajectory in phase space which other trajectories approach regardless of initial conditions. The phase space itself is a multidimensional space where the coordinates are the pertinent variables (usually chemical concentrations) of the system. This phase space has the dimension of the total number of intermediates involved in the chemical reaction. The representative point in phase space of a system approaching a steady state converges toward a fixed point, which thus is an attractor for these conditions. The attractor is a closed curve: the limit cycle, when a system is periodic. Strange attractors exist in regimes exhibiting deterministic chaos. A stange attractor looks similar to a closed curve, staying in a roughly circular volume of phase space, but it never closes on itself. The time series recorded in all of these situations are, of course, the projection of the attractor on a single response axis.

2.3.2 The Phase Portrait

The reconstruction of the exact phase portrait of an attractor in chemical dynamics requires measurement of the concentrations of all of the intermediate species involved in a chemical reaction. An impossible task, of course, because their number, nature and properties are uncertain and usually not simultaneously measurable. Fortunately, a revisitation (popularized by Ruelle prior to its publication) by Takens[14] of a 1936 theorem,[15] makes it possible to reconstruct a phase portrait from the measured time series of only a single species or response. This reconstruction becomes possible only if the experimental data are digitized. The coordinate of a point in phase space is the original value $B(t)$ at time t and, either the successive time derivatives,[16] $\dot{B}(t)$, $\ddot{B}(t)$,... or the time-shifted values[14] $B(t + T)$, $B(t + 2T)$, $B(t + 3T)$,... where T is arbitrary for noiseless data. The first method was applied to chemical data[17] as early as 1980, but the latter method is preferred because the signal to noise ratio is the same for the coordinates of all points in a times series whereas it regularly decreases in the case of successively higher derivatives. There is an optimum choice[18] for the time delay T, typically one tenth to one half of the mean orbital period. However, there has been a recent discussion[19] of the choice of T.

Both methods show that a three-dimensional phase space is sufficient to reconstruct chaotic BZ trajectories, indicating that the chaos is of low topological dimension.[14] The first attractor published (Fig. 2.2) was obtained at a high flow rate

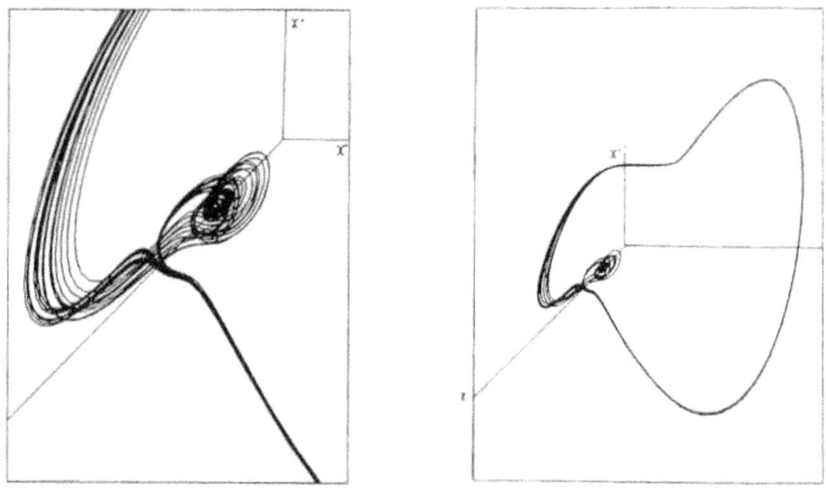

Figure 2.2. The first strange attractor obtained from the BZ reaction by the successive derivatives technique. The global view is on the left and an enlarged view of the central part of the strange attractor is shown on the right. The residence time is 5.60 minutes. Three feedstreams were used. Initial concentrations in the CSTR are: $[CH_2(COOH)_2]_0$ = 0.08 M, $[NaBrO_3]_0$ = 0.036 M, $[Ce_2(SO_4)_3]_0$ = 0.00025 M, and $[H_2SO_4]_0$ = 1.44 M. Temperature = 40 °C. Reproduced by permission from J.-C. Roux, A. Rossi, S. Bachelart and C. Vidal, *Phys. Lett.* 77A (1980) 391.

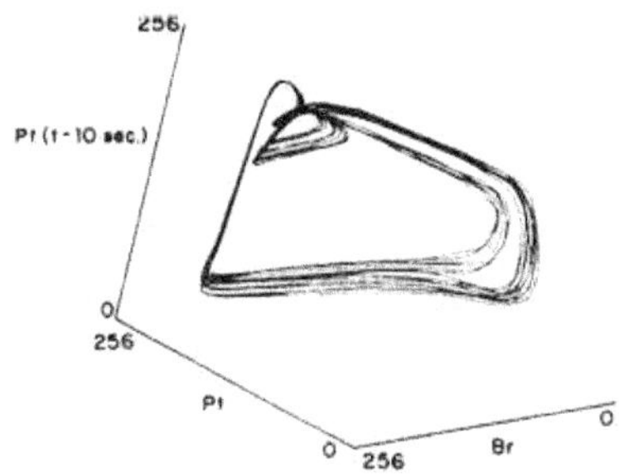

Figure 2.3. Stereoscopic plot of the strange attractor obtained by Hudson, *et al.* Variables are: Redox potential as measured at a Pt-electrode, the same signal 10 seconds later, and the potential of a Br^--selective electrode. The other conditions are the same as in Fig. 2.1 with a residence time of 6.01 minutes. Reproduced by permission from J.L. Hudson and J.C. Mankin, *J. Chem. Phys.* **74** (1981) 6171.

by the successive derivatives method. Similar attractors were independently constructed[20] from data obtained in the same flow-rate range. The 3D trajectories in the latter case were obtained from simultaneous measurements of two different responses, with the third coordinate being either the first derivative or the time-shifted value of one of these responses (Fig. 2.3).

2.3.3 *Poincaré Sections, Return maps, Lyapounov Exponents, and Fractal Dimension.*

These attractors were described as "strange" solely because of their shape, but the order in the trajectories was not clearly identified. This was done by construction of Poincaré sections independently in Virginia at high flow rates and in Texas at low flow rates.

A new periodic-chaotic sequence was discovered in Texas[21,22] on the low flow-rate side of P_{BZ}. At first sight, this (L)PCS seems to be a mirror-image of the (H)PCS obtained at high flow rates. However, as we shall see later, the (L)PCS is simpler than the (H)PCS in many aspects. The first chaotic state in the (L)PCS (referred to as C_u in the following) is illustrated in Fig. 2.4 and appears to be more chaotic than the corresponding (H)PCS state (Figs. 2.1-2.3). Actually, it has no equivalent at high flow rates, and the first chaotic regime described by Hudson *et al.*[20] corresponds to the second chaotic state of this (L)PCS. Furthermore, chaotic dynamics can be observed over a wider range of parameter values at low flow rates. A projection of the corresponding attractor obtained by the time-delay method is shown in Fig. 2.5. Its shape confirms again that BZ chaos is of low topological dimension.

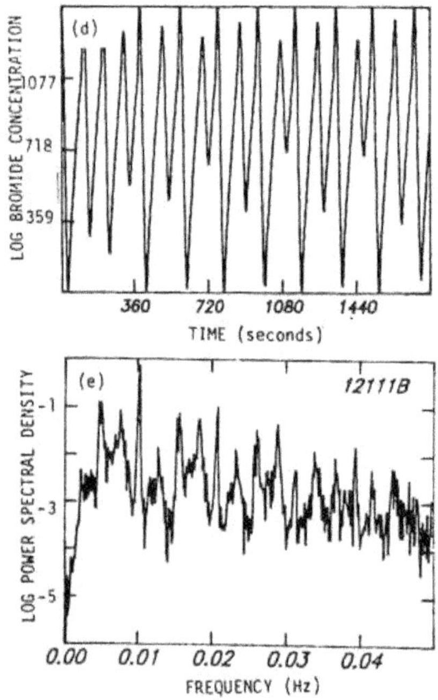

Figure 2.4. Time series and power-spectral density of a chaotic regime obtained at a low flow rate in Texas. Initial concentrations in the CSTR are: $[CH_2(COOH)_2]_0$ = 0.25 M, $[NaBrO_3]_0$ = 0.14 M, $[Ce_2(SO_4)_3]_0$ = 0.00083 M, and $[H_2SO_4]_0$ = 0.2 M. Residence time = 54 minutes. Temperature = 28.3 °C. Reproduced by permission from: J.S. Turner, J.-C. Roux, W.D. McCormick and H.L. Swinney, *Phys. Lett.* **85A** (1981) 9.

We shall illustrate with this attractor the construction of a Poincaré section formed by the intersections of the 3D phase-space orbits with a plane that passes through the dashed line in Fig. 2.5 normal to the figure. This section is not a

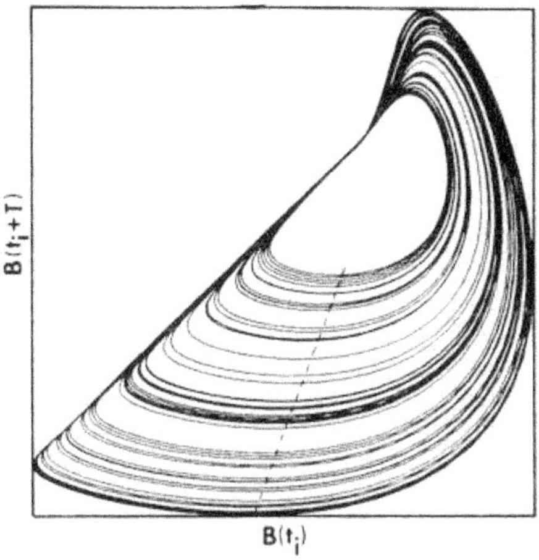

Figure 2.5. The strange attractor, C_u, obtained in Texas. $B(t_i)$ is $\log_{10}$ [Br$^-$]. The conditions are the same as in Fig. 2.4. Reproduced by permission from: J.S. Turner, J.-C. Roux, W.D. McCormick and H.L. Swinney, *Phys. Lett.* **85A** (1981) 9.

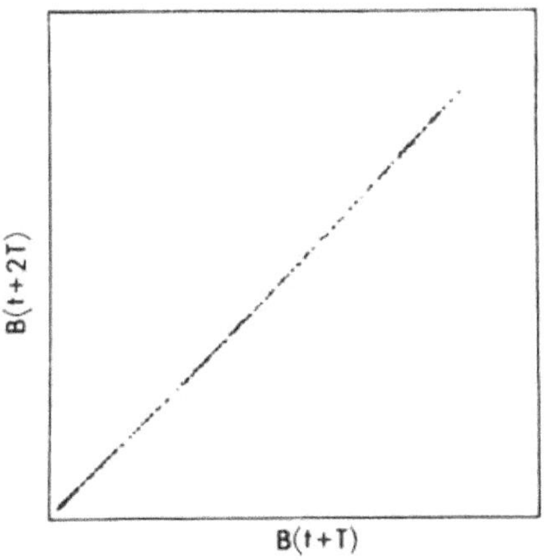

Figure 2.6. The Poincaré section of C_u. The intersecting plane passes through the dashed line in Figure 2.5. Reproduced by permission from: J.-C. Roux, R.H. Simoyi and H.L. Swinney, *Physica.* **8D** (1983) 257.

stroboscopic analysis of the attractor because the time between successive intersections is variable. The points on the Poincaré section shown in Figure 2.6 lie to a good approximation on a smooth curve. This demonstrates that there is order in the orbits because they lie approximately on a two-dimensional sheet. Further analysis of the chronology of successive intersections of the trajectories with the Poincaré section reveals an even deeper degree of order. This can be done by plotting the position of one intersection (the curvilinear abscissa, or something proportional to it, along the Poincaré section) as a function of the position of the previous intersection. This construction is shown in Fig. 2.7 and is called the (first) return map or sometimes the 1D-map. (A similar 1D-map was obtained independently at high flow rates[20]). The representative points in this construction are not scatterred as would be expected for a stochastic process, but define a smooth curve. This curve, if single-valued, in turn, allows us to **predict** the position of the next intersection, X_{n+1}, if we know the previous one, X_n. **Thus the system is deterministic.**

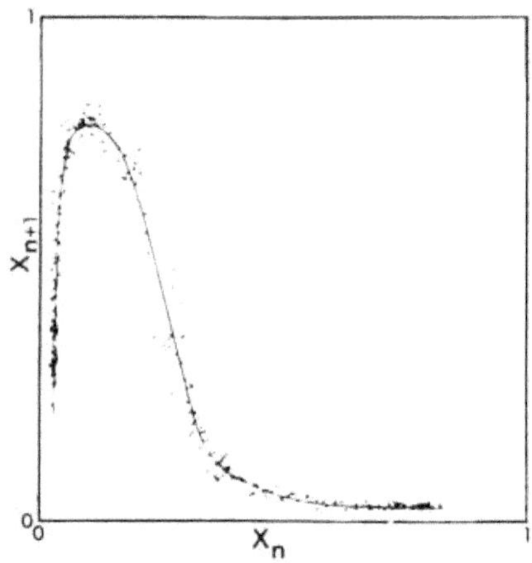

Figure 2.7. The first-return map for the Poincaré section shown in Fig. 2.6. Reproduced by permission from: J.-C. Roux, R.H. Simoyi and H.L. Swinney, *Physica* **8D** (1983) 257.

However it also is chaotic: long term prediction of the behavior of the system is impossible! This property is connected to a very sensitive dependence on initial conditions; two trajectories starting from arbitrarily close neighboring points will diverge on the average, and their relative positions are unpredictable after sufficient time has elapsed. This sensitivity to initial conditions and divergence of trajectories can be illustrated in the chaotic regime C_u (Fig. 2.8) by calculation of the Lyapounov exponent. Trajectories diverge if their Lyapounov exponent is positive. The Lyapounov exponent for the C_u trajectory in Fig. 2.8[23] is 0.3 ± 0.01,[24] and that for the attractor shown Fig. 2.3 is 0.62. The calculation of the Lyapounov exponent from experimental data has been refined[25] since these calculations were made, but its use to characterize a chaotic state or even to discriminate between chaotic and periodic dynamics cannot be generalized because of the inherent stochastic noise present in experimental data.

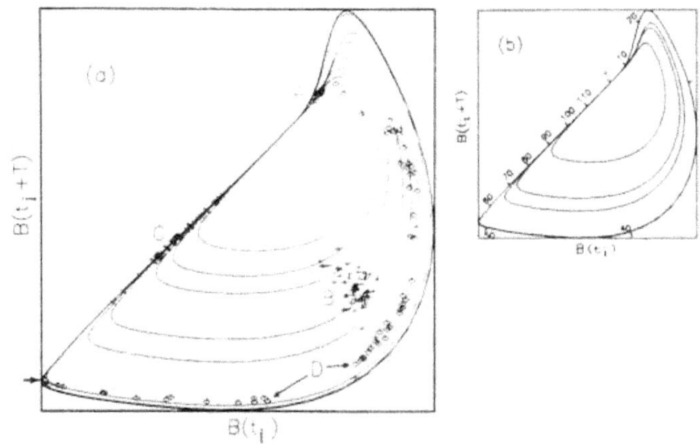

Figure 2.8. a) Sensitivity to initial conditions illustrated by the evolution of a set of points that were initially all within a very small volume located at the tip of the arrow on the left. At $\Delta T = 70.4$ s they are still grouped in A. However, at $\Delta T = 92.4$ s they are begining to disperse, as is seen at B. They are further dispersed in C at $\Delta T = 123.2$ s, and at $\Delta = 184.8$ s they are widely dispersed in D. Same conditions as in Fig. 2.4. b) Time scale on the sttractor. Reproduced by permission from: J.-C. Roux and H.L Swinney in *Nonlinear Phenomena in Chemical Dynamics*, C. Vidal and A. Pacault, Eds. (Springer-Verlag, Berlin, 1981).

However, the exponential divergence of two trajectories in a chemical reaction cannot continue forever because chemical concentrations can never become infinite. Therefore, the attractor must have many folds. This means that the curve representing the Poincaré section is not a "line" but must present some thickness, which may be impossible to resolve in either experimental or calculated data. However, these successive foldings were demonstrated in the case of C_u.[24]

This folding results in a non-integer fractal dimension (Hausdorff, information or correlation). It is possible to construct a measure of the fractal dimension from experimental data, but again, just as with the Lyapounov exponent, it is difficult to interpret because of experimental noise.[26]

However, at this point we can state that the curves in Figs. 2.3 and 2.5 are indeed strange. It also was shown that they are attractive; a trajectory ejected from them by a perturbation quickly returns[27] to the vicinity of the attractor (Fig. 2.9).

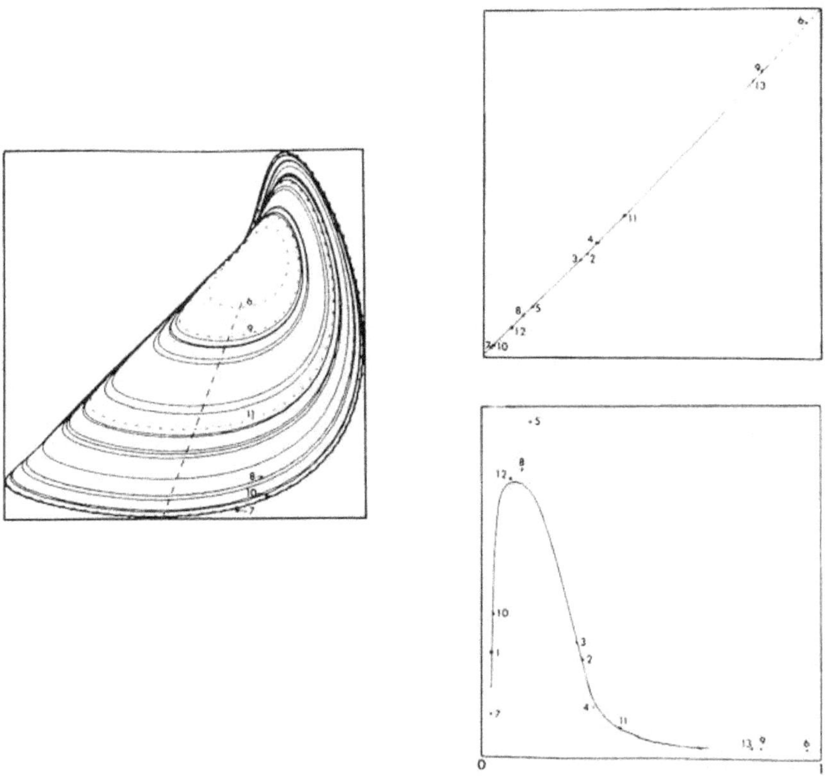

Figure 2.9. The effect of an external perturbation on an experimental trajectory viewed on the phase portrait, the Poincaré section, and the first-return map. The trajectory returns to the attractor in a few cycles. Various perturbations were used,[24] including injection of bubbles or Br^- into the CSTR, and turning off the stirrer or one of the chemical feeds for a few seconds. Reproduced by permission from H.L Swinney and J.-C. Roux in *Nonequilbrium Dynamics in Chemical Systems*, C. Vidal and A. Pacault, Eds. (Springer-Verlag, Berlin, 1984).

2.3.4 *Universal Dynamics, Part 1.*

One-hump 1D-maps similar to that shown in Fig. 2.7 reoccur in a number of physical situations, thus confirming our ideas of the universality of low-dimensional chaotic dynamics in the real word. These maps share many properties of the widely studied logistic map.[28] For instance, a continuous deformation of their shape gives birth to a succession of exotic dynamics, and most of these are, in fact, observed in the BZ reaction:

(1) The periodic dynamics preceding C_u must lose stability through the well-known period-doubling sequence,[28] and this sequence is actually observed (Fig. 2.10)[29,30,31] when the normal, periodic oscillating state bifurcates toward the

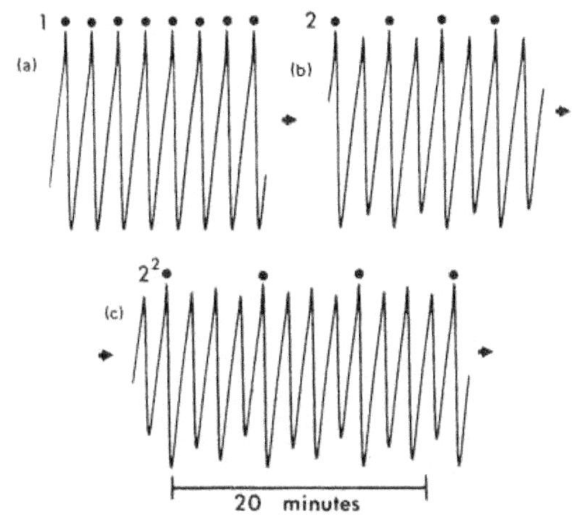

Figure 2.10. A period-doubling sequence in the BZ reaction. The trace is the potential of a Br^--selective electrode under the same conditions as in Fig. 2.4. The residence times are, respectively: period 1 - 43.5 min, period 2 - 46.38 min, period 4 - 48.18 min. Reproduced by permission from K.G. Coffman, W.D. McCormick, Z. Noszticzius, R.H. Simoyi and H.L. Swinney, *J. Chem. Phys.* **86**, (1987) 119.

chaotic state. Actually, only a few steps of this cascade are observed experimentally, which is not surprising considering the well-known rapid convergence of the period-doubling cascade; thus no attempt was made to measure the Feigenbaum constant of 4.669.[32] As we shall see later, C_u is bracketed by two simple (but different) periodic states that both bifurcate toward chaos through a period-doubling cascade.

(2) Periodic regimes (windows) are highly probable even within the chaotic regime of the logistic map. Although the waveforms of these periodic regimes can be very complex, the number of extremas, their relative height, and even their chronology in a periodic waveform are predicted by theory[33] on a very general basis. Furthermore, the order of appearance of these periodic windows also is predicted by the theory, and thus this order is called the Universal (U)-sequence. The U-sequence was observed in the C_u chaotic state.[29] The theory was later refined[34] and it was shown that this sequence could cease to be single-valued if we catch more of the detail of the 1D map; the same periodic regime could be reached for three different flow rates. This multiplicity has been demonstrated in the BZ reaction.[35]

2.4 Specifics of the Chaotic Dynamics Observed in the BZ Reaction.

2.4.1 Periodic-Chaotic Sequences: Notation

We have already mentioned the existence of two periodic-chaotic sequences (PCS) in the dynamics of the BZ reaction. One, the (L)PCS, is observed on the low flow-rate side of P_{BZ}, and the other, the (H)PCS is observed on the high flow-rate side. Both sequences are roughly symmetric, as indicated below.

We introduce a notation useful for describing a PCS. A periodic regime is termed P_L^S where L is the number of large oscillations and S the number of small oscillation per cycle. For example, the periodic regimes in Figs. 2.1-2 and 2.1-3 are labelled P_2^1 and P_1^1, respectively, and, P_{BZ} is labelled P_1^0. A chaotic state, as we shall see later, often looks like a mixture of the bracketing periodic regimes P_L^S and $P_{L_2}^{S_2}$ and is thus labelled $C_{L_1, L_2}^{S_1, S_2}$. C_u becomes $C_1^{0,1}$ and the chaotic state of Fig. 2.3 becomes $C_1^{1,2}$. We can now describe the experimentally observed BZ (H)PCS and (L)PCS.

(1) The (H)PCS sequence observed by Hudson *et al.*[6,7,8,20] at high flow rates is:

----------------------flow rate increase ------------->

$$P_1^0, P_2^1, P_1^1, C_1^{1,2}, P_1^2, C_1^{2,3}, P_1^3, C_1^{3,4}, P_1^4, C_1^n, P_0^1, \text{ steady state}$$

Other (H)PCS were observed in Bordeaux.[36]

$$P_1^0, C_m^1 \ (m>3), P_3^1, C_{3,2}^1, P_2^1, \ldots, C_2^{1,2}, P_2^2,$$

$$C_2^{2,3}, P_2^3, C_2^{3,4}, P_2^4, C_2^{4,5}, P_2^5, \ldots, C_0^m, \text{ steady state}$$

For slightly different contraint values[37] one observes:

$$P_1^0, C_{n,\ldots,n}^1 \ (n \geq 4), C_{3,4}^1, P_3^1, C_3^{1,2}, P_3^2, P_3^3, \ldots, C_3^{4,5}, P_2^1, C_2^{1,2}, P_2^2, P_2^3, \ldots,$$

$$C_2^n \ (n \approx 10), C_{1,2}^n, C_1^n, \text{ steady state}$$

However, Maselko and Swinney[38,39] used a different catalyst and observed a succession of very complicated periodic regimes that obeyed Farey arithmetic, but no chaotic states were observed.

(2) The low-flow-rate sequence shown below was initially observed in Texas and has been reproduced year after year by different experimentalists.[21,22]

------------------flow rate decrease---------------------->

$$P_1^0, C_1^{0,1}, P_1^1, C_1^{1,2}, P_1^2, C_1^{2,3}, \ldots, C_r, C_0^n, \ldots, P_0^1, \text{ steady state}$$

The C_r dynamics will be explained later (See Fig. 2.12). A similar sequence was observed in Japan.[40]

(3) It also has been shown[41] that for the same set of contraint values one can

move by continuous variation of the flow rate from a (L)PCS through P_{BZ} to a (H)PCS. The (H)PCS observed is comparable to the previous observations, whereas the (L)PCS is shorter (less rich) than the Texas (L)PCS observations. We must be aware that these sequences have nothing to do with the previously mentioned U-Sequence. The latter is observed inside the $C_1^{0,1}$ regime and is understood as the result of a continuous deformation of a one-hump, 1D-map. On the other hand, the $C_1^{1,2}$ chaotic regime, for instance, **cannot** be anticipated from such a deformation. In fact, the 1D-map corresponding to this regime shown in Fig. 2.11 is multivalued[30] and thus definitely different from the map shown Fig. 2.7.

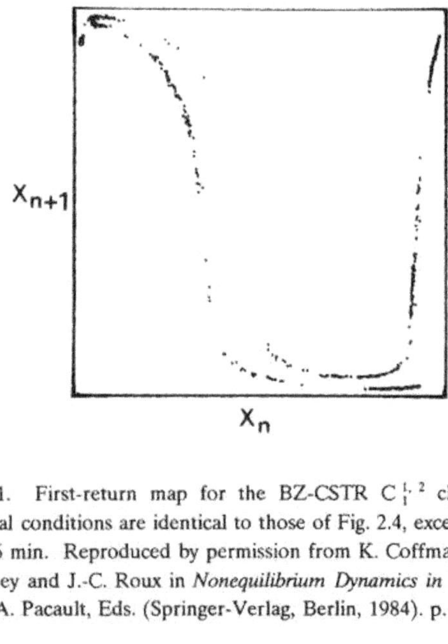

Figure 2.11. First-return map for the BZ-CSTR $C_1^{1,2}$ chaotic regime. The experimental conditions are identical to those of Fig. 2.4, except that the residence time is 66.6 min. Reproduced by permission from K. Coffman, W.D. McCormick, H.L. Swinney and J.-C. Roux in *Nonequilibrium Dynamics in Chemical Systems*, C. Vidal and A. Pacault, Eds. (Springer-Verlag, Berlin, 1984). p. 251.

2.4.2 *Periodic-Chaotic Sequences: Discussion*

As emphasized by our notation, nearly all chaotic regimes in a PCS look like a stochastic mixture of the preceding and following periodic regimes. Thus it often has been suggested that the chaotic regimes are simply the consequence of random fluctuations in the flow rate or in some other constraints which drive the system stochastically from one periodic state to the other. This assertion is reinforced by the following facts:

(a) Some chaotic windows occur in a very narrow flow-rate range that is about the size of unavoidable fluctuations in pumping speed. This is especially true for some windows in the (H)PCS.

(b) The (L)PCS is very sensitive to both impurities and to CSTR stirring rate. The entire (L)PCS disappears when sufficiently impure malonic acid[42] is used. Furthermore, a change in stirring rate may suppress a chaotic state[26,37] or transform a periodic regime into a chaotic regime.[37]

(c) Some studies in the high-flow rate domain[38,39] show sequences of only periodic states, their succession obeying Farey arithmetic. Attempts to find a chaotic state between two periodic ones only succeeds in generating new periodic states in the Farey tree. In any case, the limited time one can observe a state always leaves the possibility that an apparently chaotic state is actually a periodic state of very long period.

(d) The initial attempts to simulate BZ chaos on the basis of chemically realistic models of the BZ reaction showed only periodic regimes similar to the complex periodic state observed experimentally, but with no chaotic regimes between them.

All of these facts raise the question of the source of chaos in the BZ reaction. The question was asked, "Isn't it simply the result of **imperfect** mixing,[43] **imperfect** purification of the reactants, **imperfect** control of the flow rate, *etc*.?" Indeed, Györgyi and Field[43] showed that PCSs similar to those observed experimentally could be simulated on the basis of a non-chaotic model of the BZ chemistry, but considering incomplete mixing in the CSTR. These doubts were perfectly expressed by R.M Noyes:[44] "...There certainly are experimental systems which exhibit chaotic behavior in spite of heroic measures to control all recognized parameters. Such behavior does not prove that chaos is inherent in the mechanism itself rather than due to unavoidable stochastic fuctuations". The same ideas were later developed by Gray and Scott:[45] "..It is still an open question whether chaos arising from a homogeneous chemical mechanism has been obtained experimentally or whether it comes from the imperfect control of external features."

The last argument is no longer valid because a new version of the mechanism of the BZ reaction[46] is able to reproduce some of the chaotic states observed at both high and low flow rates, as well as most features of the (L)PCS. Actually, all these arguments concerned only the (H)PCS. It was clear that the (L)PCS could not result from imperfect experimental control because the first chaotic window, $C_1^{0,1}$, is very large compared to any experimental fluctuation. This fact, and the demonstrations of its universal properties decribed below, convinced most skeptics.

However, at high flow-rates the first three criticisms still are valid, and recent experiments[26] seem to indicate that the observed high flow-rate chaos may be caused by noise-driven transitions between neighboring orbits. With the exception of very recent[46] experiments and simulations performed in Texas and Montana, respectively, which succeeded in reproducing most of the features of the original (H)PCS observed by Hudson *et al.*,[6-8,20] the defenders of (H)PCS chaos usually do not answer directly to the above criticisms. Even though they are convinced that the experimental difficulties have been overcome, they prefer to argue instead on the basis of dynamical systems theory.

The limited width (if any) of the experimental chaotic windows is not disturbing when viewed within the context of the known behavior of dynamical systems. It is conjectured[28] that for the logistic map in the chaotic domain: (i) the set of values of the bifurcation parameter for which there is no stable periodic orbit has positive Lebesque measure, and (ii) that this set contains no intervals. If one recognizes that $C_1^{0,1}$ has a 1D-map very similar in shape to that of the of logistic map, then it may be concluded: (i) for a flow rate chosen arbitrarily in the domain of existence of $C_1^{0,1}$ there is a non-zero probability of finding a chaotic orbit, and (ii) a periodic orbit may be observed for flow rates only slightly different. It is not unreasonable that similar conclusions may hold for the complete chaotic sequence, and as defenders of chaos, "we are not necessarily deceived by a very long, but still periodic cycle."[47]

2.4.3 *Periodic-Chaotic Sequences: Universal Dynamics (Part 2): Routes to Chaos.*

Deterministic chaotic states occur in theoretical dynamical systems as the result of a bifurcation or even a cascade of bifurcations as a parameter is varied. The identification of such bifurcations in the BZ reaction puts the existence of deterministic BZ-chaos on a firmer base. These bifurcations are often called the routes to chaos, and we describe below those observed in the BZ reaction.

(1) The low-flow-rate domain. As already has been mentioned, the Feigenbaum period-doubling cascade has been unambiguously identified in the BZ

reaction. Type-I intermittency also was identified very early,[48] and more recently the type III intermittency (subharmonic intermittency) route has been identified when methanol[49] is added to the BZ reaction. The same intermittency route also has been observed at high flow rates.[50]

The final steps of the (L)PCS at very low flow rate are displayed in Fig. 2.12.[51] The C_r regime, which does not enter into our simplified labelling scheme, was shown to be a typical result of a well-known phenomenon in dynamical systems referred to as a "crisis." It occurs when the basins of attraction of a C_0^n and a $C_1^{2,3}$ chaotic attractor touch.

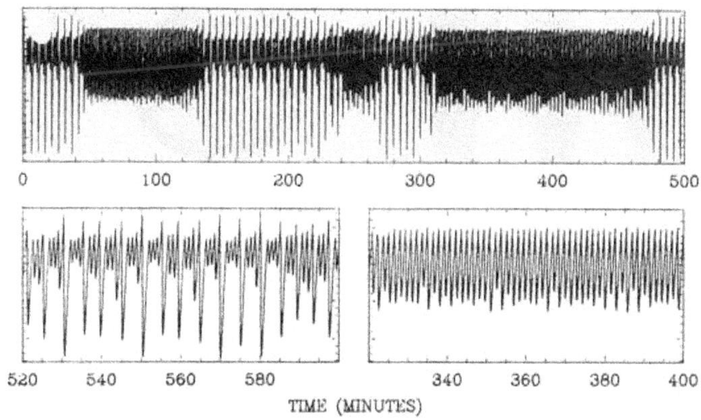

Figure 2.12. Time series of the final steps of the (L)PCS. Top = C_r time series. Bottom Left = $C_1^{2,3}$ time series. Bottom Right = C_0^n time series. Reproduced by permission from P. Richetti, P. De Kepper, J.-C. Roux and H.L. Swinney, *J. Stat. Phys.* **48** (1987) 977.

(2) The high-flow-rate domain. One of the routes to chaos recognized first is the breaking up of a torus. We first must find evidence of two-frequency dynamics in order to identify this route in the evolution of BZ chaos. Such a dynamics can be reached after a secondary Hopf bifurcation of a limit cycle. This bifurcation **has been** observed in the high-flow-rate domain of the BZ reaction, and is shown in Fig. 2.13.[52] The shape of this torus deserves a few comments. The large- and small-amplitude oscillations correspond, respectively, to trajectories on the outer part and the central hole of the torus. This hole becomes smaller and smaller as the flow rate increases, while the dynamics spend more and more time in it. Finally, in this experiment, the

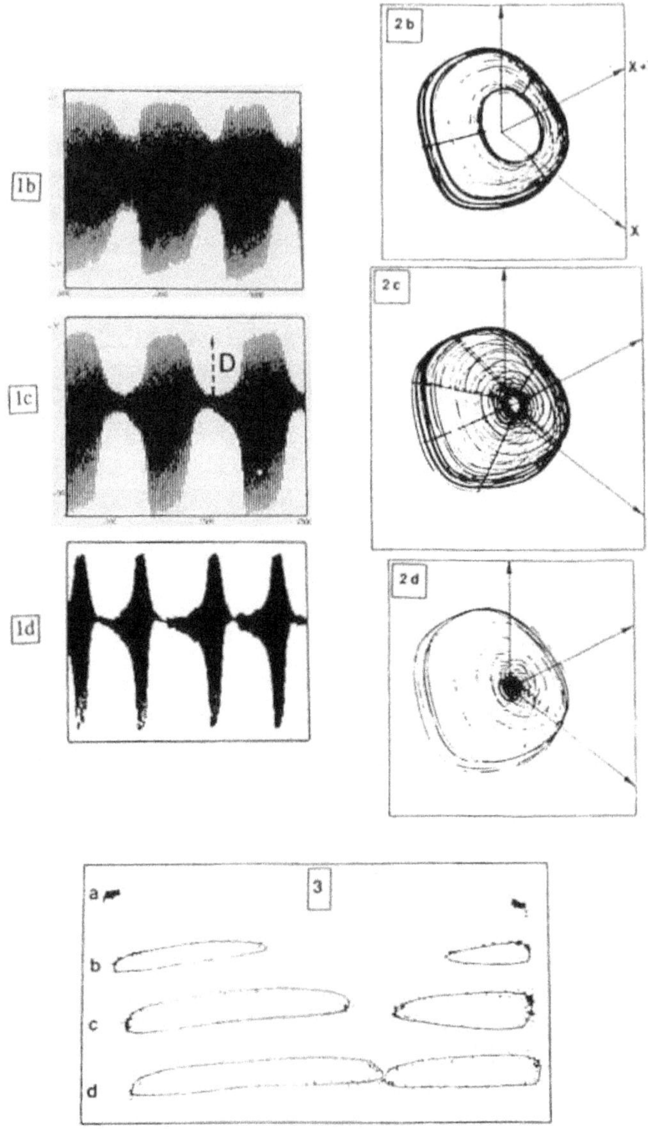

Figure 2.13. A secondary Hopf bifurcation. 1) Time Series. 2) Attractors. 3) Evolution of the Poincaré section. Flow Rates (mL/min) are: (a) 0.34, (b) 0.355, (c) 0.36, and (d) 0.43. The initial concentrations in the CSTR are: $[CH_2(COOH)_2]$ = 0.19 M, $[NaBrO_3]$ = 0.012 M, $[Ce_2(SO_4)_3]_0$ = 0.00025 M and $[H_2SO_4]_0$ = 1.0 M. T = 43 °C. Reproduced by permission from F. Argoul and J.-C. Roux, *Phys. Lett.* **108A** (1985) 426.

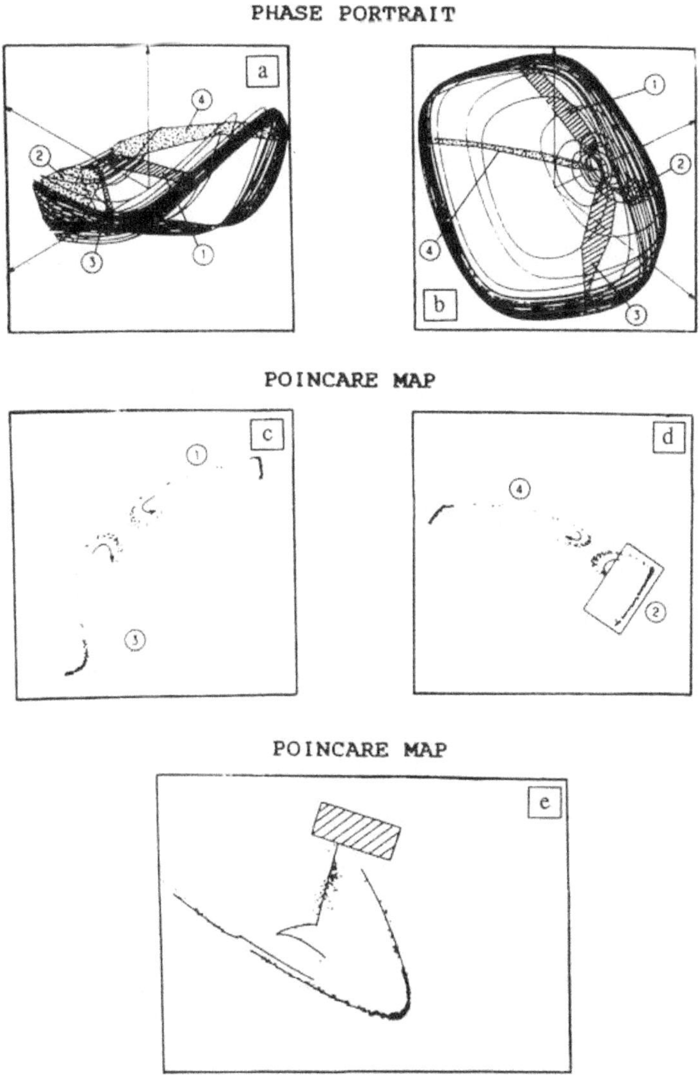

Figure 2.14. Two views (a and b) of the fractal torus observed in the vicinity of the torus shown in Fig. 2.13. The Poincaré sections (c and d) corresponding to the plane labelled 1, 2, 3, 4 in (a) and (b). e) Enlarged view of section (2) in (d) showing the folds. The initial concentrations in the CSTR are: $[CH_2(COOH)_2]_0$ = 0.18 M, $[NaBrO_3]$ = 0.012 M, $[Ce_2(SO_4)_3]_0$ = 0.00025 M, and $[H_2SO_4]_0$ = 1.0 M. T = 40 °C. Residence time = 10 min. Reproduced by permission from J.-C. Roux, *Physica* **7D** (1983) 57.

dynamics ends at a fixed point lying on the symmetry axis of the torus. No chaos was observed. However, this experiment does demonstrate that two-frequency dynamics do occur in the BZ reaction. The chemical source of the two frequencies is discussed by Györgyi and Field in Chapter 3.

The shape of the torus changes dramatically[53,54] (Fig. 2.14) for slightly different chemical parameters, although some general features are conserved. A closer analysis of the Poincaré sections shows a fractal structure which emphasizes the landmark folding and stretching of the trajectories. **The quasiperiodic route to chaos,** very well known in other nonlinear dynamical systems such as hydrodynamics, **is thus also present in the BZ reaction.**

The succession of frequency-locked periodic states observed by Maselko and Swinney[38,39] and others,[26] is in fact characteristic of trajectories on the surface of a torus. Furthermore, we know from model simulations[55] that the ordering of these frequency-locked trajectories forms a "devil staircase" and that chaotic states do exist between periodic states. However, they occur over such a small domain that experimental observation of them is impossible unless stochastic fluctuations in the flow rate drive the dynamics from neighboring periodic states into the chaotic regime. Indeed, here, the argument about the effects of external fluctuations is reversed; stochastic fluctuations may allow observation of deterministic chaos that otherwise would be unobservable.

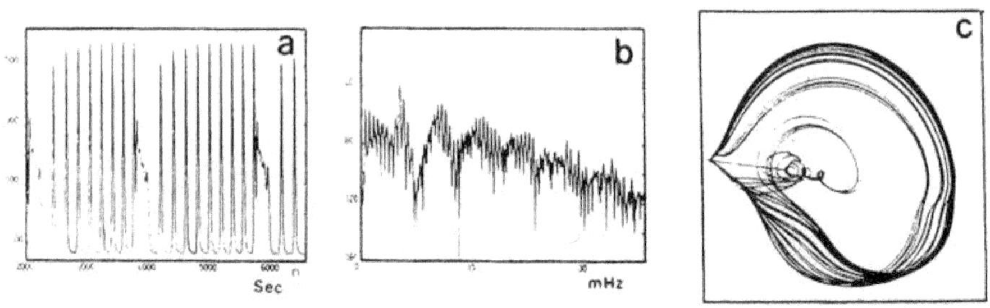

Figure 2.15. Time series, power spectrum and attractor of a $C^4_{7,8,9}$ chaotic state in the BZ (H)PCS. Initial concentrations in the reactor are: $[CH_2(COOH)_2]_0$ = 0.18 M, $[NaBrO_3]$ = 0.0025 M, $[Ce_2(SO_4)_3]_0$ = 0.00025 M, and $[H_2SO_4]_0$ = 1.0 M. T = 43 °C. Residence time = 30 min. Reproduced by permission from F. Argoul, A. Arneodo, P. Richetti, J.-C. Roux, *J. Chem. Phys.* **86** (1987) 3325.

With the quasiperiodic route to chaos in mind, the totality of the (H)PCS is understood as a variety of dynamics on a fractal torus. This torus is easy to distinguish in phase portraits of chaotic states such as is shown in Fig. 2.15. The general features of the torus remain even though it is partly broken. Trajectories stay for a while on the exterior of the torus, and then visit the center hole via the two reverse spirals that are clear in the figure. The sensitive dependence on initial conditions seems to develop near two points that are near the two ends of the inner tube, where the trajectories are more scattered.

2.5 Toward a Global Understanding

These experimental hints, as well as arguments from the theory of dynamical systems, allow us to put forward a general understanding of the complex dynamics of the BZ reaction in the high-flow-rate range. This is not the place to develop these arguments. The interested reader will find the major details elsewhere.[56,57] We shall simply underline here the main assumptions and results of this approach.

The main instabilities in the BZ reaction in a CSTR are the Hopf bifurcation (one of the generic routes to oscillatory behavior observed in the BZ reaction), and the hysteresis bifurcation that is responsible for the bistability observed in many open chemical systems[58]). The simplest nonlinear equations that describe the interaction of these instabilities are the "normal form," a set of three ordinary differential equations containing arbitrary parameters. These equations are valid only in the vicinity of a surface where both instabilities occur simultaneously. However, a variety of behaviors can be inferred by continuity from the study of numerical simulations based on this normal form.

It has been shown[55] that the result of interaction of these two instabilities is that the oscillatory regime may undergo a secondary Hopf bifurcation to yield a torus very reminiscent to that observed experimentally[52] (Fig. 2.13). Furthermore, it also has been shown[55] that a continuous change in the adjustable parameters may lead to a fractal torus similar to that observed experimentally.[54] Simulations at some distance from the polycritical surface show a periodic-chaotic sequence reminiscent of the (H)PCS described above.

Further studies of chaos in the BZ reaction following the same normal-form approach seem to indicate that the periodic-chaotic sequence observed at low flow rates {(L)PCS} could correspond to a scenario close to homoclinic conditions.[59] Orbits and even multivalued 1D-maps similar to those of the (L)PCS are predicted on the way to homoclinicity by the normal-form studies. Furthermore, a characteristic "homoclinic chaos" has been observed experimentally in the high flow-rate domain.[60,61] Finally, it has been shown[61] that the nonlinear interaction of three elementary instabilities seems to be sufficient to describe the periodic-chaotic sequences observed at high **and** low flow rates.

2.6 Conclusions

Chemical dynamics are described by systems of often **nonlinear** ordinary differential equations. It is thus not surprising that deterministic chaos, a characteristic behavior of such dynamical systems, is observed in the dynamics of the BZ reaction. The interpretation of these observations as examples of deterministic chaos is strongly supported by the theory of dynamical systems. The succession of periodic and chaotic states, the shape of the chaotic and periodic orbits, and the transitions between them when chaos occurs in the BZ reaction are reproduced perfectly by **simple**, paradigmatic dynamical systems.

The variety of dynamical behaviors observed in the BZ reaction makes this experimental system a good standard, if not **the** prototype for illustration of the properties of new area of unpredictable but still deterministic dynamics.

Aknowledgement. I deeply thank Dr. P. De Kepper for a critical reading of this manuscript.

Literature Cited

1. D. Ruelle, *Trans. N.Y. Acad. Sci.* **35** (1973) 66.
2. I.R. Epstein, *Physica* **7B** (1983) 47.
3. P. De Kepper, A. Rossi and A. Pacault, *C.R. Acad. Sci.* (Paris) **C283** (1976) 371.
4. P.G. Sorensen, *Faraday Symp. Chem. Soc.* **9** (1974) 88.
5. P.G. Sorensen, *Ann. N.Y. Acad. Sci.* **316** (1979) 667.
6. K.R. Graziani, J.L. Hudson and R.A. Schmitz, *Chem. Eng. J.* **12** (1976) 9.
7. R.A. Schmitz, K.R. Graziani and J.L. Hudson, *J. Chem. Phys.* **67** (1977) 3040.
8. J.L. Hudson, M. Hart and D. Marinko, *J. Chem. Phys.* **71** (1979) 1601.
9. O.E. Rossler and K. Wegmann, *Nature* (London) **271** (1978) 89.
10. K. Wegmann and O.E. Rossler, *Z. Naturforsch*, **A33** (1978) 1179.
11. J. Maselko, *Chem. Phys. Lettr.* **73** (1980) 194.
12. C. Vidal, J.-C Roux, A. Rossi and S. Bachelart, *C.R. Acad. Sci.* (Paris) **C289** (1979) 73.
13. C. Vidal, J.-C. Roux, S. Bachelart and A. Rossi, *Ann. N.Y. Acad. Sci.* **357** (1980) 377.

14. F. Takens, *Lect. Notes Math.* **898** (1981) 366.
15. H. Whitney, *Ann. Math.* **37** (1936) 645.
16. N.H. Packard, J.P. Crutchfield, J.D. Farmer and R.S. Shaw, *Phys. Rev. Lett.* **45** (1980) 712.
17. J.-C. Roux, A. Rossi, S. Bachelart, C. Vidal, *Phys. Lett.* **77A** (1980) 391.
18. A.M. Fraser, H.L. Swinney, *Phys. Rev.* **33B** (1986) 1134.
19. A.M. Albano, A. Passamante and M.E. Farrel, *Physica* **54D** (1991) 85.
20. J.L. Hudson and J.C. Mankin, *J. Chem. Phys.* **74** (1981) 6171.
21. J.S. Turner, J.-C. Roux, W.D. McCormick, H.L. Swinney, *Phys. Lett.* **85A** (1981) 9
22. J.-C. Roux, J.S. Turner, W.D. McCormick, H.L. Swinney, in *Nonlinear Problems: Present and Future*, A.R. Bishop, D.K. Campbell and B. Nicolaenko, Eds. (North Holland Press, Amsterdam, 1982) p. 409.
23. J.-C. Roux and H.L. Swinney, in *Nonlinear Phenomena in Chemical Dynamics*, C. Vidal and A. Pacault, Eds. (Springer-Verlag, Berlin, 1981) p. 38.
24. J.-C. Roux, R.H. Simoyi and H.L. Swinney, *Physica* **8D** (1983) 257.
25. A. Wolf, J.B. Swift, H.L. Swinney and J.A. Vastano, *Physica* **16D** (1985) 285.
26. F.W. Schneider and A.F. Münster, *J. Phys. Chem.* **91** (1991) 2130.
27. H.L. Swinney and J.C. Roux, in *Nonequilibrium Dynamics in Chemical Systems*, C. Vidal and A. Pacault, Eds. (Springer-Verlag, Berlin, 1984) p. 124.
28. P. Collet and J.P. Eckmann, *Iterated Maps of the Interval as Dynamical Systems*, (Birkäuser, Basel, 1980).
29. R.H. Simoyi, A. Wolf and H.L. Swinney, *Phys. Rev. Lett.* **49** (1982) 245.
30. K. Coffman, W.D. McCormick, H.L. Swinney and J.-C. Roux in *Nonequilibrium Dynamics in Chemical Systems*, C. Vidal and A. Pacault, Eds, (Springer-Verlag, Berlin, 1984) p. 251.
31. K.G. Coffman, W.D. McCormick, Z. Noszticzius, R.H. Simoyi and H.L. Swinney, *J. Chem. Phys.* **86** (1987) 119.
32. M.J. Feigenbaum, *J. Stat. Phys.* **19** (1978) 25.
33. N. Metropolis, M.L. Stein and P.R. Stein, *J. Combinatorial Theory Ser.* **A15** (1973) 25.
34. W.A. Beyer, R.D. Mauldin and P.R. Stein, *J. Math. Anal. Appl.* **115** (1986) 305.
35. K. Coffman, W.D. McCormick and H.L. Swinney, *Phys. Rev. Lett.* **56** (1986) 999.
36. C. Vidal, in *Nonequilibrium Dynamics in Chemical Systems*, C. Vidal and A. Pacault, Eds. (Springer-Verlag, Berlin, 1984) p. 49.
37. F. Argoul, A. Arneodo, P. Richetti, J.-C. Roux, *J. Chem. Phys.* **86** (1987) 3325.
38. J. Maselko and H.L. Swinney, *Physica Scr.* **T9** (1985) 35.
39. J. Maselko and H.L. Swinney, *J. Chem. Phys.* **85** (1986) 6430.
40. M. Hourai, Y. Kotake, K. Kuwata, *J. Phys. Chem.* **89** (1985) 1760.
41. P. Ibison, S.K. Scott, *J. Chem. Soc. Faraday Trans.* **86** (22) (1990) 3695.
42. Z. Noszticzius, W.D. McCormick and H.L. Swinney, *J. Phys. Chem.* **91** (1987) 5129.
43. L. Györgyi and R.J. Field, *J. Phys. Chem.* **92** (1988) 7079; *J. Chem. Phys.* **91** (1989) 2159.
44. R.M. Noyes. in *Stochastic Phenomena and Chaotic Behavior in Complex Systems*, P. Schuster, Ed. (Springer, Berlin, 1984) p. 106.
45. P. Gray and S.K. Scott, *J. Phys. Chem.* **89** (1985) 22.
46. L. Györgyi, S.L. Rempe and R.J. Field, *J. Phys. Chem.* **95** (1991) 3159; L. Györgyi and R.J. Field, Z. Noszticsius, W.D. McCormick and H.L. Swinney, *J. Phys. Chem.* **96** (1992) 1228; L. Györgyi and R.J. Field, *Nature* **355** (1992) 808.
47. R.V. Jensen, *American Scientist*, **75** (1987) 168.

48. Y. Pomeau, J.-C. Roux, A. Rossi, S. Bachelart and C. Vidal, *J. Phys. (Paris) Lett.* **L271** (1981) 42.
49. N. Kreisberg, W.D. McCormick and H.L. Swinney, *Physica,* **50D** (1991) 463.
50. G. Baier, K. Wegmann and J.L. Hudson, *Phys. Lett.* **141A** (1989) 340.
51. P. Richetti, P. De Kepper, J.C. Roux and H.L. Swinney, *J. Stat. Phys.* **48** (1987) 977.
52. F. Argoul and J.-C. Roux, *Phys. Lett.* **108A** (1985) 426.
53. J.-C. Roux and A. Rossi, in *Nonequilibrium Dynamics in Chemical Systems*, C. Vidal and A. Pacault, Eds. (Springer-Verlag, Berlin, 1984) p. 141.
54. J.C. Roux, *Physica*, **7D** (1983) 57.
55. P. Richetti, J.-C. Roux, F. Argoul, A. Arneodo, *J. Chem. Phys.* **86** (1987) 3339.
56. W.F. Langford, in *Nonlinear Dynamics and turbulence,"* G. Iooss and D.D. Joseph, Eds. (Pitman, New York, 1982).
57. W.F. Langford in *International Series of Numerical Mathematics #70*, T. Kupper, H.D. Mittelmann and H. Weber, Eds, (Birkhäuser, Basel, 1984). p. 285.
58. P. De Kepper and J. Boissonade, in *Oscilllations and Traveling Waves in Chemical Systems*, R.J. Field and M. Burger, Eds. (Wiley-Interscience, New York, 1985). p. 223.
59. P. Richetti and A. Arneodo, *Phys. Lett.* **109A** (1985) 359.
60. F. Argoul, A. Arneodo and P. Richetti, *Phys. Lett.* **A120** (1987) 269.
61. F. Argoul, A. Arneodo and P. Richetti, in *A Chaotic Hierarchy*, G. Baier and M. Klein, Eds. (World Scientific Press, Singapore, 1991). p. 79.

CHAPTER 3: MODELING AND INTERPRETATION OF CHAOS IN THE BELOUSOV-ZHABOTINSKY REACTION

LÁSZLÓ GYÖRGYI
Institute of Inorganic and Analytical Chemistry
Eötvös Loránd University
PO Box 32, Budapest-112
H-1518 Hungary

and

RICHARD J. FIELD
Department of Chemistry
University of Montana
Missoula, MT 59801

ABSTRACT

Deterministic chaos has unequivocally been observed and has been well characterized when the Belousov-Zhabotinsky (BZ) reaction is run in a continuous-flow, stirred tank reactor (CSTR). Its presence has been confirmed by the construction of a number of different types of maps and by observation of the classic routes from periodicity to aperiodicity. We discuss here several phenomenological models of BZ-CSTR chaos as well as a number of models based on the detailed chemistry of the BZ reaction. The chemical models reproduce very well the vast majority of the observed chaotic and complex periodic behavior of the BZ reaction.

3.1. Introduction

3.1.1. Background

We review here some of the efforts that have been made to model chemical systems exhibiting deterministic chaos. Our particular focus is on the temporal behavior of *isothermal, homogeneous solution-phase systems*, especially the Belousov-Zhabotinsky (BZ) reaction in a continuous-flow, stirred tank reactor (CSTR). A brief account of early work is given, but the focus is on recent developments. Much early work that is mentioned only briefly here is treated in more depth in a recent monograph by Scott.[1]

The bulk of experimental work on chaotic chemical systems has involved the Belousov-Zhabotinsky (BZ) reaction. Chaos has been observed in other chemical systems, mostly by Epstein and coworkers,[2-5] but BZ chaos is by far the best characterized. Very careful, quantitative experiments have determined the exact form of the chaos, identified the types of transitions between periodicity and aperiodicity, and the way in which chaos is embedded within other complex but periodic behavior. Thus most modeling efforts have focused on the reproduction and interpretation of chaos observed in the BZ reaction. One exception is the work of Citri and Epstein[6] on modeling the BrO_3^--ClO_2^--I^- system, where chaotic waveforms were seen but proved to be a numerical artifact.

Initial efforts to understand chemical chaos used simple models not necessarily related to real systems in order to determine whether chaotic dynamics is possible in differential equations of polynomial form and to provide simple objects for study. The most important of these efforts are those of Willamowski and Rössler,[7] Hudson and coworkers,[8-10] Kawczyńsky *et al.*[11] and more recently Peng *et al.*[12] Furthermore, Samardzija *et al.*[13] have shown that several simple chaotic models not originally of chemical form can be transformed into chemical mechanisms whose associated differential equations show the same dynamics. Unfortunately, the resulting mechanisms usually contain more autocatalytic steps than found in any known chemical system. Nevertheless, this work with simple chemical models showed that there is no reason why chemical reactions should escape from chaotic behavior. We will not discuss this work any further as our interest is in real chemical systems, particularly the BZ reaction, which is the only real, isothermal, solution-phase, chaotic system for which several models are available for comparison with a large body of experimental data.

3.1.2. Chemical Models

The temporal evolution of reacting chemical systems is described by models based on reaction kinetics. These models are formed from detailed chemical mechanisms, and the behavior of other than the simplest of them is studied mainly by numerical simulation. A chemical mechanism is an often complex set of elementary chemical reactions that leads to the observed overall chemical change. The model is either the full mechanism itself or a simpler reaction set containing its essential dynamics. The behavior of the model is investigated using the set of ordinary differential equations obtained from it as described below.

The general form of the chemical reactions in a homogeneous model can be written as below.

$$\sum_j \alpha_{i,j} X_j \rightarrow \sum_k \beta_{i,k} X_k, \quad i=1,..., NR \tag{3.1}$$

In Eq. 3.1 X_j ($j=1,..., NC$) are the NC components involved in the NR reactions of the model and $\alpha_{i,j}$ and $\beta_{i,k}$ (i=1,..., NR; j, $k=1,..., NC$) are the non-negative stoichiometric coefficients of the i-th reaction. An NR x NC matrix of stoichiometric coefficients can be constructed via Eq. 3.2.

$$\nu_{i,j} = \beta_{i,j} - \alpha_{i,j} \tag{3.2}$$

The temporal behavior of a model can be studied if the rates of its component reactions (r_i, $i=1,..., NR$) are known. It is then described in a closed, homogeneous system by the set of ordinary differential equations

$$x_j' = \sum_i \nu_{i,j} r_i, \qquad j=1,..., NC, \tag{3.3}$$

where the prime indicates differentiation with respect to time, and x denotes the concentration of component X.

There are two ways of obtaining the reaction rates, r_i. If the rate constants of the NR reactions, k_i, are known or can be estimated, then the Law of Mass Action can be applied to the model. This approximation assumes that the rate, r, of an elementary chemical reaction is proportional to the product of the concentration of its reactants, each raised to the power of its stoichiometric coefficient, $\alpha_{i,j}$ in Eq. 3.1.

$$r_i = k_i \prod_j x_j^{\alpha_{i,j}}, \tag{3.4}$$

In many cases, the use of net stoichiometries and empirical rate laws is preferred. It is often true that the net stoichiometry resulting from a group of elementary chemical reactions may be directly measured and its rate law determined experimentally, even though the rate constants or even the identity of the individual component reactions are not known. The net stoichiometry and rate law of a group of component reactions can be predicted in other cases using techniques such as the equilibrium, pseudo-steady-state or rate-determining-step approximations. In either case, the rate, r_i, of the net stoichiometry i may be expressed as Eq. 3.5,

$$r_i = f_i(\underline{x}; \underline{p}), \tag{3.5}$$

where f_i is a function of the concentration array of the chemical species, $\underline{x}$, and the array of empirical kinetic parameters, $\underline{p}$.

Chaos has been seen in the BZ reaction only in an open system, *e.g.*, a continuous-flow, stirred tank reactor (CSTR). The CSTR feedstreams maintain the system far from equilibrium, a requirement for the occurrence of chaos. However, at least in the BZ reaction, the appearance of chaos also is directly related to the kinetics of the flow. Inflow and outflow terms must be added to the differential equations describing the behavior of the model in a CSTR, resulting in Eq. 3.6,

$$x_j' = \sum_i \nu_{i,j} r_i + k_o (x_{j,in} - x_j), \quad j=1,..., NC, \tag{3.6}$$

where k_o is the inverse of the mean residence time of the reactor, and $x_{j,in}$ is the mixed-feed concentration of X_j, *i.e.*, the concentration of X_j in the reactor in the hypothetical case where no chemical reaction occurs. The concentration variables must always be non-negative (assuming non-negative initial conditions) if the reaction rates are defined according to Eq. 3.4 because the negative terms in Eq. 3.6 are proportional to the variable itself and will vanish as the concentration approaches zero. Although not guaranteed, this usually also holds when empirical rate laws are used.

3.1.3. *Mechanism of the Belousov-Zhabotinsky Reaction*

The name Belousov-Zhabotinsky[14,15] (BZ) reaction covers a number of chemical systems. The so-called classical system has been used in most chaotic BZ experiments. It consists of an aqueous, 0.1-1.5 M H_2SO_4 medium containing BrO_3^-, a Ce(III)/Ce(IV) catalyst and malonic acid, $HOOCCH_2COOH$. The overall chemical process is the simultaneous oxidation and bromination of malonic acid by BrO_3^-. The first detailed mechanistic rationalization of the BZ oscillations is due to Field, Kőrös and Noyes[16] and is referred to as the FKN mechanism. According to more recent studies,[17] there are 50-100 chemical reactions occurring among 20-30 species in this system. However, the models to be discussed in the next sections are relatively simple derivatives of the FKN mechanism, with the 3-variable Oregonator model[18] used to demonstrate the basic features of the dynamics of the BZ reaction.

The chemical identification of the species in the Oregonator (Table 3.1) are: A = BrO_3^-, H = H^+, X = $HBrO_2$, Y = Br^-, Z = 2Ce(IV), P = HOBr. The species

P is treated as an inert product in the first approximation. The components A and H are present in large excess. Thus a and h are usually taken to be constant, which leaves three species of variable concentration: X, Y and Z. The first four steps of the model are key reactions of the FKN mechanism. The fifth step is a simple stoichiometric representation of a complex set of reactions involving the oxidation of organic material by Ce(IV) with the simultaneous production of Br^-. The parameters k_{O5} and f are adjustable.

The corresponding differential equations can be constructed using Eq. 3.3 and the reactions rates given in Table 3.1. Thus the model exhibits true steady states and undamped limit cycles. The model is essentially open even though flow terms usually are not included in the governing differential equations because a and h are held constant. However, important kinetic effects of the flow observed experimentally, *e.g.*, chaos, are not reproduced by this simple model.

Table 3.1. The Oregonator Model of the BZ Reaction.

	Reaction	Rate
(O1)	$A + Y + 2H \rightarrow X + P$	$k_{O1} a h^2 y$
(O2)	$X + Y + H \rightarrow 2P$	$k_{O2} h x y$
(O3)	$A + X + H \rightarrow 2X + Z$	$k_{O3} a h x$
(O4)	$2X \rightarrow A + P + H$	$k_{O4} x^2$
(O5)	$Z \rightarrow f Y$	$k_{O5} y$

The chemical interpretation of the behavior of the Oregonator model, and thus that of the BZ reaction, involves relaxation oscillations resulting as control of the system switches back and forth between either reactions O1 and O2 or reactions O3 and O4, depending on $[Br^-]$. The critical controlling competition is that for $HBrO_2$ between reactions O2 and O3. Reaction O2 dominates reaction O3 when Br^- is high, causing reactions O1 and O2 to be dominant. However, these reactions lead to the net consumption of Br^-, and when $[Br^-]$ falls below a critical value, the autocatalysis of reaction O3 becomes the dominant fate of $HBrO_2$; reactions O3 and O4 then gain control of the system. There is, though, a negative feedback loop present such that after a delay while Ce(IV) accumulates, reaction O5 regenerates sufficient Br^- for reaction O2 to again become the dominant fate of $HBrO_2$. Reactions O1 and O2 then regain control of the system, and the cycle is reinitialized as Z is consumed. It should not be surprising given this structure that bistability is a common feature of the BZ reaction in a CSTR. The critical importance of the parameters k_{05} and f is apparent as they control the efficiency of the feedback loop. At sufficiently inefficient Br^- production, reactions O3 and O4 remain in control, and a high [Ce(IV)]-low [Br^-] steady state is approached. At very efficient Br^- production, reactions O1 and O2 remain in control, and a low [Ce(IV)]-high [Br^-] steady state is approached. Oscillations and bistability occur at intermediate efficiencies.

The detailed chemistry represented by reaction O5 is still not completely understood, and the processes involved are so complex that this may always be the case. It was originally suggested that it is oxidation of brominated organic compounds (mainly bromomalonic acid, BrMA) by Ce(IV) that yields Br^-. Later Varga *et al.*[19] found, that a significant amount of Br^- comes from the reduction of HOBr or Br_2. Organic radicals rather than Br^- were recently proposed to be the control intermediate under some circumstances.[20-22] Nevertheless, the basic concept of the Oregonator model, *i.e.*, the autocatalytic production of $HBrO_2$ with simultaneous production of some species that completes a delayed, negative feedback loop by consuming $HBrO_2$, seems universally accepted.

Field *et al.*[16] also suggested a set of rate constants for use with the FKN mechanism. These later proved to be incorrect,[23,24] and the presently accepted values were compiled by Field and Försterling,[25] following a suggestion due to Tyson.[24] Some were further modified by Kshirsagar and Field.[26] These new values are usually referred to as the Field-Försterling (FF) rate constants. Accordingly, BZ systems of different composition are usually modeled using the FKN or more recently the FF rate constants for the first four reactions, and by choosing appropriate values for k_{O5} and f.

The Oregonator has been the subject of many detailed analytical and numerical investigations.[24] It shows true steady states, bistability between a steady state and a limit cycle at a subcritical Hopf-bifurcation, excitability and simple limit-cycle oscillations. Simple means here that every variable exhibits only one minimum and one maximum per period, so that the limit cycle can be embedded in a two-dimensional space.

3.1.4 Chaos in the Belousov-Zhabotinsky Reaction in CSTR Mode

The major experimental characteristics of BZ-CSTR chaos have been discussed in a previous chapter. We briefly summarize its most important characteristics. The various dynamical regimes are usually mapped out using the CSTR flow rate as the bifurcation parameter. The system is nonoscillatory and approaches a reduced, low [Ce(IV)]-high [Br^-] steady state at very low flow rates, presumably because the accumulation of brominated organic materials causes Br^- production to be very efficient. This behavior is referred to as the thermodynamic branch because chemical equilibrium is approached in the limit of zero flow rate. The system approaches an oxidized, high-[Ce(IV)], low-[Br^-] steady state at sufficiently high flow rate, presumably because the rapid flow washes out brominated organic material so rapidly that the efficiency of the Br^--producing processes remains very low. This behavior is referred to as the flow branch. Bistability and relaxation oscillations appear at intermediate flow rates. Small-amplitude, sinusoidal oscillations appear near the points where the thermodynamic- and flow-branch steady states loose their stability at a Hopf bifurcation. The relaxation oscillations are clearly related to switching back and forth between the high [Ce(IV)]-low [Br^-] and low [Ce(IV)]-high [Br^-] quasi-steady states. The flow-rate scan shown in Fig. 3.1 reveals the following sequence of dynamics. The reduced steady state is dominant at the lowest flow rates.

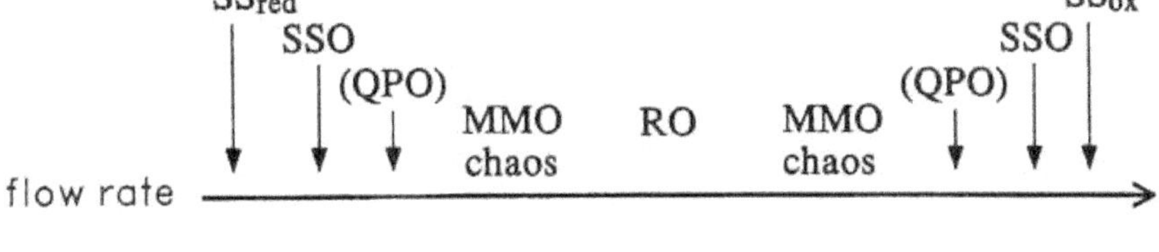

Figure 3.1. Schematic representation of typical behaviors of the BZ reaction as the CSTR flow rate is varied. Notation: SS – steady state; SSO – small-amplitude, sinusoidal oscillations; QPO – quasiperiodic oscillations; MMO – mixed-mode oscillations; RO – relaxation oscillations. The residence times range from ≈ 100-200 min to ≈ 2-5 min, depending on the particular system studied.

It loses stability at a Hopf bifurcation as the flow rate is increased, and small-amplitude, sinusoidal oscillations (SSO) are observed. A secondary-Hopf bifurcation (also called a Naimark-Sacker bifurcation) may follow as the flow rate is increased, resulting in quasiperiodicity associated with the presence of toroidal dynamics. The torus becomes wrinkled at even higher flow rates, leading to phase-locking and chaos. The same dynamics also can be reached without seeing the torus in the case where the SSO loses stability in a manner probably associated with a subcritical secondary-Hopf bifurcation. The phase-locked and chaotic oscillations in this region consist of relaxation oscillations in which several small, quasi-sinusoidal oscillations around the reduced steady state are superimposed on each excursion to its vicinity. The intermittency and period-doubling routes from a periodic window to chaos are both observed in this region. There may be several relaxation oscillations with different numbers of cycles around the reduced steady state before the pattern is repeated. These are often called "mixed-mode" oscillations (MMO). Several types of notation have been used to characterize these complex patterns. We use the L^S notation, where L is the number of relaxation oscillations and S is the number of quasi-sinusoidal excursions per cycle. Some very complex periodicities may occur, and notations such as $1^3 1^4$ are occasionally seen. A note is added here in the figure caption if a different notation is used in work reproduced from other authors.

The ratio of small-amplitude oscillations to relaxation excursions per period decreases as the flow rate is increased. The so-called low-flow-rate complexity ends when simple, relaxation oscillations are reached at higher flow rates approximately in the middle of the flow-rate scan.

Another area of complex behavior is observed at the high-flow-rate end of the relaxation oscillation region. It contains a sequence of dynamic behaviors very similar to and roughly symmetrical to the low-flow-rate complexity. Mixed-mode periodicities and chaos are found when the simple relaxation-oscillation limit cycle loses stability, although the quasi-sinusoidal oscillations occur around the oxidized rather than the reduced steady state. The ratio of quasi-sinusoidal to relaxation oscillations increases as the flow rate increases. This is the "high-flow-rate" region of complexity. Sinusoidal, small-amplitude oscillations (SSO) appear just before the oxidized steady state is reached at the highest flow rates.

An important feature of these phenomena is the presence of a torus, *i.e.*, two frequencies, in the dynamics. As will be seen later, some successful modeling efforts have been guided by this observation.

Various models of BZ-CSTR chaos will be discussed in the next sections. We focus on the chemical processes generating the complexity, so models that are not of chemical origin will be mentioned only briefly. Chemical models are divided into two major groups, those where the old FKN rate constants were used, and more recent models using the FF rate constants. The complexity in these models is the result of different chemical processes in the two cases.

3.2. Phenomenological Models of BZ-CSTR Chaos

3.2.1. Introduction

Phenomenological models are not based on the chemistry of the BZ reaction. They are instead mathematical constructs designed to reproduce the form of the experimental dynamics. Two types of phenomenological models have been used: mappings and continuous-time (ODE) models. Both approaches aim at revealing the mathematical essence of the experimental aperiodicity. This may be useful when comparing BZ-CSTR chaos to similar phenomena in other fields, but it is of limited use for understanding the chemical mechanism that generates the complexity.

3.2.2. Discrete-Time Models

The basic assumption of these approaches is that the complex periodicities and chaos exhibited by the BZ reaction can be described by 1-dimensional dynamics. This seems to be true in most cases. First we look at those attempts where experiments were used to construct a 1-dimensional map. We follow the general practice in this field and use the term "map" in an non-rigorous, informative way, since we do not alway require these constructions to be single-valued.

There have been a number of efforts to describe the high-flow-rate periodic-chaotic sequences using iterative methods. Tomita and Tsuda[27] used the experimental data of Hudson *et al.*[28] to construct a return map of the minimum values of a Br^--selective electrode (Br^--ISE) potential (corresponding to $\log[Br^-]_{max}$) recorded in two different chaotic states. This map was assumed to be single-valued and fitted by a mixture of polynomial and exponential functions, $f(x)$, that has a single extremum. The dynamics of the system was described by the recursion formula below.

$$x_{n+1} = F(x_n) = f(x_n) + b. \tag{3.7}$$

An increase in flow rate corresponds to decreasing b, *i.e.*, to shifting the map downward. A sequence of behaviors similar to that observed by Hudson *et al.*[28] was obtained.

Hudson and Mankin[29] themselves used data from the same dynamical regime to construct a return map from the potential minima of the Br^--ISE. The data was fitted with a discontinuous function very similar to that used by Tomita and Tsuda[27] over most of the interval, but with a small, detached, linear part at high minimum values. They did not study the effect of changing the flow rate but calculated the Lyapounov exponent of the mapping, and looked at its dependence on the shape of the map. The Lyapounov exponent was positive in most cases, and it was concluded that the data was from a chaotic time series.

Hourai *et al.*[30] modeled their own experiments at low CSTR flow rates. This attempt was aimed at reproducing the full range of low-flow-rate complexity. It is not apparent from their report how they constructed their return maps, but probably successive maxima of the potential of a platinum electrode were used. They then combined two return maps, one obtained in a chaotic region consisting mainly of relaxation oscillations and the other obtained in a region where quasi-sinusoidal excursions were observed. This combined experimental return map was fitted by a series of hyperbolic functions. They defined their iterative dynamical system in the form of Eq. 3.7, but in this case $f(x)$ had two extrema. The control parameter b is increased with increasing flow rate, and the simulated sequence of periodic and chaotic states is in good agreement with the experiment, indicating that a wide range of complexity can be described by a map of two extrema and suggesting the universal nature of the form of the BZ-CSTR chaos.

Coffmann *et al.*[31] considered in great detail the complexity occurring in a narrow range of low flow rates. Their experiments belong to the group of "Texas-experiments," which were performed with great care and high-quality instrumentation in the Center for Nonlinear Dynamics at the University of Texas in Austin. Coffman *et al.*[31] attempted to model a sequence of periodic states found beyond a cascade of period-doubling bifurcations. These periodic states have the same symbolic dynamics as the universal sequence (U-sequence)[32] characteristic of certain unimodal maps. They also found the reoccurrence in some experiments of certain states (multiplicity) as the flow rate was varied. They approximated the return map of Br^--ISE potential minima by a single-valued, unimodal function on the interval [0,1], $g(x)$. The discrete-time dynamical system below was defined by,

$$x_{n+1} = F(x_n) = \lambda g(x_n) , \tag{3.8}$$

where the parameter λ increases as the flow rate is decreased. The sequence of complex periodic states of interest was studied as a function of λ. The function $g(x)$ was approximated both by a cubic spline and by the piecewise-linear indented trapezoid shown in Fig. 3.2. Even this very simple, indented-trapezoid map is capable of reproducing the experimental sequence of states including the observed multiplicity.

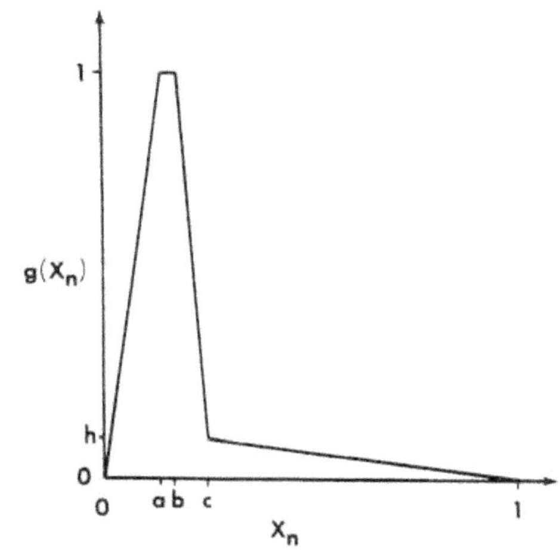

Figure 3.2. The indented-trapezoid map used by Coffman *et al.*[31] to simulate the symbolic dynamics of the Texas-experiments at low flow rates. The following parameter values reproduce the experimental sequence of periodic states found between the 1^0 and 1^1 periodicities: $a = 0.13$, $b = 0.16$, $c = 0.25$, $h = 0.09$. Reproduced by permission from K.G. Coffman, W.D. McCormick, Z. Noszticzius, R.H. Simoyi and H.L Swinney, *J. Chem. Phys.* **80**, (1987) 119.

The other approach to modeling the BZ-CSTR complexity with a 1D dynamical system is to use simulations based on a continuous-time system to derive a 1D mapping. The goal here is to gain a better understanding of the continuous-time model by simplifying its dynamics to that of a discrete-time system. A drawback of this approach, however, is that the mapping is only as good a model of the BZ system as is the original system of ODE's. To our knowledge, only the high-flow-rate complexity has been studied this way.

Pikovsky[33] studied a map with two extrema derived from a set of ODE's not related to the chemistry of the BZ system. The map was constructed from the successive maxima of one of the three variables of the model, and no analytical expression was given for it. The effect of increasing the flow rate was modeled by slowly tilting the transfer function of the map toward the diagonal from above. The model successfully simulates the increasing number of small oscillations appearing together with one large, relaxation oscillation as the flow rate is increased as well as the character of the chaotic states appearing between two neighboring mixed-mode periodicities as a stochastic mixture of the two neighboring limit cycles.

Rinzel and Schwartz[34] studied a map derived at high CSTR flow rates from the Showalter-Noyes-Bar-Eli[35] (SNB) model of the BZ reaction. This model consists of seven ODE's and will be discussed in detail later. Numerical integration of the SNB ODEs shows mixed-mode periodicities, but no chaos is observed at high flow rates. Rinzel and Schwartz constructed a Poincaré-map from the first returns of the variable P ([HOBr]) to a plane defined by Y ($[Br^-]$) = constant. As will be discussed in the next section, P changes very little during oscillations, and its average value determines the ratio of small to large excursions in a limit cycle. The stepwise map shown in Fig. 3.3 was obtained from the first returns of P over a wide range of initial values. Every plateau (or sheet) of this function belongs to a certain type of periodicity, and since P changes very little, at most two plateaus are visited by the system. The system has a fixed point on the sheet that represents the mixed-mode periodicity with three small-amplitude oscillations for each relaxation oscillation. The two types of periodicities are mixed to give a more complex limit cycle if two sheets are visited. The map is shifted downwards as the flow rate is increased. The discrete system describes very well the behavior of the original set of ODE's. The model did not exhibit chaos. However, Schwartz showed in a separate work[36] that the 1-dimensional dynamics is very sensitive to random fluctuations in the flow rate (random shift of the map up or down) if the derivative of the map at its intersection with the diagonal is just below -1. Small random fluctuations of the flow rate in such a case may result in an aperiodic mixture of the periodicities represented by the sheets on either side of the intersection. The resulting aperiodicity is similar to that observed experimentally.

The mixed-mode oscillations are characteristic of phase-locking on the surface of a torus and can be modeled by the iteration of circle maps. Some experiments[37-38] and numerous model simulations[39] showing toroidal behavior also exhibit a complete Farey sequence of mixed-mode oscillations with non-overlapping states. The lack of both quasiperiodicity and chaos is characteristic of a two-parameter path in the

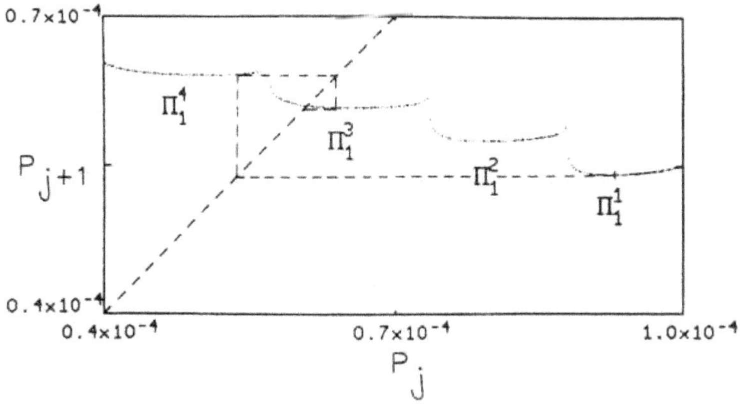

Figure 3.3. Global return map of P (HOBr) in simulations with the SNB model from Rinzel and Schwartz.[34] The notation is analogous to that used by us: $\Pi_L^S \equiv L^S$. The 1^3 periodicity is the stable behavior at this particular flow rate (k_f = 0.0079 s^{-1}). Reproduced by permission.

parameter space of a circle map, and it is surprising why so many experiments and simulations where only one parameter was varied obtain it. Barkley[40] argues in an elegant work that this behavior also is characteristic of another parameter path, where the corresponding circle map has an almost infinite slope. Since this path is asymptotic, relatively large changes in the values of parameters cause very small changes in the distance from the asymptotic limit, and various one-parameter scans will result in very similar near-critical behavior.

Some recent work of Ringland *et al.*[41] is relevant here. They showed that a two-extrema map can be used to model a very wide variety of dynamical behavior, much of which has been seen in BZ-CSTR experiments. Their map is the form of

$$x_{n+1} = Z(x_n),$$

$$\text{where } Z(x) \equiv (c + ax) R_s^-(x) + (d + bx) R_s^+(x),$$

$$\text{and } R_s^\pm(x) \equiv 0.5 \pm 0.5 \tanh(sx). \tag{3.9}$$

The function $Z(x)$ approximates a smooth cubic map for small values of s and a piecewise linear map as $s \to \infty$. Figure 3.4 shows the bifurcation diagram as d is varied with $s = 20$. The structure for values of d near zero (right-hand side of Fig. 3.4) is very similar to that seen in the low-flow-rate Texas-experiments with the flow rate and d being changed in the same direction.

Ringland *et al.*[41] point out that the dynamics of the Z map can range from the universal sequence to a Farey-sequence of periodic states. The latter, however, does

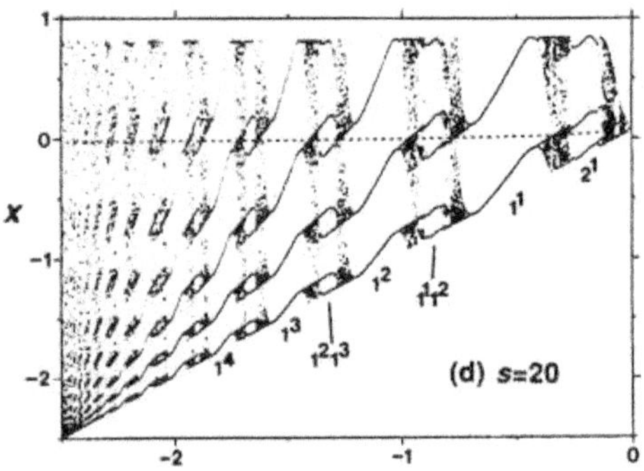

Figure 3.4. Orbit diagram of map Z as a function of d from Ringland *et al.*[41] at $s = 20$. The major periodic orbits are indicated. Reproduced by permission.

not result from phase-locked motion on a torus, and differences from the dynamics of circle maps is also discussed. Chaos is also commonplace in the dynamical zoo of the Z-map, with tangent or period-doubling bifurcations being the route from periodicity to chaos.

3.2.3 *Continuous-Time Models*

Besides the ODE's used above by Pikovsky[33] to generate a 1-dimensional map, two major attempts have been made to understand the dynamics of the BZ reaction with a continuous-time phenomenological model.

Richetti *et al.*[42] constructed what seemed to be the simplest continuous-time model that could account for the quasiperiodicity, mixed-mode oscillations and periodic-chaotic sequences observed in the BZ reaction. These phenomena occur if the attractor of the system is a torus (quasiperiodicity and phase-locking) or a fractal torus (chaos and phase-locking). They used a simple, 3-dimensional set of

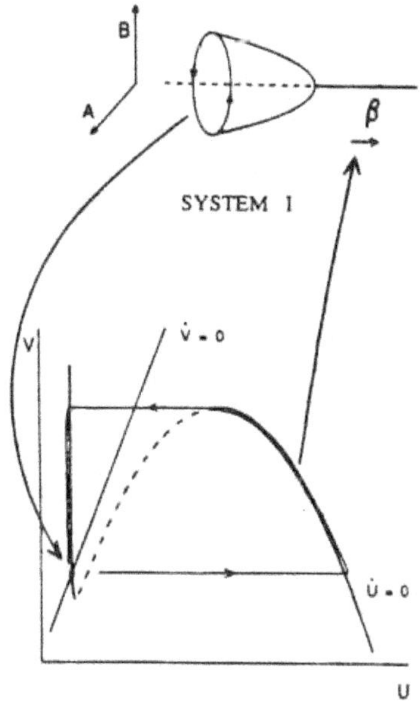

Figure 3.5. Schematic representation of the two subsystems of Barkley's[43] Model A. System I is represented by its bifurcation diagram showing a supercritical Hopf bifurcation as the control parameter β is varied. For system II the nullclines are shown. Their intersection is on the stable part of the u nullcline, thus this subsystem is excitable. The trajectory shown is an excursion of system II following a sufficiently large perturbation of its locally stable steady state. Reproduced by permission.

differential equations derived from the so-called hysteresis-Hopf normal form (HHNF), which has for certain parameter values a steady state with a zero and a pair of purely imaginary eigenvalues. Of more interest to us, following a Hopf bifurcation, this set of ODE's shows a secondary-Hopf bifurcation that introduces a second frequency. This leads to toroidal dynamics. Richetti *et al.*[42] showed in simulations using this simple model that the HHNF equations show quasiperiodicity when they are very close to being axysimmetric as well as the wrinkling of the torus with the development of mixed mode oscillations under the influence of a non-axysimmetric perturbation.

Barkley[43] focused on possible slow-manifold models describing the mixed-mode oscillations and contrasted the HHNF with the SNB model[35] and experiments. While it does not seem proper to discard the HHNF as a possible phenomenological model of BZ chaos, there are some disturbing discrepancies (See the next section) between its dynamics and that of the chemically based SNB model. Barkley thus developed a four-dimensional model (called Model A) that is built up from two, two-dimensional subsystems. As is shown in Fig. 3.5, the first subsystem (System I) exhibits a Hopf bifurcation as its control parameter β is varied. This is coupled to System II, which has a slow manifold and may show relaxation oscillations between the two sheets of this manifold. System II is actually only excitable for this particular situation because the steady state at the intersection of the u and v nullclines is stable. The coupling is such that increasing v in System II increases the effective bifurcation parameter β for System I, and the small oscillations at low values of β perturb the stable steady state of System II, eventually exciting it. Model A shows mixed-mode oscillations very similar to those seen in the SNB model. Its weakness is that supercritical secondary-Hopf bifurcation occurs far too infrequently, and certain stability properties of the steady states do not match that of the SNB model.

3.3. Modeling BZ-CSTR Chaos with the FKN Mechanism.

3.3.1. With the Original FKN Rate Constants

Various Oregonator-type models have been used in attempts to interpret BZ-CSTR chaos. The simple Oregonator[18] (see Table 3.1) does not exhibit complex oscillations. Nevertheless, it has been so successful in simulating other experimental BZ reaction phenomena that several attempts at modeling BZ-CSTR chaos have used various modifications of it.

Janz *et al.*[44] recognized that some complexity seen in BZ-CSTR experiments can be rationalized by periodically moving the Oregonator dynamics back-and-forth across a subcritical Hopf bifurcation. They parameterized the Oregonator stoichiometric factor, f, with P, which they identified as bromomalonic acid (BrMA) because HOBr is eventually converted into BrMA. They set $f = FP^2/(K+P^2)$ and studied the high-f behavior of the Oregonator. The parameter K is very close to the average value of P, and $F = 4$. The value of f oscillates because the production of P during relaxation oscillations is faster than its removal by the CSTR flow. This allows f to increase to a level where the Hopf bifurcation border is crossed. The reduced steady state then becomes stable and is approached by the system. However, the rate of production of P is lower when the model is near the reduced steady state than when oscillation is occurring. Thus, for suitable flow rates, P is washed out of the reactor faster than it is produced, and f decreases back to a level where the bifurcation border is recrossed, and oscillations restart. The simulations showed complex periodic bursting patterns, but an extensive search failed to find a parameter region where aperiodic oscillations occurred. Troy[45] studied the same bursting behavior using 1-dimensional maps generated from a simplified version[46] of this model. We made an attempt[47] to find chaos in this model when the FKN rate constants were updated, but again only complex periodic oscillations were seen.

Turner and coworkers[48-51] studied the reversible Oregonator originally introduced by Field.[52] All of the reactions of the original Oregonator are assumed to be reversible in this model. This is a reasonable approximation in the case of reactions O1-O4, but it is not valid for reaction O5, which represents a very complex set of essentially irreversible reactions. A wide variety of complex periodic and chaotic waveforms were found in one series of calculations[49] showing remarkable qualitative similarities to the Texas-experiments. However, the control parameters of the flow system (inflow concentrations, flow rate, *etc.*) used in the simulations were quite dissimilar to those used experimentally.

Although the reversible Oregonator provided good qualitative agreement with experiments, another Oregonator-type mechanism due to Showalter, Noyes and Bar-Eli (SNB model)[35] has attracted much more attention because it is more satisfying

Table 3.2. The SNB[35] Model of the BZ Reaction.

	reaction	rate
(S1)	A + Y + 2H ⇄ X + P	$k_{S1} a h^2 y \; - k_{S-1} x p$
(S2)	X + Y + H ⇄ 2P	$k_{S2} h x y \; - k_{S-2} p^2$
(S3)	A + X + H ⇄ 2W	$k_{S3} a h x \; - k_{S-3} w^2$
(S4)	W + C + H ⇄ X + Z	$k_{S4} w c h \; - k_{S-4} x z$
(S5)	2X ⇄ A + P + H	$k_{S5} x^2 \; - k_{S-5} a h p$
(S6)	Z → gY + C	$k_{S6} z$

chemically. It was originally proposed to account for the complex oscillations found in various BZ-CSTR experiments and is shown in Table 3.2. The chemical identifications of species are the same as in the simple Oregonator with the addition of C ≡ Ce(III), W ≡ $BrO_2\cdot$ and Z ≡ Ce(IV). It is an extension of the reversible Oregonator in which the autocatalytic production of $HBrO_2$ (reaction O3) is split into two steps by introducing the additional variable, $BrO_2\cdot$. The model also explicitly includes the reduced form of the catalyst and identifies Z as one Ce(IV) ion, not two. This latter change causes the stoichiometric factor, g, to be around 0.5, half the value of f used in the Oregonator. Although there are 8 variable concentrations in the model, H and A are usually held constant, and the mass-conservation relationship $c + z = [Ce]_{tot}$ allows reduction of the dimensionality of the system to five. Complex waveforms appear in this model at both high and low flow rates with the addition of appropriate CSTR flow terms. Extensive simulations by Showalter *et al.*[35] and by Ganapathisubramanian and Noyes[53] found only periodic waveforms and caused some skepticism that the source of the BZ-CSTR chaos was in the chemical kinetics itself. However, a further search of the parameter space by the Texas group has shown differently.

Barkley *et al.*[54] observed the formation of a torus in a secondary-Hopf bifurcation. Furthermore, the torus became wrinkled as the control parameter ($c + z = [Ce]_{tot}$) was varied. The resulting fractal attractor supported the experimentally observed mixed-mode, periodic-chaotic sequences, although a detailed description of this behavior was not given. This mixed-mode behavior is related to the high-flow-rate complexity because the quasi-sinusoidal oscillations occur in the neighborhood of the oxidized steady state. Although chaos appeared in these simulations, the parameter values used did not correspond to any real experiments, and the periodic-chaotic sequences are too hard to find to claim that the SNB model is an appropriate description of the high-flow-rate BZ-CSTR complexity, where chaos is easily found.

Barkley[40] in further studies of the high-flow-rate regime looked at the mixed-mode oscillations and their relationship to simple phenomenological models (See Section 2.3). He concluded that the most characteristic features of this slow-manifold dynamics are: (1) the funneling of the spiraling trajectories approaching the saddle-focus of the system (oxidized steady state), (2) the fast change of the reinjection variable, p, during the decaying small oscillations, and (3) the following slow approach of the reinjection variable to the saddle-focus. Because neither the slow-manifold model of Rössler, nor the HHNF reproduces this behavior, he suggested the four-variable Model A (See Section 2.3) as an alternative. We stress, however, that his conclusions are based on the assumption that the SNB model with the incorrect FKN rate constants is a proper representation of the chemistry of the BZ reaction. We discuss below several ways in which the SNB model can be improved.

There has been some recent success using the SNB model to reproduce the low-flow-rate complexity. Lindberg *et al.*[55] observed robust, large-scale chaos in simulations based on the SNB model with the FKN rate constants at low CSTR flow rates. A periodic-chaotic sequence was observed at appropriate flow rates but with feedstream concentrations quite different from the experimental. Their results (Fig. 3.6) showed considerable similarity to the experiments shown in Fig. 3.1. They also report that an extensive search using the Field-Försterling rate constants failed to find chaos.

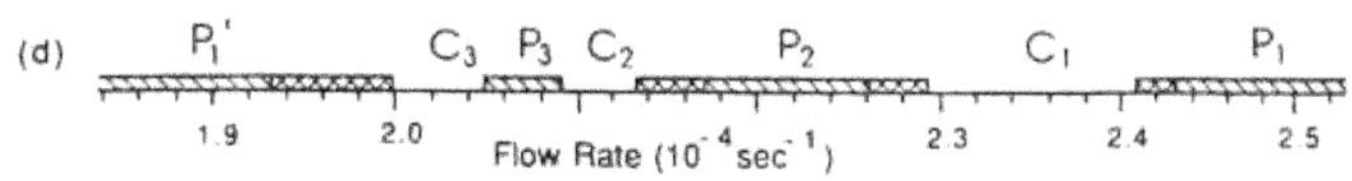

Figure 3.6. Low-flow-rate complexity as simulated by Lindberg *et al.*[55] with the SNB model. Periodic regions are indicated by diagonal hatching. Cross-hatching shows the approximate regions of period-doubling cascades. Notation: $P_1 \equiv 1^0$ (RO); $P_2 \equiv 1^1$; $P_3 \equiv 1^2$; $P_1' \equiv 0^1$ (SSO). C_x shows chaotic regions. Reproduced by permission.

A slightly modified version of the SNB model was used by Bar-Eli and Noyes.[39] It differs from that in Table 3.2 in that malonic acid (MA) is considered to be a reactant in reaction S6, and reaction Q1,

$$W + Z \rightleftarrows A + C, \qquad \text{(Q1)}$$

is added to the model. They used the FKN rate constants and carried out simulations along cross-sections of the $[Y]_0$-$[MA]_0$ plane (the zero subscript denotes mixed-feed concentrations) and found Farey sequences of mixed-mode oscillations as $[Y]_0$ is increased that seem to be related to the low-flow-rate complexity. This sequence of periodic states qualitatively agrees with the experimental findings of

Maselko and Swinney[37] in the Mn^{2+}-catalyzed BZ system using flow rate as the bifurcation parameter, although they seem to have monitored the high-flow-rate complexity because the small-amplitude oscillations circle the oxidized steady state. Bar-Eli and Noyes[39] also suggest that there is a hint of chaos in their computations, but they do not elaborate.

A particularly significant recent contribution to the study of the SNB model is that of Ringland.[56] The models discussed so far all have several adjustable parameters, in the SNB model usually g, k_6, and $[Y]_o$. Therefore, it has been difficult to carry out sufficiently exhaustive searches to ascertain that the parameters used are actually those that give the best agreement with experiment. Ringland has developed a rapid-reconnaissance method for the SNB model in the 4-dimensional space of k_o, $[Y]_o$, g, k_6. He followed the crossing of two, codimension-2 bifurcation lines (in the k_o-$[Y]_o$ plane) across the parameter space of the remaining two parameters, allowing k_6 to vary from 0 to ∞ and g to vary from 0 to 2. Then he compared the results with bench experiments with the BZ reaction with similar (although not identical) conditions. The important conclusions are that it is possible with the FKN rate constants to qualitatively reproduce the experimentally observed crossing, although quantitatively there is at least an order of magnitude deviation in its position in the parameter space. However, when Tyson's LO rate constants[24] were used (these triggered the Field-Försterling values and are quite similar to them), the model is unable to reproduce the experimental crossing even qualitatively for physically reasonable values of the parameters.

A chemical explanation can be given[57] for the ability of the SNB model to reproduce experimental complexity with the FKN rate constants[16] but not with the FF values.[25] Two important reactions of the FKN mechanism, the Br_2-hydrolysis equilibrium followed by the bromination of malonic acid (MA), reactions Q2 and Q3

$$P + Y + H \rightleftarrows Br_2 + H_2O \qquad (Q2)$$

$$Br_2 + MA \rightarrow BrMA + Y \qquad (Q3)$$

are missing from all Oregonator-derived models discussed above. This allows P (HOBr) to accumulate to unrealistically high concentrations. In the case of the FKN rate constants, P then plays an artifactual role in the dynamics through the reversibility of step (S1).

$$A + Y \rightleftarrows X + P \qquad (S1)$$

The FKN value of k_{S-1} is $\approx 10^4$ $M^{-1}s^{-1}$, while the correct value is on the order of 1. This makes reaction (−S1) an important Br^--producing step (it completes the negative feedback loop) and [P] effectively a bifurcation parameter[34] for the core of the model. That is, the model oscillates without reaction (−S1) for appropriate parameter values, but a Hopf bifurcation at which the steady state becomes stable is passed when the rate of reaction (−S1) becomes sufficiently rapid. The concentration of P oscillates in this model in the same way that it does in the Janz *et al.*, parameterized-*f* model described above. The complexity occurs when the SNB core oscillations interact near the bifurcation point with the oscillations in *P*. The oscillations in *P* decouple from the SNB oscillations because their frequency is determined by both the flow and the SNB kinetics. This was verified mathematically by Barkley[43] who found that the rate of the slow approach to the saddle-node fixed point in the mixed-mode SNB oscillations is determined by the smallest eigenvalue of the fixed point, which is $-k_o$. The chaos in this model results from two problems with the assumed chemistry, and it disappears if these problems are corrected.

The last modeling effort with the FKN rate constants is the work of Richetti *et al.*[42] who studied the quasiperiodicity and periodic-chaotic sequences of mixed-mode oscillations using both the HHNF and a chemical mechanism called Model M2. The mechanism is a somewhat arbitrary subset of the FKN mechanism. The main difference between M2 and the Oregonator-type models discussed above is that reaction (S1) is not reversible, thus preventing *P* from playing a role in Br^- production. Bromide ion is produced in their mechanism by the oxidation of BrMA (formed from malonic acid and HOBr) by Ce(IV). They include reaction Q2 left out of the SNB model, but ignore its reversibility and fail to specify the fate of the Br_2 formed. These modifications make it questionable[47] to what extent this model represents the chemistry of the BZ reaction. There are three important results of their simulations. There is qualitative agreement between the simulated and Texas experimental waveforms. However, the simulated sequence of states that appears as the flow rate is varied is opposite to the experimental. Good qualitative agreement was obtained in simulations of the sequence of dynamical regimes observed by Argoul *et al.*[58] The flow rate was changed in the same direction in both experiments and simulations. However, there is again a discrepancy in that the simulated complexity occurs close to the reduced steady state (low-flow-rate complexity), while the experiments were done in the high-flow-rate region where the small oscillations are close to the unstable oxidized steady state. Richetti *et al.*[59] showed in a separate work that Model M2 reproduces qualitatively the experimentally observed collapse of a strange attractor in an interior crisis.

The differences between experiments and simulations using Model M2 are due to the incomplete chemistry of the model. The simulated $[Br^-]$ is very low, about 10^{-9} M, and Br^- is pumped into the CSTR modeled by Richetti *et al.* at a mixed-feed concentration $[Br^-]_0 > 10^{-7}$ M. This way the flow acts as a major Br^- source that removes HOBr without the production of any other bromine-containing species that would be the normal precursor of Br^- in the kinetics of the SNB and related models. It is because of this difference that the Model-M2 sequence of mixed-mode states is sometimes opposite to that of the SNB model and experiments. Furthermore, when the ordering is correct, the complexity occurs near the wrong steady state.

3.3.2. *With the Field-Försterling Rate Constants*

3.3.2.1 *Chaos Resulting from Inhomogeneities and from Flow-Rate Fluctuations*

The difficulty of simulating the robust experimental BZ-CSTR chaos using the homogeneous BZ kinetics with the correct rate constant values prompted studies of mixing and stirring effects in a CSTR, especially because of reports of significant stirring effects associated with experimental BZ-CSTR chaos. We showed[60,61] that very small concentration heterogeneities occurring near or in the tips of the CSTR inlets can cause a wide variety of complex phenomena, including chaos, as they couple with the bulk of the reactor contents. The model for heterogeneity-induced chaos in the BZ-CSTR experiments is shown in Fig. 3.7. This is a phenomenological macro-mixing model[62] of a sort often used to simulate stirring effects on complex chemical reactions occurring in a CSTR.[63,64] The chemical model used in both the main reactor and the heterogeneity is a partially reversible Oregonator similar to the SNB model in Table 3.2, except that only S3 and S4 are treated as reversible. This model exhibits steady states and relaxation oscillations in a perfectly mixed CSTR, but no complexity or chaos. The coupling between the reactor and the heterogeneity has a symmetrical part (diffusion) and an asymmetrical part (flow). The oscillations in the heterogeneity have a frequency and amplitude different from those in the bulk because its chemical composition is different. A perturbation of the bulk of the reactor (30 mL) by a very small heterogeneity (0.1 mL) is sufficient to reproduce much of the complexity seen in BZ-CSTR experiments. All experimentally observed routes to chaos (intermittency, a cascade of period-doubling bifurcations, secondary-Hopf bifurcation), an interior crisis of a strange attractor, birhythmicity, tristability among a steady state, a limit cycle and a strange attractor were observed in a single flow-rate scan at experimentally realistic values of control parameters. The observed

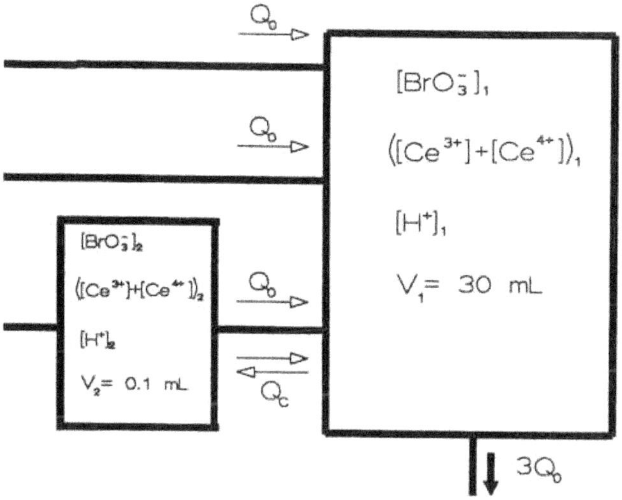

Figure 3.7. The coupled reactor model of Györgyi and Field[60,61] used to model the effect of heterogeneities resulting from incomplete mixing on the dynamical behavior of the BZ reaction. Both the large (30 mL) and the small (0.1 mL) reactors are assumed to be perfectly mixed. They are coupled by diffusion, and the smaller reactor perturbs the larger one by affecting the concentrations in one of its inflows. The concentrations of the major reactants in the small reactor are chosen to be between those in the large reactor and those in the feedstream that flows through the small reactor. Reproduced by permission.

sequence of dynamics, however, did not relate to any single BZ-CSTR experiment, although no exhaustive search of the space of the control parameters was performed to obtain such an agreement. These simulations do clearly demonstrate that very small heterogeneities in a CSTR can drastically alter the behavior of an otherwise simple nonlinear chemical system and suggest that the BZ-CSTR chaos might be readily rationalized this way.

Schneider *et al.*[38] considered the possibility that the BZ-CSTR chaos results from stochastic environmental perturbation. They suggested that the high-flow-rate chaos seen, *e.g.*, in the experiments of Hudson and coworkers,[28,29,65] is a stochastic mixture of neighboring Farey states resulting from fluctuations in the CSTR flow rate. They investigated this region experimentally and with the SNB model using the FF rate constants but with k_{-1} increased to 200 $M^{-1}s^{-1}$, thus allowing P to play the role of a Br^- source as discussed above. They found experimental periodic-chaotic sequences, but the largest Lyapounov-exponent calculated from their data was positive even for some of the mixed-mode periodic states. The unperturbed SNB model shows only complex periodic states at high flow rates with this parameterization. However, when 3-7% gaussian noise was superimposed on the SNB-CSTR flow rate, the largest Lyapounov-exponent and the Fourier spectra of the computed aperiodic oscillations matches those seen in the unperturbed experiments. It was thus suggested that it is not necessary to explain the high-flow-rate, mixed-mode periodic-chaotic sequences by intrinsic deterministic chaos in the homogeneous BZ kinetics. We note here the similarity of the approach with that of Schwartz[36] discussed in Section 2.2.

Following this work it was discovered that in fact BZ-CSTR chaos can be very accurately simulated using chemically complete and reasonable, homogeneous unperturbed models.

3.3.2.2. Chaos in Homogeneous Models of the BZ Reaction

Ibison and Scott[66] recently proposed a new model of the BZ reaction designed for simulation of their high- and low-flow-rate CSTR experiments.[67] The novel feature of this model is that Br^- is generated by reactions of HOBr with organic material, in accordance with the experimental findings of Varga *et al.*[19] Although they find sequences of behavior in their model similar to experiments, there are some problems. This mechanism is complete with reactions Q2 and Q3. Thus, in order to generate enough Br^- to complete the negative feedback loop, either the equilibrium constant of reaction Q2 or the rate constant of Q3 (they appear in the model as terms in a composite rate constant) must be decreased from their experimental values by five orders of magnitude. This allows enough Br_2 to accumulate so that its hydrolysis keeps [HOBr] at a high level. This agrees with our modeling experience that [HOBr] is never large enough under normal BZ conditions to provide enough Br^- to complete the negative feedback loop. There is also a significant amount of Br^- ($\approx 10^{-5}$ M) in their CSTR feedstream. This value seems especially large when compared with the maximum simulated CSTR [Br^-] of $\approx 10^{-11}$ M, which is in turn very low compared to experiment. The CSTR feedstream may contain ≈ 10^{-7} M Br^- as a contaminant of BrO_3^-, and usually the experimental $[Br^-]_{max}$ is greater than 10^{-6} M. Nevertheless, this was the first BZ model using the FF rate constants that exhibited low- and high-flow-rate complexity and chaos separated by relaxation oscillations.

Another model based on the homogeneous kinetics of the BZ reaction was suggested by Györgyi *et al.*[68] at essentially the same time and is shown in Table 3.3. It is a SNB model augmented by a subset of reactions from a detailed mechanistic model[17] of the BZ reaction. The differential equations were set up using the Law of Mass Action with appropriate flow terms and the rate constant values given by Györgyi *et al.*[68] The concentration of BrO_3^-, H^+ and H_2O were assumed to be constant, leaving 11 species with variable concentration. The identifications of the organic species are as follows: MA ≡ $CH_2(COOH)_2$, MA· ≡ ·$CH(COOH)_2$, BrMA≡ $BrCH(COOH)_2$. The model also contains the Br_2-hydrolysis equilibrium (reaction A1). The Br^- regeneration route in the model is very similar to the original FKN mechanism,[16] *i.e.*, Ce(IV) oxidizes malonic acid (MA) and bromomalonic acid (BrMA) to yield Br^-. The reaction of Ce(IV) with MA produces a MA·, that eventually may produce Br^-, while the reaction of Ce(IV) with BrMA results immediately in Br^-. Most of the rate constants in the model are experimentally determined quantities, although the values of k_9, k_{10} and k_{11} were treated as somewhat adjustable parameters to fit experiments. It was necessary to do this because the organic chemistry in the model is clearly incomplete.[17] The value of k_{11} must be assumed to be 300 times larger than the experimental value. We interpret this to mean that products of the reaction of Ce(IV) with MA (or of the reaction of HOBr or Br_2 with brominated organics) go on to release Br^-, but that reaction A11 is probably not the actual route by which this occurs. It is very difficult to separate the mechanistic effects of Ce(IV) and HOBr because they are formed simultaneously and in a fixed stoichiometric ratio. The experimental conditions of the Texas-experiments[31,69,70] were used in the simulations.[68] Figure 3.8 shows a comparison of the experimental and simulated low-flow-rate complexity in the region just beyond the large-amplitude relaxation oscillations (1^0 periodicity). The agreement between the dynamics of the model and that of the real system is also good in the rest of the region of complexity. The number of small-amplitude peaks per period grows to roughly 15 per relaxation oscillation as the flow rate is further decreased.

There are chaotic regions between the complex limit cycles. They usually are approached via cascades of period-doubling bifurcations as the flow rate is varied. Sinusoidal, small-amplitude oscillations are found between the mixed-mode behavior

Table 3.3. A Homogeneous, 11-Variable, Chaotic Model of the BZ Reaction due to Györgyi *et al.*[68]

(A1)	Br^-	$+ HOBr$	$+ H^+$	$\rightleftarrows Br_2$	$+ H_2O$		
(A2)	BrO_3^-	$+ Br^-$	$+ 2H^+$	$\rightleftarrows HBrO_2$	$+ HOBr$		
(A3)	$HBrO_2$	$+ Br^-$	$+ H^+$	$\rightarrow 2HOBr$			
(A4)	BrO_3^-	$+ HBrO_2$	$+ H^+$	$\rightleftarrows 2BrO_2\cdot$			
(A5)	$BrO_2\cdot$	$+ Ce(III)$	$+ H^+$	$\rightleftarrows HBrO_2$	$+ Ce(IV)$		
(A6)			$2HBrO_2$	$\rightarrow BrO_3^-$	$+ HOBr$	$+ H^+$	
(A7)		MA	$+ Br_2$	$\rightarrow$ BrMA	$+ Br^-$	$+ H^+$	
(A8)		MA	$+ HOBr$	$\rightarrow$ BrMA	$+ H_2O$		
(A9)		MA	$+ Ce(IV)$	$\rightarrow MA\cdot$	$+ Ce(III)$	$+ H^+$	
(A10)		BrMA	$+ Ce(IV)$	$\rightarrow Br^-$	$+ Ce(III)$	$+ \{Q\}$	
(A11)		$MA\cdot$	+ BrMA	$\rightarrow$ MA	$+ Br^-$	$+ \{Q\}$	
(A12)		$MA\cdot$	$+ Br_2$	$\rightarrow$ BrMA	$+ Br\cdot$		
(A13)		$MA\cdot$	$+ HOBr$	$\rightarrow Br\cdot$	$+ \{Q\}$		
(A14)			$2MA\cdot$	$\rightarrow$ MA	$+ \{Q\}$		
(A15)		$Br\cdot$	+ MA	$\rightarrow Br^-$	$+ MA\cdot$	$+ \{Q\}$	

and the reduced steady state at the lowest flow rates used. Quasiperiodic transients observed suggest that this SSO limit cycle loses stability in a secondary-Hopf bifurcation and that the resulting periodic-chaotic sequence develops on the "surface" of a fractal torus.

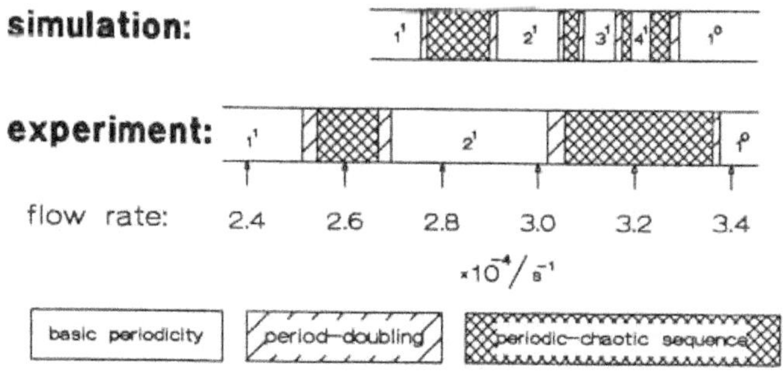

Figure 3.8. Comparison of the dynamics of the model[68] in Table 3.3 with experiments[69,70] done under the same conditions. The cross-hatched regions contain further narrow periodic windows in both experiments and simulations.

This model is very successful in reproducing the low-flow-rate complexity, but it is too complicated mathematically (11 variables) and chemically for its dynamic essence to be discerned. We have thus used reaction rate sensitivity methods[71] and the classical approximations of chemical kinetics to develop several simpler models[72-74] that are less cumbersome to work with numerically and whose dynamic structure is more apparent than that of the 11-variable model. Two four-variable and one three-variable models will be discussed. We demonstrate the chemical mechanism generating the complex periodic-chaotic sequences using one of the four-variable models.

The four-variable models of BZ-CSTR chaos are shown in Table 3.4. Some new notations are: $M \equiv$ MA, $V \equiv$ BrMA, c_{tot} = [Ce(III)] + [Ce(IV)] = c_o. The quantities $\hat{w}$ and $\hat{r}$ are estimated concentrations of $BrO_2\cdot$ and $MA\cdot$, respectively. The concentrations of A, H and M are held constant. The concentration of $MA\cdot$ was estimated using a quasi-steady-state approximation that ignores the effect of the CSTR flow and results in Eq. 3.10.

$$\hat{r} = \frac{-k_{B8}v + \sqrt{(k_{B8}v)^2 + 2.4\times 10^{10}k_{B6}mz}}{1.2\times 10^{10}} \qquad (3.10)$$

Table 3.4. Four-Variable Models of the BZ reaction.[72]

	Reaction	Rate	Rate Constant
(B1)	A + Y + 2H → X + V	$k_{B1} a h^2 y$	2.0 $M^{-3}s^{-1}$
(B2)	X + Y + H → 2V	$k_{B2} h x y$	2.0x10^6 $M^{-2}s^{-1}$
(B3)	0.5X + A + H → X + Z	$k_{B3} h(c_{tot}-z)\hat{w}$	6.2x10^4 $M^{-2}s^{-1}$
(B4)	X + Z → 0.5X	$k_{B4} x z$	7.0x10^3 $M^{-2}s^{-1}$
(B5)	2X → V + A + H	$k_{B5} x^2$	3.0x10^3 $M^{-1}s^{-1}$
(B6)	Z + M →	$k_{B6} m z$	0.3 $M^{-1}s^{-1}$
(B7)	Z + V → Y	$k_{B7} z v$	30.0 $M^{-1}s^{-1}$
(B8)	V → Y	$k_{B8} v f$	2.4x10^4 $M^{-1}s^{-1}$

There are two ways of approximating $[BrO_2\cdot]$. Equation 3.11 results if it is assumed that reaction A4 in Table 3.3 is always at equilibrium.

$$\hat{w} = \sqrt{K_{A4} a h}\sqrt{x} \tag{3.11}$$

K_{A4} = 7.9 x 10^{-7} M^{-1} for reaction A4. This approximation leads to a square-root dependence of the rate of the autocatalytic reaction on $[HBrO_2] \equiv x$. It is also possible to approximate $\hat{w}$ using the quasi-steady-state approximation for $[BrO_2\cdot]$. The removal of W by the CSTR flow is again ignored. Equation 3.12 results from this approximation.

$$\hat{w} = \frac{-k_{B3} h c + \sqrt{(k_{B3} h c)^2 + 3.36\times10^8\, x\,(66.0 a h + k_{B4} z)}}{1.68\times10^8} \tag{3.12}$$

The concentration of [Ce(III)] in the reactor is given by $c = c_{tot} - z$. The constants in Eqs. 3.10-3.12 are derived from the rate constants given by Györgyi *et al.*[68,72] Two four-variable (x, y, z, v) models of the BZ reaction, Model B_{QSS} and Model B_{EQ}, are obtained depending on whether Eq. 3.11 or Eq. 3.12 is used for $\dot{w}$. Both models were investigated in both the high- and low-flow-rate regions of complexity.[72,73] They reproduce well the behavior of the 11-variable model in the low-flow-rate region, although Model B_{QSS} is a somewhat better approximation than Model B_{EQ}. Model B_{EQ} shows[72] periodic-chaotic sequences at low flow rates, but the flow rate range where the complexity occurs is shifted to values twice as large as are observed experimentally. This is not a large discrepancy. On the other hand, the low-flow-rate bifurcation sequence in Model B_{EQ} agrees better with experiments in the region between the 1^1 and the 0^1 periodicities than does Model B_{QSS}. Simulations[72] with Model B_{QSS} (and with the model in Table 3.3) reveal more complexity and limit cycles with more quasi-sinusoidal oscillations than are seen in the Texas-experiments. Fig. 3.9 shows the bifurcation diagram obtained with Model B_{QSS} at flow rates just below the region of relaxation oscillations (1^0 limit cycle). The points indicate log z values of the intersections of trajectories with a 3-dimensional hyperplane (Poincaré plane). Chaos or very complex periodicity is indicated by many intersections. Few intersections indicate simple limit cycles. Figure 3.9 (A) was calculated with increasing flow rates and Fig. 3.9 (B) with decreasing flow rates. Several regions of bistability were found between two limit cycles where one contained twice the number of large and small excursions as the other. There also is a bistability (not shown) at lower flow rates between the small-amplitude, sinusoidal limit cycle (0^1) and mixed-mode behavior.

Both four-variable models exhibit complex oscillations and chaos at high flow rates as well. Conditions close to those used by Schneider and Münster[38] and by Hudson and Mankin[29] were used in one simulation,[72] while in another case[73] we experimentally and in simulations reproduced the chaotic experiments of Hudson and Mankin.[29]

Although the four-variable models are quite simple and successful, it is of interest to derive a chemically based three-variable model that exhibits CSTR chaos. Model B_{EQ} was used as a starting point in order to obtain the simplest possible three-variable model. It was simplified by replacing reactions B6-B8 by two alternative Ce(IV) consuming pathways, one that produces Br^- and one that does not. The resulting four-variable chemical scheme is shown in Table 3.5.

A three-variable model can be obtained from the mechanism in Table 3.5 if a fast variable can be identified and removed by setting its time derivative equal to zero. Such perturbation techniques have been successful in reducing the FKN

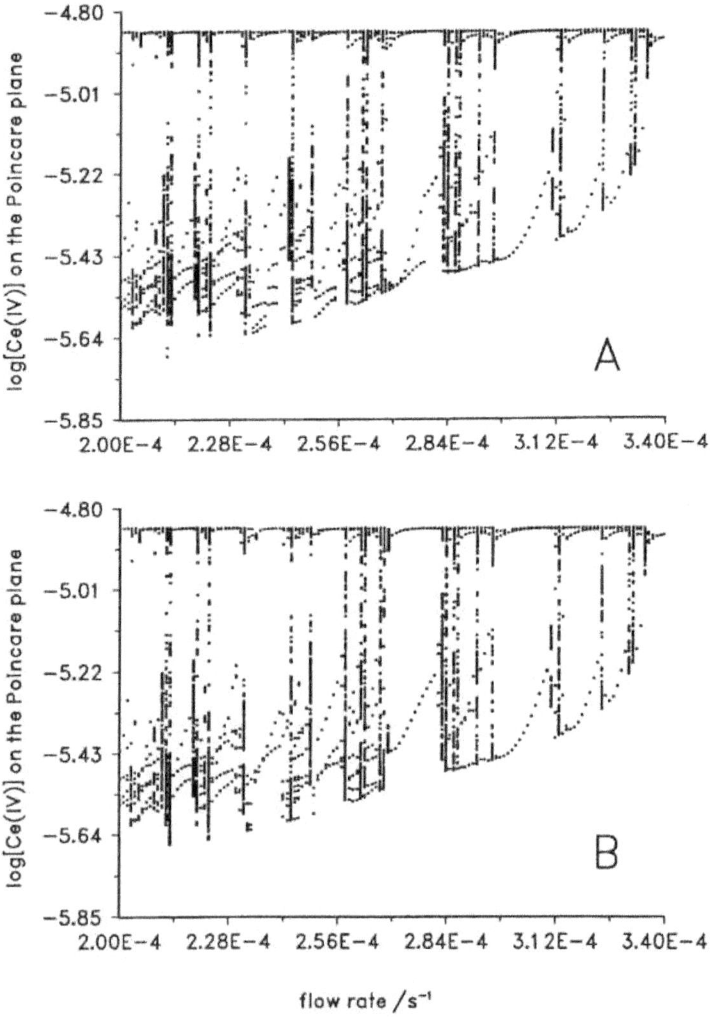

Figure 3.9. Bifurcation diagram[72] of Model B_{OSS} at low CSTR flow rates. Two scans are shown because of the existence of bistable regions, one with increasing flow rate (A) and one with decreasing flow rate (B). Reproduced by permission.

mechanism[75] and the Oregonator.[18,76] The differential equations resulting from Model C with appropriate flow terms were first scaled to yield variables with magnitudes on the order of one. Experience[72,76] suggests that $[Br^-]$ may be eliminated as a fast variable with appropriate scaling, and this is our finding as well. The scaled differential equation for Y is set equal to zero, and the scaled concentration of Y is expressed as an algebraic function of the remaining variables. This reduces the dimension of the problem to three. Table 3.6 shows the scaled differential equations of this three-variable model of the BZ reaction.

The rate constants of reactions C1-C5 in Table 3.5 are all experimentally determined quantities. Although experimental values of k_{C6} and k_{C7} exist, these

Table 3.5. A simple four-variable scheme of the BZ reaction.[74]

	Reaction	Rate	Rate Constant
(C1)	A + Y + 2H → X + V	$k_{C1} a h^2 y$	$2.0\ M^{-3}s^{-1}$
(C2)	X + Y + H → 2V	$k_{C2} h x y$	$4.0 \times 10^6\ M^{-2}s^{-1}$
(C3)	0.5X + A + H → X + Z	$k_{C3} h^{1.5} a^{0.5} (c_{tot}-z)\, x^{0.5}$	$55.2\ M^{-2.5}s^{-1}$
(C4)	X + Z → 0.5X	$k_{C4} x z$	$7.0 \times 10^3\ M^{-2}s^{-1}$
(C5)	2X → V	$k_{C5} x^2$	$3.0 \times 10^3\ M^{-1}s^{-1}$
(C6)	Z + V → Y	$k_{C6} z v$	adjustable
(C7)	Z + M →	$k_{C7} z m$	adjustable

quantities were treated as expendable parameters in order to account for the inhibitor (Y) production lost by ignoring that the products of reaction C7 may react with V to produce Br^- and that other intermediates from BrMA may contribute to inhibition of the positive feedback loop represented by reaction C3.

This model reproduces a number of phenomena found in BZ-CSTR experiments despite its chemical simplicity and low dimensionality. In particular, complexity is found at both high and low flow rates and is embedded between small-amplitude, sinusoidal oscillations and large amplitude, 1^0-type relaxation oscillations. The experimentally observed low-flow-rate 1^1 and 1^2 periodicities and their period-

Table 3.6. Scaled differential equations of a three-variable BZ model.

$$\mathrm{d}\bar{x}/\mathrm{d}\tau = T_o\{k_{C1}AH^2Y_o/X_o\bar{y} - k_{C2}HY_o\bar{y}\bar{x} + 0.5k_{C3}A^{0.5}H^{1.5}X_o^{-0.5}(C - Z_o\bar{z})\bar{x}^{0.5} - 0.5k_{C4}Z_o\bar{x}\bar{z} - 2k_{C5}X_o\bar{x}^2 - k_f\bar{x}\}$$

$$\mathrm{d}\bar{z}/\mathrm{d}\tau = T_o\{k_{C3}A^{0.5}H^{1.5}X_o^{0.5}(C/Z_o - \bar{z})\bar{x}^{0.5} - k_{C4}X_o\bar{x}\bar{z} - k_{C6}V_o\bar{z}\bar{v} - k_{C7}M\bar{z} - k_f\bar{z}$$

$$\mathrm{d}\bar{v}/\mathrm{d}\tau = T_o\{k_{C1}AH^2Y_o/V_o\bar{y} + 2k_{C2}HX_oY_o/V_o\bar{y}\bar{x} + k_{C5}X_o^2/V_o\bar{x}^2 + k_{C6}Z_o\bar{z}\bar{v} - k_f\bar{v}\}$$

where,

$$\tau \equiv t/T_o,\ \bar{x} \equiv x/X_o,\ \bar{z} \equiv z/Z_o,\ \bar{v} \equiv v/V_o$$

$$\bar{y} = \{k_{C6}Z_oV_o\bar{z}\bar{v}/(k_{C2}HX_o\bar{x} + k_{C1}AH^2 + k_f)\}/Y_o$$

and,

$$T_o = (10k_{C1}AHC)^{-1},\ X_o = k_{C1}AH^2/k_{C4},\ Y_o = 4k_{C1}AH^2/k_{C4},$$

$$Z_o = CA/(40M),\ V_o = 4AHC/M^2.$$

doubling bifurcations to chaos are also reproduced. The bifurcation diagram of the low-flow-rate complexity obtained with $k_{C6} = 60\ M^{-1}s^{-1}$ and $k_{C7} = 0.08\ M^{-1}s^{-1}$ is shown in Fig. 3.10 for simulated conditions equivalent to the experiment modeled.[69,70] The simulated appearance and sequence of periodic-chaotic windows at high flow rates also agrees with experiment.

Figure 3.11 shows a bifurcation diagram obtained with $k_{C6} = 30\ M^{-1}s^{-1}$ and $k_{C7} = 0.06\ M^{-1}s^{-1}$ in the neighborhood of two complex periodicities, where there appears to be an intermittency transition from periodicity to chaos as seen in experiments. The conditions are those of the high-flow-rate experiments of Hudson

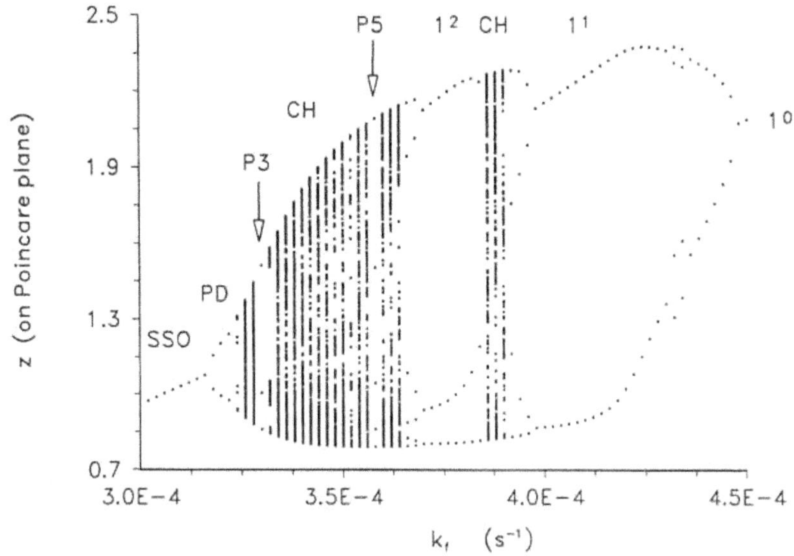

Figure 3.10. Bifurcation diagram of the low-flow-rate complexity obtained from the three-variable version of Model C [74] with increasing flow rate. The chemical conditions for these simulation are:[69,70] A = 0.1 M; M = 0.25 M; H = 0.26 M; C = 0.000833 M. Notation: SSO: small-amplitude sinusoidal oscillations; PD: period-doubling cascade; PX: period-X oscillations (X=3,5); CH: chaos; L^S: periodic states. Reprinted by permission from *Nature* (355, 808) copyright (1992) Macmillan Magazines Limited.

and Mankin[29] and Györgyi *et al.*[73] There are two differences between experiment and simulations at low flow rates with the parameters used.[74] The experimental small-amplitude, sinusoidal oscillations at low flow rates appear to lose their stability via a secondary-Hopf bifurcation rather than by period-doubling as in the simulations. Furthermore, chaos does not appear in the simulations between the low-flow-rate 1^0 and 1^1 periodicities where most of the experimental chaos was found. These shortcomings may be overcome in future studies because we did not exhaustively explore the k_{C6} - k_{C7} parameter plane in order to optimize agreement with experiment. It should also be noted that the elimination of Br^- from the model assumes that it is a fast variable. However, the system spends much time at low flow rates in the neighborhood of the high-$[Br^-]$, reduced steady state, suggesting that this approximation may not be very accurate in this case, resulting in some discrepancies.

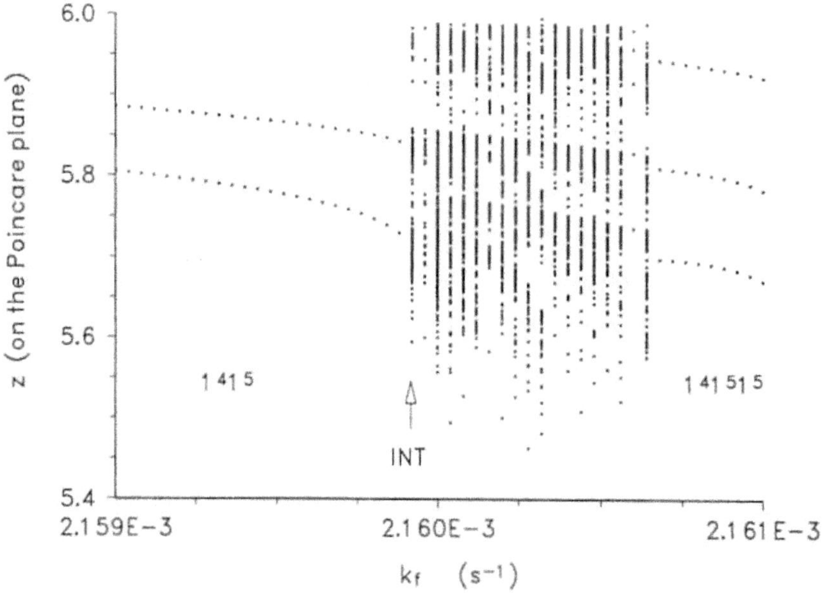

Figure 3.11. Bifurcation diagram of the high-flow-rate complexity obtained with the three-variable version of Model C between two complex periodic states[74] obtained with increasing flow rate. The chemical conditions for these simulations are:[29,73] A = 0.14 M; M = 0.30 M; H = 0.26 M; C = 0.001 M. Notation: INT: intermittency route to chaos; L^S: complex periodic states.

3.4. Chemical Interpretation of BZ-CSTR Chaos

A lot of effort has been expended on the phenomenological interpretation of BZ-CSTR chaos. Most of that work has been reviewed here. This approach is useful for understanding its underlying dynamics and for drawing analogies to other chaotic systems. However, no phenomenological model is able to reproduce the complete body of experimental fact. Furthermore, various phenomenological models designed to reproduce particular experiments often are incompatible with each other.

Here we look for the chemical basis of BZ-CSTR chaos. The success of the chemically devised 11-variable model (Table 3.3) and the four-variable models derived from it (Table 3.4) in simulating both low- and high-flow-rate complexity

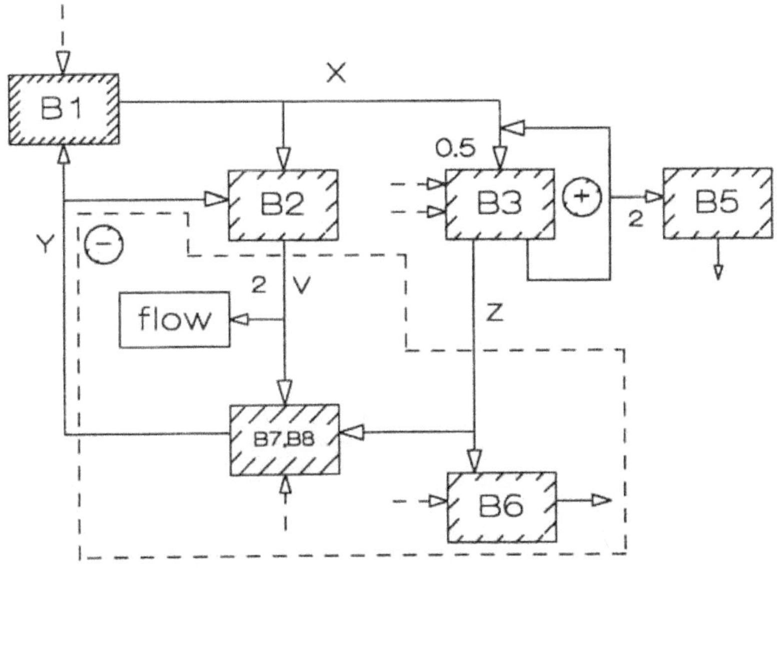

Figure 3.12. Schematic representation of the dynamic structure of Model B. Only the interactions important for our discussion are indicated. See text for details.

enables us to look for the chemical source of this exotic behavior. Figure 3.12 shows the most important reactions in Model B and their major interactions. Boxes represent chemical reactions, and solid lines represent the intermediates of variable concentration. The numbers give the stoichiometric coefficients if different from one. Dashed arrows indicate components of constant concentration. Light shading marks processes active during the oxidized, low-[Br^-] state. Heavy shading indicates that reaction B1 is active mainly during the reduced, high-[Br^-] state. No shading indicates that the CSTR flow is always an important process.

The mechanism can be broken into two parts. Reactions B1-B5 are the bistable part of the mechanism, and the reactions in the dashed box complete the negative feedback loop. Reactions B1-B5, together with a negative feedback loop like step O5 of Table 3.1, forms the "Oregonator-core" of the model. It is capable of exhibiting only steady states and simple (mostly relaxation) oscillations. However, the reactions in the dashed box additional to the Oregonator-core all participate in the delayed negative feedback loop, and they are the source of the observed complexity. Some of these reactions involve an added variable species, bromomalonic acid (V), which is not part the Oregonator core. Its effect is essentially to parameterize f and k_{O5} of the Oregonator step O5.

The role of BrMA in Model B is similar[57,68,72] to that of HOBr (P) in the SNB model when the FKN rate constants are used (see Section 3.1). Its concentration is large compared to the other species and hardly changes during a single oscillation. That is, there are only shallow oscillations of [BrMA] on the top of a very large concentration. Thus [BrMA] is a slow variable, and it can be treated[77] as a bifurcation parameter for the Oregonator-core through its effect on the effective values of f and k_{O5}, which control the rate and stoichiometry of Br^- regeneration from Ce(IV). Similarly to the interpretation of Janz *et al.*[44] for the parameterized-f Oregonator and Rinzel and Schwartz[34] for the SNB model, complex oscillations arise if [BrMA] oscillates across a subcritical Hopf bifurcation point of the Oregonator core between a stable focus (steady state) and large-amplitude oscillations.

Simulations and experiments indicate that this complexity is characterized by two frequencies. Bromomalonic acid participates in a second negative feedback mechanism in the vicinity of the high- and low-flow-rate Hopf bifurcations, and this generates a second frequency different from that of the fundamental frequency of the Oregonator-core. This occurs because BrMA is mainly produced by reaction B2 during the low-[Br^-] phase of the Oregonator-core oscillations, but its removal is primarily due to the CSTR flow. An oxidative excursion when the system is near the low-flow-rate bifurcation may generate enough BrMA to temporarily stabilize the reduced state of the Oregonator-core, where BrMA production is minimal.

Bromomalonic acid is removed by the flow while the system spirals towards the steady state, and oscillation may restart if the bifurcation border is recrossed in the other direction. The second negative feedback loop thus occurs because an increase in [BrMA] causes it to decrease. The second frequency is at least partially independent of the Oregonator-core frequency because it is governed by the flow rate. Thus the two frequencies do not necessarily entrain.

The situation is analogous at the high-flow-rate bifurcation. Here the intermittent BrMA production during the oxidized excursions is not enough to balance its removal by the flow, and the system ultimately starts to spiral toward the oxidized steady state of the Oregonator-core. As opposed to the events at low flow rates, this spiraling is characterized by continuous production of BrMA, and if the flow rate is not too large, [BrMA] may increase to a level where the Oregonator-core becomes oscillatory again. Thus a decrease in [BrMA] leads to an increase in [BrMA].

The importance of the CSTR flow rate is apparent from the above discussion, and it is clear why chaos only appears when the BZ reaction is run in a CSTR rather than a batch reactor. It also emphasizes that chemical chaos is essentially a phenomenon of coupled frequencies, as is discussed in Chapter 4. This analysis provides a useful guideline for both interpretation of chaos in other chemical reactions and the design of new chaotic chemical systems.

The question remains of what the effect of imperfect mixing is on the appearance of BZ-CSTR chaos. We believe that BZ-CSTR chaos almost certainly arises from the homogeneous chemical kinetics of the BZ reaction. However, there are stirring effects on BZ-CSTR chaos, and it seems equally certain that the exact form of chaotic behavior observed may in some cases be perturbed by mixing effects. We know of no work combining any of the chaotic models described here with mixing effects.

Acknowledgement. This work was supported by the National Science Foundation under grant CHE-8822886, by the OTKA of Hungary and by the Magyar Tudományért Alapitvány.

Literature cited

1. S.K. Scott, *Chemical Chaos* (Clarendon Press, Oxford, 1991).
2. M. Orbán and I.R. Epstein, *J. Phys. Chem.* **86** (1982) 3907.
3. J. Maselko and I.R. Epstein, *J. Chem. Phys.* **80** (1984) 3175.
4. M. Alamgir and I.R. Epstein, *Int. J. Chem. Kinet.* **17** (1985) 429.
5. J. Maselko, M. Alamgir and I.R. Epstein, *Physica* **19D** (1986) 153.

6. O. Citri and I.R. Epstein, *J. Phys. Chem.* **92** (1988) 1865.
7. A.D. Willamowski and O.E. Rössler, *Z. Naturforsch.* **35a** (1980) 317.
8. J. Hudson and O.E. Rössler, in *Modeling of Patterns in Space and Time*, eds. W. Jäger and J.D. Murray (Springer, Berlin, 1984).
9. H.C. Killory, O.E. Rössler and J.L. Hudson, *Phys. Lett.* **A122** (1987) 341.
10. J.L. Hudson, O.E. Rössler and H.C. Killory, *IEEE Trans. Circ. Syst.* **35** (1988) 902.
11. A.L. Kawczński, M. Misiurewicz and K.W. Leszczinski, *Polish J. Chem.* **63** (1989) 239.
12. B. Peng, S.K. Scott and K. Showalter, *J. Phys. Chem.* **94** (1990) 5243.
13. N. Samardzija, L.D. Greller and E. Wasserman, *J. Chem. Phys.* **90** (1989) 2296.
14. B.P. Belousov, in *Sb. Ref. Radiat. Med.* (Medgiz, Moscow, 1958) p. 145.
15. A.M. Zhabotinsky, *Biofizika* **9** (1964) 306.
16. R.J. Field, E. Kőrös and R.M. Noyes, *J. Am. Chem. Soc.* **94** (1972) 8649.
17. L. Györgyi, T. Turányi and R.J. Field, *J. Phys. Chem.* **94** (1990) 7162.
18. R.J. Field and R.M. Noyes, *J. Chem. Phys.* **60** (1974) 1877.
19. M. Varga, L. Györgyi and E. Kőrös, *J. Am. Chem. Soc.* **107** (1985) 4780.
20. H.-D. Försterling and Z. Noszticzius, *J. Phys. Chem.* **93** (1989) 2740.
21. H.-D. Försterling, S. Murányi and Z. Noszticzius, *J. Phys. Chem.* **94** (1990) 2915.
22. H.-D. Försterling and S. Murányi, *Z. Naturforsch.* **45a** (1990) 1259.
23. Z. Noszticzius, E. Noszticzius and Z.A. Schelly, *J. Phys. Chem.* **87** (1983) 510.
24. J.J. Tyson, in *Oscillations and Traveling Waves in Chemical Systems*, eds. R.J. Field and M. Burger (Wiley, New York, 1985) p. 93.
25. R.J. Field and H.-D. Försterling, *J. Phys. Chem.* **90** (1986) 5400.
26. G. Kshirsagar and R.J. Field, *J. Phys. Chem.* **92** (1988) 7074.
27. K. Tomita and I. Tsuda, *Prog. Theor. Phys.* **64** (1980) 1138.
28. J.L. Hudson, M. Hart and D. Marinko, *J. Chem. Phys.* **71** (1979) 1601.
29. J.L. Hudson and J.C. Mankin, *J. Chem. Phys.* **74** (1981) 6171.
30. M. Hourai, Y. Kotake and K. Kuwata, *J. Phys. Chem.* **89** (1985) 1760.
31. K.G. Coffman, W.D. McCormick, Z. Noszticzius, R. Simoyi and H.L. Swinney, *J. Chem. Phys* **86** (1987) 119.
32. N. Metropolis, M.L. Stein and P.R. Stein, *J. Comb. Theory* **15A** (1973) 25.
33. A.S. Pikovsky, *Phys. Lett.* **A85** (1981) 13.
34. J. Rinzel and I.B. Schwartz, *J. Chem. Phys* **80** (1984) 5610.
35. K. Showalter, R.M. Noyes and K. Bar-Eli, *J. Chem. Phys.* **69** (1978) 2514.
36. I.B. Schwartz, *Phys. Lett.* **A102** (1984) 25.
37. J. Maselko and H.L. Swinney, *J. Chem. Phys* **85** (1986) 6430.
38. F.W. Schneider and A.F. Münster, *J. Phys. Chem* **95** (1991) 2130; R. Bittersdorf, A.F. Münster and F.W. Schneider, *J. Phys. Chem.* **96** (1992) 5893.
39. K. Bar-Eli and R.M. Noyes, *J. Chem. Phys.* **88** (1988) 3646.
40. D. Barkley, *Phys. Lett.* **A129** (1988) 219.
41. J. Ringland, N. Issa and M. Schell, *Phys. Rev.* **A41** (1990) 4223.
42. P. Richetti, J.C. Roux, F. Argoul and A. Arneodo, *J. Chem. Phys.* **86** (1987) 3339.
43. D. Barkley, *J. Chem. Phys.* **89** (1988) 5547.
44. R.D. Janz, D.J. Vanacek and R.J. Field, *J. Chem. Phys* **73** (1980) 3132.
45. W.C. Troy, in *Oscillations and Traveling Waves in Chemical Systems*, eds. R.J. Field and M. Burger (Wiley, New York, 1985) p. 145.

46. J. Rinzel and W.D. Troy, *J. Chem. Phys.* **76** (1982) 1775.
47. L. Györgyi and R.J. Field, *J. Phys. Chem.* **92** (1988) 7079.
48. J.S. Turner, in *Kinetics of Physicochemical Oscillators*, Discussion meeting held by the Deutsche Bunsengesellschaft für Physikalische Chemie (Aachen, 1979) preprints of submitted papers, p. 61.
49. J.S. Turner, J.C. Roux, W.D. McCormick, Swinney, H.L., *Phys. Lett.* **A85** (1981) 9.
50. J. Ringland and J.S. Turner, *Phys. Lett.* **A105** (1984) 93.
51. J.S. Turner, *Adv. Chem. Phys.* **LV** (1984) 205.
52. R.J. Field, *J. Chem. Phys.* **63** (1975) 2289.
53. N. Ganapathisubramanian and R.M. Noyes, *J. Chem. Phys.* **76** (1982) 1770.
54. D. Barkley, J. Ringland and J.S. Turner, *J. Chem. Phys.* **87** (1987) 3812.
55. D. Lindberg, J.S. Turner and D. Barkley, *J. Chem. Phys.* **92** (1990) 3238.
56. J. Ringland, *J. Chem. Phys.* **95** (1991) 555.
57. L. Györgyi and R.J. Field, *J. Chem. Phys.* **93** (1990) 2159.
58. F. Argoul, A. Arneodo, P. Richetti and J.-C. Roux, *J. Chem. Phys.* **86** (1987) 3325.
59. P. Richetti, P. DeKepper, J.-C. Roux and H.L. Swinney, *J. Stat. Phys.* **48** (1987) 977.
60. L. Györgyi and R.J. Field, *J. Phys. Chem.* **93** (1989) 2865.
61. L. Györgyi and R.J. Field, *J. Chem. Phys.* **91** (1989) 6131.
62. J. Villermaux, in *Spatial Inhomogeneities and Transient Behaviour in Chemical Kinetics*, Eds. P. Gray, G. Nicolis, F. Baras, P. Borckmans, S.K. Scott (Manchester University Press, Manchester, 1990) p. 119.
63. E. Kumpinsky and I.R. Epstein, *J. Chem. Phys.* **82** (1985) 53.
64. K. Bar-Eli and R.M. Noyes, *J. Chem. Phys.* **85** (1986) 3251.
65. R.A. Schmitz, K.R. Graziani and J.L. Hudson, *J. Chem. Phys.* **67** (1977) 3040.
66. P. Ibison and S.K. Scott, *J. Chem. Soc. Faraday Trans.* **87** (1991) 223.
67. P. Ibison and S.K. Scott, *J. Chem. Soc. Faraday Trans.* **86** (1990) 3695.
68. L. Györgyi, S.L. Rempe and R.J. Field, *J. Phys. Chem.* **95** (1991) 3159.
69. Z. Noszticzius, W.D. McCormick and H.L. Swinney, *J. Phys. Chem.* **91** (1987) 5129.
70. Z. Noszticzius, W.D. McCormick and H.L. Swinney, *J. Phys. Chem.* **93** (1989) 2796.
71. T. Turányi, *New. J. Chem.* **14** (1990) 795.
72. L. Györgyi and R.J. Field, *J. Phys. Chem.* **95** (1991) 6594.
73. L. Györgyi, R.J. Field, Z. Noszticzius, W.D. McCormick and H.L. Swinney, *J. Phys. Chem.* **96** (1992) 1228.
74. L. Györgyi and R.J. Field, *Nature* **355** (1992) 808.
75. J.J. Tyson, *J. Phys. Chem.* **86** (1982) 3006.
76. J.J. Tyson, in *Nonlinear Phenomena in Chemical Kinetics*, eds. C. Vidal and A. Pacault (Springer, Berlin, 1981) p. 222.
77. J.D. Murray, *Mathematical Biology* (Springer, Berlin, 1989) p. 153.

CHAPTER 4: CHAOS IN FORCED AND COUPLED CHEMICAL OSCILLATORS AND EXCITATORS

MILOS MAREK
Department of Chemical Engineering
Prague Institute of Chemical Technology
166 28 Prague 6, Czechoslovakia

and

IGOR SCHREIBER
Department of Chemistry
Stanford University
Stanford, CA 94305, U.S.A.

ABSTRACT

Forced and/or interacting chemical reactions occurring in separate reactors can display a variety of dynamical behaviors, including chaos. Here we focus on systems which, if uncoupled, are either oscillatory or excitable. Forcing of a chemical oscillator by an external periodic constraint produces the well-known hierarchy of resonance responses and quasiperiodic oscillations, which frequently bifurcate into chaos. Such a behavior is explicable in terms of dynamics on an invariant torus and its subsequent disintegration. Similar behavior also occurs when an excitable system is periodically forced or coupled to an oscillator. We overview experimental work in this area and give some insight into the underlying theory by calculating bifurcation diagrams and by showing typical dynamics of forced/coupled chemical oscillators and excitators.

4.1. Introduction

Forced chemical oscillators and excitators can occur both as finite-dimensional, lumped-parameter systems, *e.g.*, a continuous-flow, stirred tank reactor or CSTR, described by a set of non-autonomous, nonlinear, ordinary differential equations, and as infinite-dimensional, distributed-parameter systems, *e.g.*, reaction-diffusion systems forced at the boundary, described by partial-differential or integral equations.

Similarly, chemical reactors coupled in arrays can be modelled by networks of coupled cells with each cell described by appropriate differential equations.

Motivation for the study of simple forced and coupled chemical systems stems, on one hand, from their significance to an interpretation and understanding of observed spatio-temporal patterns in chemical systems as such, and on the other hand, from their importance for the interpretation of the mechanism of generation of spatio-temporal patterns in often far more complex biological systems. Experiments with model chemical systems usually can be performed under well-controlled conditions and interpreted on the basis of relatively well-founded and documented kinetic (mechanistic) models.

Studies of chaos in forced and coupled chemical systems have been reviewed in two recent books[1,2] and are the subjects of frequent publications in specialized journals.[3-5]

We first overview briefly the types of dynamic behavior arising in forced oscillatory and excitatory systems and discuss specific features of dynamics in coupled systems (Section 4.2). In Section 4.3 we review currently available experimental observations and briefly discuss typical modelling studies. The review of experimental studies is focused on homogeneous reactions, but even there it is not comprehensive, reflecting the personal preferences and involvements of the authors. Section 4.4 is devoted to more detailed demonstration and discussion of the principal features of the origin of chaos in forced and coupled systems using low-dimensional dynamical systems as examples. A parallel treatment of oscillatory and excitatory systems differentiates this review from earlier ones.

4.2. Phenomenology

4.2.1 Forced Systems

The term oscillator denotes a dynamical system which, in the absence of external forcing, evolves toward a stable limit cycle, *i.e.*, oscillations are asymptotically periodic. On the other hand, the asymptotically approached limit set of an excitator is a stable steady state. However, disturbance of this steady state by a small but finite pulse perturbation leads to a response which substantially amplifies the perturbation before it eventually damps back to the steady state. The transient dynamics following the perturbation is called an excitable cycle, *i.e.*, a trajectory in the phase space to which other trajectories are attracted before final approach to the steady state. It is analogous in some sense to the limit cycle of an oscillatory system. In fact, the same chemical system can exhibit both oscillatory and excitable dynamics, depending on the external constraints imposed such as CSTR inlet concentrations or flow rate. Often only a moderate change of constraints is needed to switch from one dynamical mode to the other one.

Intuitively, due to its sensitivity to perturbation, the dynamics of a periodically perturbed excitator should share at least some of the features of the dynamics of a periodically perturbed oscillator. Our parallel treatment of both types of dynamical systems reveals some of similarities but points out some differences as well.

One most often uses the forcing amplitude A and frequency ω as externally controlled parameters in forced oscillator and excitator studies. It is somewhat more convenient for the purpose of this review to use the forcing period, $T = 2\pi/\omega$, instead of the frequency. The partition of the A - T plane into regions corresponding to various dynamical regimes is usually referred to as a *resonance diagram* for oscillatory systems and an *excitation diagram* for excitable systems, see Figs. 4.1 and 4.7, respectively. There is a particular mode of dynamics for any chosen value of A and T, which may, in general, include various kinds of periodic oscillations, quasiperiodicity and chaos. The observed dynamical behavior changes mode abruptly via bifurcations at discrete values of a varying parameter. The loci of bifurcations are curves dividing the A - T plane into sets, within each of which the dynamical behavior is qualitatively the same. Considering all of the bifurcation curves would produce a highly complex partition of the parameter plane because the bifurcations tend to accumulate in a small area, particularly in the presence of chaos. Neither experimental measurements nor computer modelling are capable of resolving all of the fine details of the bifurcation structure. However, knowledge of even only a few of the bifurcation curves facilitates the understanding of the possible dynamics of the system.

The structure of the resonance diagram when forcing is weak (A is small) is well known. Let T_0 be the period of the unforced oscillator. Periodic regimes are confined to a set of non-overlapping cusp-shaped regions, called *entrainment bands* or *Arnold tongues*, emanating from the T-axis at every value of T such that T/T_0 is rational, say p/q, where p, q are integers. There is a simple interpretation of the number p/q. The system will oscillate p times during q cycles of the external forcing before the temporal pattern is repeated. Such a phenomenon is also referred to as *phase locking* or *resonance*. Thus the overall period is qT. Mathematically, p/q is the rotation number, ρ. A geometrical interpretation of the dynamics with $\rho = p/q$ is a closed orbit that winds an invariant, two-dimensional torus p times in one direction and q times in the other. The entrainment bands are well-ordered in the A - T plane. If there are two bands with the rotation numbers p/q and r/s, then a band with the rotation number $(p+r)/(q+s)$ is located between them. Combining rational numbers in this way generates a Farey tree which defines an infinite ordering of the entrainment bands. The complementary set to the entrainment bands is a set of curves, each one corresponding to an irrational rotation number. When ρ is irrational, then the invariant torus is covered densely by a single orbit and the dynamics are quasiperiodic. A consequence of this structure is that ρ as a function of T is continuous, constant on intervals and increasing at points. Its graph contains an infinite hierarchy of plateaus called a *devil's staircase*.

However, this well-ordered structure may disintegrate with increasing forcing strength and lead to chaos if the coupling of the external to the internal oscillator is able to produce appropriate bifurcations, such as break up of the invariant torus followed by cascades of period doublings. Such a behavior is indicated by an overlap of some of the Arnold tongues. Regardless of the presence or the absence of chaos at intermediate values of A, the resonance diagram becomes simple again as A increases further. The invariant torus reappears but eventually shrinks to a periodic orbit which is stable for sufficiently large values of A. Simultaneously, Arnold tongues close up and disappear. There are several different manners of disappearance. Weakly resonant tongues (those with $q > 4$) vanish at tips where the torus degenerates into a periodic orbit. The remaining strongly resonant tongues display more complex bifurcation structure.[6]

Forced excitators display many of the features outlined above for oscillators. Their excitation diagram contains entrainment bands as well as sets of parameters corresponding to quasiperiodicity and chaos. The rotation number can be defined in a similar fashion. However, unlike for a perturbed oscillator, ρ is not expected to exceed the value of 1. In other words, a forced excitator normally cannot oscillate faster than the forcing term. This is a simple consequence of the fact that the unforced excitator does not possess a non-zero intrinsic frequency. There is, however, an exception giving the possibility of having an oscillatory response with a higher frequency than that of the forcing. This may happen when a relatively low frequency forcing of sufficiently high amplitude parametrically switches the excitator into an oscillator during a part of the cycle. Consequently, the response during one cycle consists of a series of spikes followed by a phase of quiescence. This phenomenon is called bursting[7,8] and has been studied mostly in systems with no external forcing.[8] Another implication of the oscillator/excitator difference is that entrainment bands exist only above a threshold value of A (which itself is dependent on T). A stable periodic regime is observed for small sub-threshold forcing amplitudes. Depending on the type of excitator, this regime is either 0/1-phase locked or it is not a resonance in the sense that no rotation number can be assigned to it. However, the picture for super-threshold forcing amplitudes when $\rho < 1$ is quite similar to a forced oscillatory system.

Poincare's time-discretizing method is a convenient way of looking at the dynamics of forced systems. This is done by stroboscopically recording data only once during a forcing cycle. This method reduces the dimension by one. For example, the dynamics on the torus is converted into a circle mapping. One can go even further in certain cases and reduce a multi-dimensional dynamical behavior to a one-dimensional mapping. This holds true in particular for systems which are perturbed by a periodic sequence of brief pulses. The transient resulting from the delivery of a pulse typically relaxes rapidly back to the limit or excitable cycle. A 1D-mapping approximating the dynamics can be derived if this is the case using the phase-transition curve (PTC)[9] or phase-excitation curve (PEC).[10,11] The idea is to define a phase on the limit (or excitable) cycle and find a relationship between the phases at which two consecutive pulse perturbations are applied. Such a mapping can be readily obtained from experiments by looking at the response to single perturbations delivered at different phases and subsequently used to model the periodically forced system.

The rich spectrum of dynamical behavior of forced systems was reviewed long ago.[12] A detailed analysis of typical responses can be based on:

1) Analytic studies of simplified systems close to bifurcation points, mostly using the small-forcing-amplitude assumption as in reference 13 and references quoted therein.

2) Detailed numerical studies most often including calculation of the boundaries of entrainment bands and other bifurcation sets.[7,14-19]

3) The study of corresponding low-dimensional mappings, which can be directly obtained from experimental time series in the form of Poincaré mappings or reduced 1D-maps. See references 21-28 and citations contained in references 1, 2, 19, 20.

4.2.2 *Coupled Systems*

Numerical studies of ergodicity in a nonlinear chain of coupled oscillators are classical.[29] It has been demonstrated among other results that continuum systems admit stable, localized excitations.[30] The term soliton is applied in a rough sense to any stable elementary excitation. Such solitons can decompose in a chaotic or turbulent fashion. Interaction and propagation of elementary excitations is associated with many elementary phenomena in solid-state physics and chemistry. The phenomenology of spatio-temporal patterns observed in coupled elementary cells has been reviewed recently.[2,31]

Let us assume that there are n oscillatory or excitatory units coupled in such a way that the coupling strength is represented by a single parameter. This parameter is analogous to the forcing amplitude. Clearly, the complexity of dynamics resulting from such coupling will depend on n and on the connectivity of the network. One can expect a high degree of complexity for two- or three-dimensional networks consisting of a large number of cells. However, even the simplest arrangement of two units can produce complex chaotic oscillations as a result of mutual coupling. The situation is analogous to forced systems except that the coupling is bi-directional. For an oscillator-oscillator system the dynamics is again phase locked or quasiperiodic for low coupling strengths. Increasing the intensity of interaction may break up the torus and cause the onset of chaos. However, the analogy with forced systems may be lost at an intermediate coupling strength as stationary states may appear. Obviously, periodically forced systems cannot display stationary states.

The case of an oscillator-excitator pair is analogous to the periodically forced excitator and, again, one may expect all the bifurcation features to be carried over, except for stationary states. The third possible combination, excitator-excitator, is not expected to be analogous, though, since none of the coupled units shows an intrinsic rhythmicity.

New factors come into the play in the case of $n > 2$. If all the units are oscillators, then for weak coupling there is an n-dimensional invariant torus. The number of various phase-locked states, quasiperiodic regimes and their combinations grows rapidly as n is increased. A new phenomenon is the possibility of chaos even without breaking up of the n-torus.[32,33] The disintegration of the torus may still be the dominant mode of chaos generation for intermediate coupling strengths. The dynamics does not significantly depend on the connectivity of the network when the coupling is weak enough. This universal behavior derives from structural stability arguments, but a strong enough coupling is likely to destroy the torus in one way or another, and steady-state Turing structures strongly dependent on the network connectivity are likely to emerge.

As in the case of two units, mixed oscillator-excitator networks are expected to preserve some of the phase locking-quasiperiodicity-chaos structure, whereas all-excitator networks are not.

So far we have discussed complexity arising from mutual interaction. There is, in addition, the possibility of chaos generation due to the coupling of a chaotic oscillator with periodic oscillators or excitators. A single chaotic cell can induce chaotic oscillation in the entire network.

Similarly, as in studies of forced systems, the use of analytic techniques, numerical methods and studies of simplified model systems facilitates the understanding of the mechanisms of formation and development of chaos. Systems with weak coupling or phase-related descriptions often are used in analytic approaches.[34,35] The use of numerical methods for the studies of coupled cells has been reviewed.[2] Recently, globally coupled dynamical systems, *e.g.*, globally coupled maps, have been the subject of intense study.[35,36,37] Periodically forced oscillatory and excitatory reaction-diffusion systems described by coupled sets of PDEs have been studied,[38] and chaos resulting from spatially periodic forcing of a spatially periodic array of oscillators has been investigated.[39]

4.3. Experimental Observations

4.3.1 Forced Systems

The Belousov-Zhabotinsky (BZ) reaction has been used most often in studies of periodic forcing of chemical oscillators. Quasiperiodic responses were observed in the 1960s in a cerium-catalyzed batch system perturbed by ultraviolet radiation.[40,41] The response of the classical BZ reaction running in a CSTR to periodic forcing by variation of the feedstream [Ce^{4+}] has been investigated, and quasiperiodic and aperiodic dynamics observed.[42]

Various periodic and quasiperiodic patterns also have been observed in a CSTR when the Briggs-Rauscher reaction is subjected to periodic illumination.[43] Stimulations of both oscillatory and excitable states were performed, and phase transition curves (PTC) constructed. However, no chaotic behavior was found.

Resetting of the phase of BZ oscillations by injection of Br^- and Ce^{4+} also has been studied, and the transition of a PTC of topological degree one to one of topological degree zero reported.[44] The phase transition curves were used later in an interpretation of more detailed forcing studies[22,24,45] in which the effect of the amplitude and the period of forcing by injection of pulses of Br^- was determined. Phase-locked periodic and chaotic dynamics were observed. Both strong (*e.g.*, 7/3) and weak (*e.g.*, 12/5) periodic resonance regimes were found to alternate with chaotic regimes characterized by positive Lyapounov exponents. The PTC was determined from phase resettings following single stimulations at different phases. It then was used in computer simulations to confirm the chaotic response to periodic stimulation. The effects of additive noise were analyzed on the basis of numerical solution of the corresponding Langevin equation. It was found that random noise causes narrowing of the resonance windows and some of the positive Lyapounov exponents to change sign. It was concluded that both true chaos and noise-induced randomness were present in the experimental data.

Schneider and coworkers[46,48] demonstrated with the BZ reaction that sinusoidal variation of the CSTR flow rate causes entrainment in several strong resonance regimes and the occurrence of quasiperiodic oscillations. Hudson and coworkers[49,50] varied the CSTR feedsteam [Br^-] periodically and observed both entrained periodic and chaotic regimes. Quasiperiodic oscillations were observed with smaller forcing amplitudes. A period-doubling cascade leading to chaos was found at higher forcing strengths as the period of the Br^- forcing was varied. The broad-band power spectra and stroboscopic maps confirmed the presence of deterministic chaos. The character of the entrainments in the forced BZ reaction in oscillatory and excitable regimes in a CSTR have been compared.[27,28]

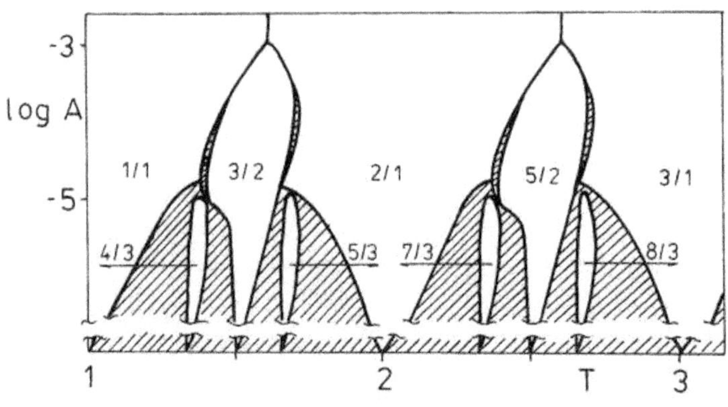

Figure 4.1. Resonance diagram in the parameter plane A - T for the BZ reaction in a CSTR forced by pulse perturbations of Ag^+: A - amplitude of the Ag^+ pulses (amount of added Ag^+); T - forcing period; • - periodic regimes; ∘ - aperiodic regimes. Reproduced by permission from M. Dolnik, J. Finkeova, I. Schreiber and M. Marek, *J. Phys. Chem.* **93** (1989) 2764.

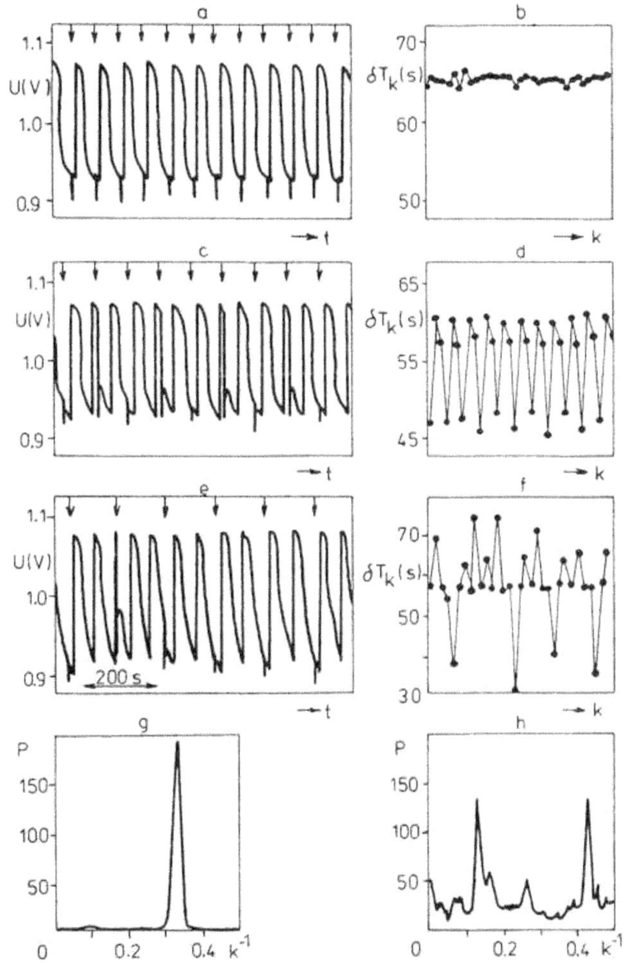

Figure 4.2. Experimental periodic and aperiodic time series with corresponding "periodograms" and power spectra. Reproduced by permission from M. Dolnik, J. Finkeova, I. Schreiber and M. Marek, *J. Phys. Chem.* **93** (1989) 2764.

A typical experimental resonance diagram is shown in Fig. 4.1. An intermingling of p/q-phase locking with aperiodic regimes is seen. Typical redox potential time series illustrating 1/1 and 3/2 resonance regimes and an aperiodic regime with corresponding "periodograms" (*i.e.*, sequences of interspike time intervals) and power spectra are shown in Fig. 4.2. Simulation of experimentally observed dynamical responses and confirmation of the of the presence of chaos often can be based on experimentally determined PTCs.[22,24] Two types of PTCs for experiments on the Br^--forced BZ reaction in a CSTR are shown in Fig. 4.3. The presence of chaos is indicated by sufficiently developed extrema of the PTC. A direct verification can be done by calculation of the Lyapounov exponent. Another indication of the presence of chaotic behavior in experiments is its simulation on the basis of appropriate kinetic models. Study of the "expanded Oregonator" model[27] used in the simulation of a Br^--forced CSTR confirms the presence of chaotic behavior. (See Fig. 4.4).

Both entrained periodic and aperiodic (chaotic) regimes have been observed in experiments with Ag^+-pulse forcing of an excitable BZ reaction mixture in a CSTR.[28] An example of an aperiodic experimental time series is shown in Fig. 4.5. The stroboscopic map, Fig. 4.5c, suggests that the dynamics can be described as random jumps between regimes with firing numbers 1/1 and 1/2. The firing number is defined as the average number of large-amplitude responses per one forcing period. It is an experimental quantity corresponding to the rotation number. The

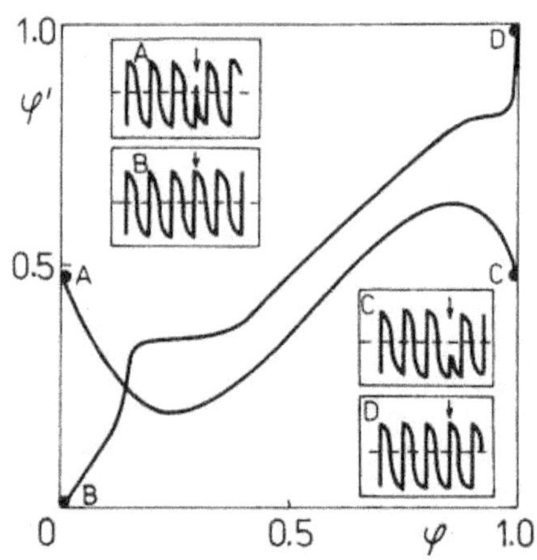

Figure 4.3. Phase transition curves: curve AC - PTC of type "0"; curve BD - PTC of type "1"; φ - old phase of oscillations; φ' - new phase of oscillations. Reproduced by permission from M. Dolnik, J. Finkeova, I. Schreiber and M. Marek, *J. Phys. Chem.* **93** (1989) 2764.

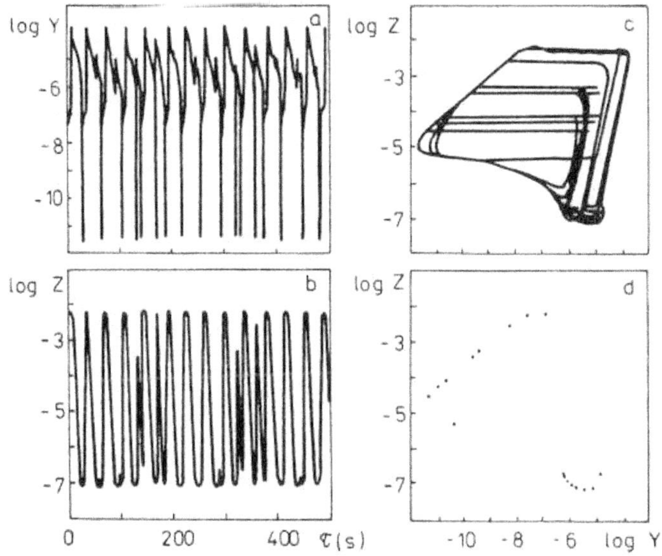

Figure 4.4. Chaotic oscillations found numerically in the forced "expanded Oregonator" model: $T = 1.35$; $A = 1.3 \times 10^{-5}$ M; a), b) - time evolution of $[Br^-]$ and $[Ce^{4+}]$, respectively; c) - phase-plane log $[Br^-]$ - log $[Ce^{4+}]$; d) - stroboscopic map. Reproduced by permission from M. Dolnik, J. Finkeova, I. Schreiber and M. Marek, *J. Phys. Chem.* **93** (1989) 2764.

devil's staircase-like structure of the plot of firing number *vs* forcing period presented for the excitable system in Fig. 4.6 is quite similar to the analogous plot for the oscillatory system. The aperiodic regimes are denoted by dotted lines. The observed aperiodic regimes appear to be random combinations of $1/q$ and $1/(q+1)$ periodic resonance regimes. The excitation diagram is shown in Fig. 4.7. The entrainment bands indicate occurrence of periodic resonances with specified firing numbers. Aperiodic regimes can be found in the regions where periodic resonances overlap. The presence of an excitation threshold differentiates the excitation diagram from the resonance diagram for the forced oscillatory system (*cf.* Fig. 4.1).

Both the PTC for an oscillatory system and the PEC for an excitable system determined on the basis of single stimulation experiments have been used to construct resonance and excitation diagrams.[27,28] The construction of excitation

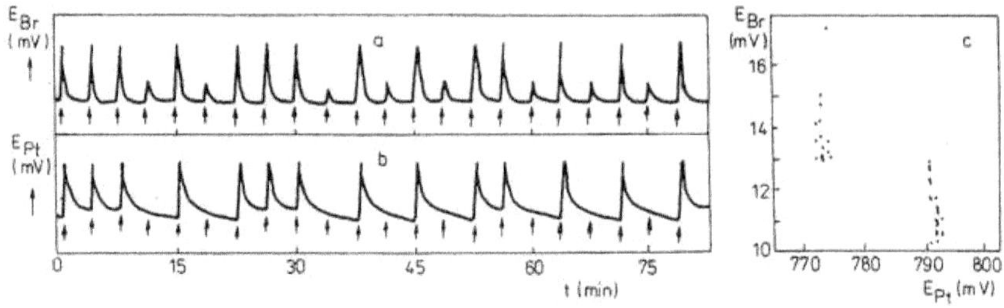

Figure 4.5. Pulse forcing of an excitable BZ reaction in a CSTR by Ag^+: a) - potential of a Br^--selective electrode; b) redox potential at a Pt electrode; c) - stroboscopic map $E_{Br^-} - E_{Pt}$. Reproduced by permission from J. Finkeova, M. Dolnik, B. Hrudka and M. Marek, *J. Phys. Chem.* **94** (1990) 4110.

diagrams using various types of PECs determined from experiments and Oregonator-type models also has been discussed.[10,11]

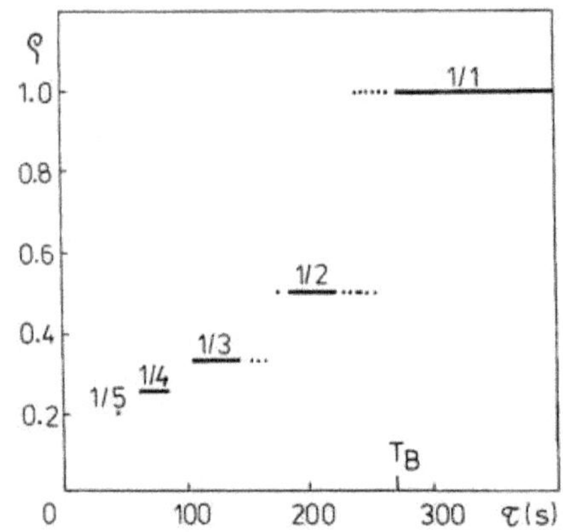

Figure 4.6. Dependence of the firing number, ρ, on the forcing period T for an excitable BZ reaction mixture: $A = 1 \times 10^{-5}$ M $AgNO_3$. Reproduced by permission from J. Finkeova, M. Dolnik, B. Hrudka and M. Marek, *J. Phys. Chem.* **94** (1990) 4110.

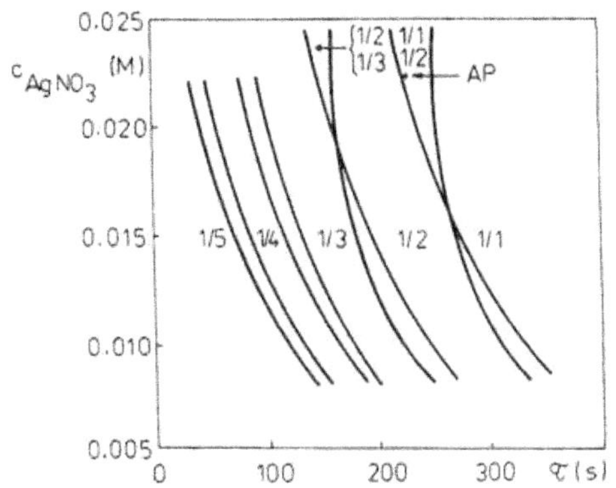

Figure 4.7. Excitation diagram for the BZ reaction in a CSTR forced by $AgNO_3$ pulses: AP - aperiodic regime shown in Fig. 4.5. Reproduced by permission from J. Finkeova, M. Dolnik, B. Hrudka and M.Marek, *J. Phys. Chem.* **94** (1990) 4110.

Dynamical patterns of traveling pulse waves have been studied experimentally in an excitable, modified BZ reaction mixture contained in a narrow capillary tube perturbed periodically by decreasing the local $[Br^-]$ at one end.[27] Both phase-locked periodic regimes and chaotic regimes were observed. A complete excitation diagram showing entrainment bands in the amplitude-period plane has been obtained from computer simulations of a related reaction-diffusion problem.[51] Two types of chaotic wave patterns were observed:

a) Waves generated irregularly directly at the boundary and propagated along the spatial coordinate.

b) A chaotic pattern arising during the propagation of an originally periodic wave pattern due to the interaction of transport and reaction somewhere along the spatial coordinate.

Bursting dynamics, often observed in various neurophysiological systems,[8] can be periodic or chaotic as well. Both types of bursting patterns have been observed recently as a result of periodically forcing the BZ reaction in a CSTR by a square wave variation of the inlet $[Br^-]$. The amplitude and period of the forcing were varied in such a way that the high value of the inlet concentration corresponded to an excitable regime and the low value to an oscillatory one.[52]

Probably the first *in vitro* biochemical system to be subjected to periodic perturbation in a controlled manner was a glycolytic system. Hess and coworkers[53] investigated the response of autonomous glycolytic oscillations in a yeast extract to a periodic square-wave input of the substrate. In addition to subharmonic entrainment and quasiperiodic behavior, an irregular behavior was reported for a perturbing frequency close to the entrainment edge. Chaos also was observed for sinusoidal glucose input into extracts of *Saccharomyces cerevisiae* in a CSTR.[54]

Chaotic behavior of forced biological systems has been the subject of many studies.[55] For example, phase-locked, quasiperiodic and chaotic sequences of impulse production (firing rates) resulting from periodic current stimulation of squid giant axon have been reported.[56] Three routes to chaos (successive period-doubling, intermittency, and collapse of a torus) were described. It is interesting that the sequence of rotation numbers of the phase-locked states satisfies a one-sided Farey sequence similar to that observed in a periodically forced, excitable BZ reaction in a CSTR.[10]

4.3.2 Coupled Systems

Three types of CSTR coupling have been used to this point for experimental studies of coupled reactors:

a) Coupling by a direct convective exchange of mass between CSTRs through a porous wall.[57-68]

b) Coupling by an indirect convective exchange of mass between CSTRs via controlled pumping between the reactors.[69,72]

c) Electrical coupling.[70,71]

The reactors in a) are coupled through a porous wall of variable area. The exchange coefficients characterizing the intensity of coupling between the cells are evaluated from the results of independent tracer experiments. Thermostatting at various temperatures, use of different stirring rates and/or different compositions of inlet reaction mixture to the CSTRs can be used for setting up different oscillation frequencies in the reactors. Various stationary and oscillatory regimes are observed as the coupling strength is varied. For example, non-homogeneous concentration profiles along the array of reactors ("dissipative structures") have been recorded in

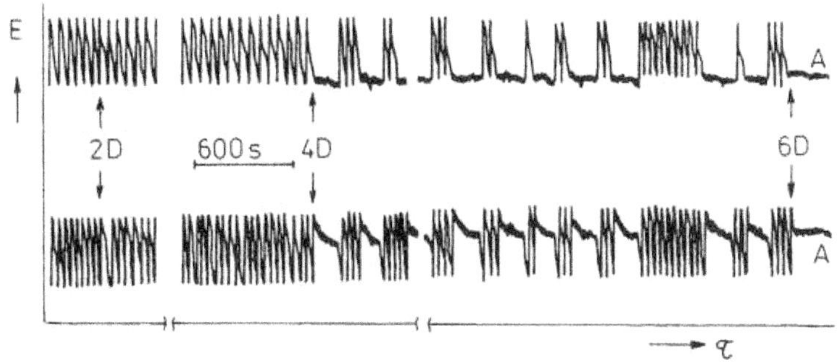

Figure 4.8. Temporal evolution of the redox potential showing chaotic bursting and an asymmetric steady state in two coupled CSTRs with the classical BZ reaction: A - asymmetric steady state; D - relative unit of the coupling intensity. Reproduced by permission from M. Marek, in *Lecture Notes in Biomathematics, Modelling of Patterns in Space and Time*, (Springer, Berlin 1984). p. 214.

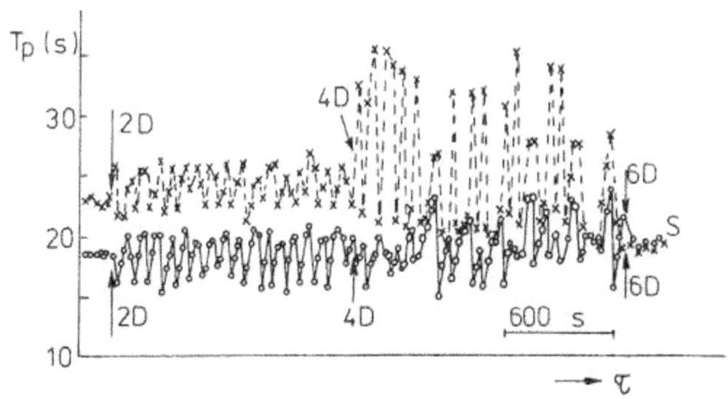

Figure 4.9. Sequence of interspike time intervals (a "periodogram") with varying coupling strengths showing a progression of periodic, quasiperiodic and chaotic regimes in two coupled CSTRs with the classical BZ reaction. Uncoupled oscillation periods, T_{P1} = 22.8 s; T_{P2} = 18.4 s; S - 1/1 synchronized regime. Reproduced by permission from M. Marek, in *Lecture Notes in Biomathematics, Modelling of Patterns in Space and Time*, (Springer, Berlin, 1984). p. 214.

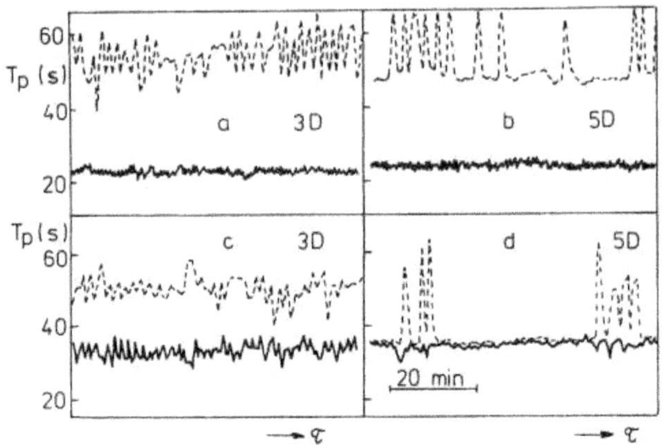

Figure 4.10. "Periodogram" of chaotic oscillations and intermittency in two coupled CSTRs with the classical BZ reaction: a, c) chaotic oscillations, coupling intensity 3D; b, d) intermittency, coupling intensity 5D; Uncoupled oscillation periods: a,c) T_{P1} = 24.1 s and T_{P2} = 49.4 s; c,d) T_{P1} = 35.0 s, T_{P2} = 50.0 s. Reproduced by permission from M. Marek, in *Lecture Notes in Biomathematics, Modelling of Patterns in Space and Time*, (Springer, Berlin 1984). p. 214.

a system of seven, identical, linearly coupled reactors.[73] A chaotic bursting regime observed in two coupled CSTRs with the BZ reaction is shown in Fig. 4.8. For an intensity of mutual mass exchange equal to 4D (D is a chosen value of the mass-exchange coefficient used as a relative unit) an irregular alternation of periods of quiescence with bursts of oscillations of various length can be seen. When the intensity of mass coupling is increased to 6D, oscillations in both cells disappear and an asymmetric steady state with different levels of the redox potential in the two cells (simplest Turing structure) develops

The "periodogram" (*i.e.*, the sequence of interspike time intervals) recorded for different values of the exchange coefficient in two coupled reactors with the BZ reaction is shown in Figs. 4.9 and 4.10. If the ratio of periods in the uncoupled reactors is close to one, then noisy quasiperiodic oscillations, chaotic oscillations and 1/1-phase locking are observed for values of the exchange coefficient of 2D, 4D, and 6D, respectively (See Fig. 4.9). An interesting dynamics is observed when the uncoupled periods differ significantly (See Fig. 4.10). If the ratio is close to 1/2, then at intermediate interaction intensities (*e.g.*, 3D) there are large irregular fluctuations in the reactor with slower oscillations. However, only small fluctuations are observed in the reactor with faster oscillations (See Fig. 4.10a). The behavior is intermittent for larger coupling intensities (*e.g.*, 5D) in the reactor with slower oscillations (See Fig. 4.10b). As before, oscillations are nearly periodic in the faster reactor. The interaction between the cells seems to be effectively one-directional as if the faster reactor were independently driving the slower one. If the ratio of uncoupled periods is close to 5/7 the behavior is again chaotic for lower interaction intensities (*e.g.*, 3D), (See Fig. 4.10c), and intermittent for a higher interaction intensities (*e.g.*, 5D) (See Fig. 4.10d). However, the fluctuations in the faster reactor are much larger in this case.

Thus the intermittency and torus-breaking routes to chaos apparently have been observed in the above experiments. The period-doubling route, which often occurs over a narrow interval of the control parameter has not been found. Both the period-doubling and the torus-breaking routes to chaos have been found in simulations with two coupled Brusselators.[2,74,75]

Sawada and coworkers[62,63] performed experiments with the BZ reaction in closed, coupled reactors. They claim the presence of chaotic behavior in addition to phase-locked regimes .

Crowley and Field[70,71] electrically coupled two CSTRs containing the BZ reaction with acetylacetone replacing malonic acid. A large-area platinum electrode was placed in each of the reactors and connected through an external circuit and an internal ion bridge. If the $[Ce^{4+}]/[Ce^{3+}]$ ratio is different in the two reactors, the resulting potential difference drives a current through the system, causing the electrical coupling. The current is carried by reaction of Ce^{4+} and Ce^{3+} at the electrodes, which causes $[Ce^{4+}]/[Ce^{3+}]$ to approach the same value in both CSTRs. A large resistance introduced into the external electrical circuit corresponds to uncoupled reactors. A decrease of this resistance then represents an increase in coupling strength. Weak coupling leads to mutual entrainment or quenching of the oscillations. Similarly to the case of convective coupling, it was observed that if the uncoupled cells have different frequencies, then at low coupling strength the slow reactor is driven by the faster one, but not *vice versa*. Subharmonic entrainment, quasiperiodicity, bursting oscillations and chaos also were observed.

Experiments have been reported[64-67] with the BZ reaction in which two reactors are coupled, and one of them is in addition periodically forced. The reactors are coupled directly via mass exchange through a porous wall. Both single and periodic pulse perturbations with Br^- or Ag^+ were used to study the dynamics of the

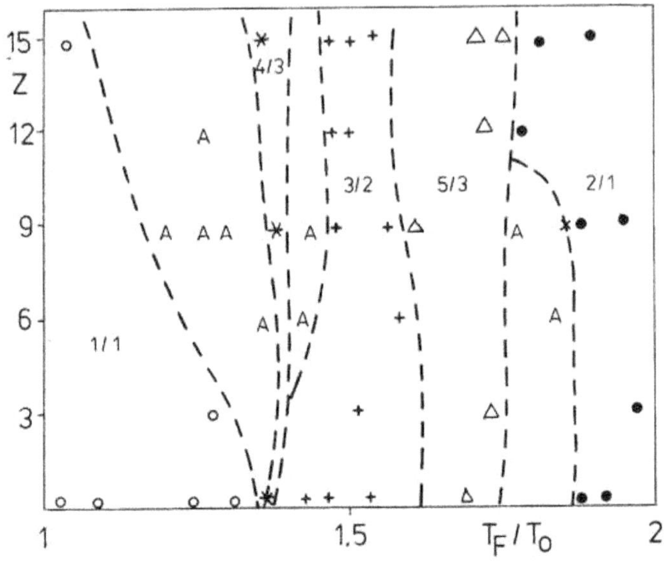

Figure 4.11. Resonance diagram for the BZ reaction in a CSTR when periodically stimulated by Br^- ions. The intensity of coupling, Z, *vs* relative period of stimulation, T_F/T_0, is shown: stimulation amplitude, $a = 5 \times 10^{-5}$ M Br^-; o - 1/1 phase locking; * - 4/3 phase locking; + - 3/2 phase locking; ▵ - 5/3 phase locking; • - 2/1 phase locking; x - 7/4 phase locking; A - aperiodic regime. Reproduced by permission from M. Dolnik, E. Padusakova and M. Marek, *J. Phys. Chem.* **91** (1987) 4407.

response to perturbations of various amplitudes and frequencies. If the unforced and uncoupled reactors are oscillatory, then forcing by Br^- causes the resonance periodic regimes corresponding to the first two rows of the Farey tree to intermingle with aperiodic dynamics[64] (See the resonance diagram in Fig. 4.11.) A dynamical feature specific to this system is a *transient extinction of oscillations* which also could be called *chaotic bursting* or *intermittent chaos.*[67] Similar dynamics are observed when uncoupled cells are excitable and Ag^+ is used as the perturbant.[66] Various phase-locked regimes have been found, *cf.* Fig. 4.12, as well as aperiodic oscillations. Two qualitatively different parts of a time series from one experimental run are compared in Fig. 4.13. One type of dynamics, shown in Fig. 4.13a-c, exhibits synchronous aperiodic oscillations of the redox potentials in both cells. Another behavior, shown

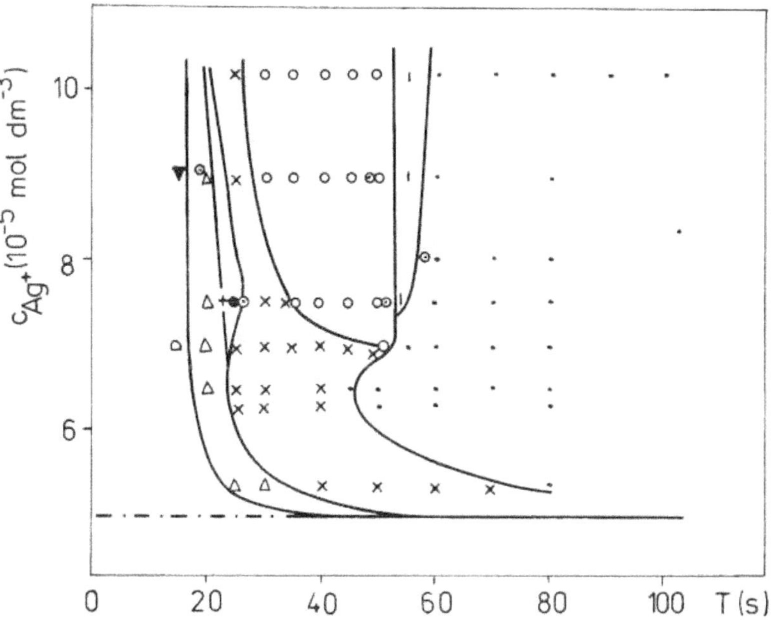

Figure 4.12. Excitation diagram for the BZ reaction when periodically perturbed by Ag^+ in a CSTR. The concentration of added Ag^+ is shown *vs* period of stimulation, *T*. Different phase-lockings are characterized by their rotation vector, $\rho_{1,2}$ = (*m*/*n*, *r*/*s*), indicated by the symbols: • - (1/1, 1/1); " - (1/1, 1/2); ○ (1/1, 0/1); × - (1/2, 1/2); ● - (1/2, 1/4); + - (1/2, 0/1); △ - (1/3, 1/3); ▽ - (1/4, 1/4) and ▲ - (1/4, 0/1). The curves separating entrainment bands are approximate. The relative coupling strength, *z*, is 10. Reproduced by permission from J. Kosek and M. Marek, *J. Phys. Chem.*, submitted (1992).

in Fig. 4.13d-f, can be understood as an irregular switching between periodic regimes with rotation vectors $\rho_{1,2}$ = (1,1/2) and $\rho_{1,2}$ = (1,0). Here the two components of $\rho_{1,2}$ are phase-locking ratios in both cells with respect to the external forcing. The difference between the two types of dynamics also is reflected in the phase-plane of measured voltages, E_{Pt}, E_{Br}, of two independent electrodes. See Figs. 4.13c and f. The aperiodic regime found at a higher stimulation amplitude is shown in

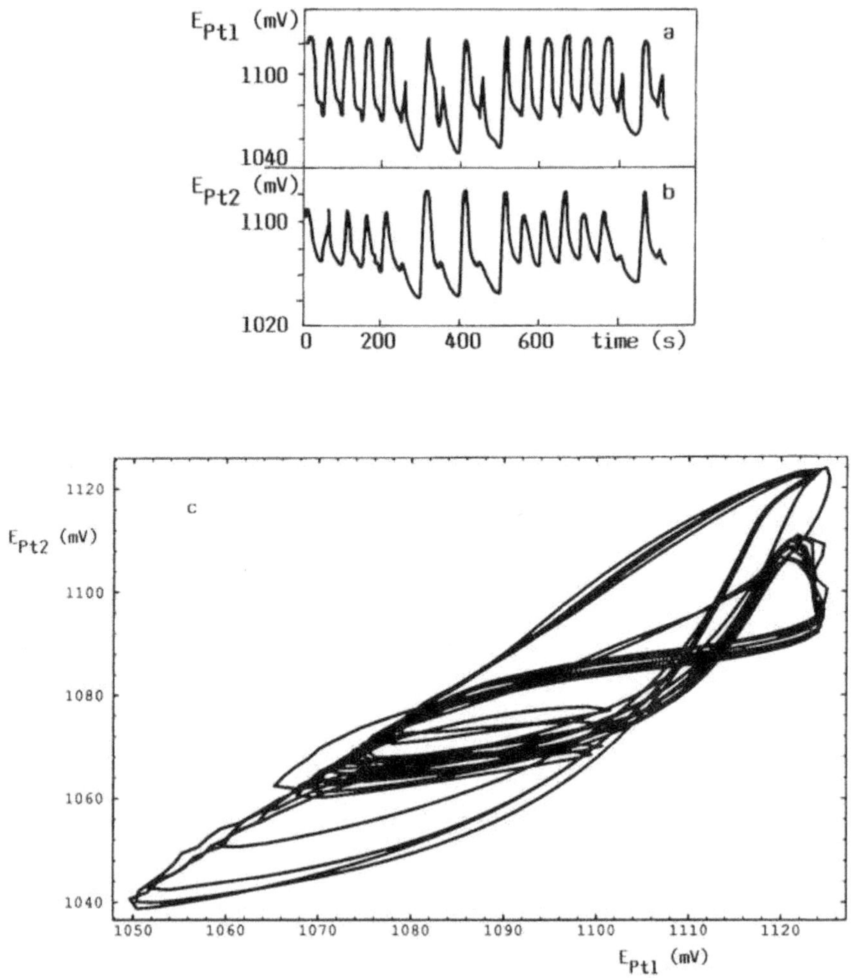

Figure 4.13. (a-c) Aperiodic dynamical response of two coupled CSTRs subjected to periodic perturbations of the first cell: relative coupling strength, z, is 10; a)-c) $T = 50$ s; d)-f) $T = 58$ s; $[Ag^+] = 7.5 \times 10^{-5}$ M. The $E_{Pt\,1}$, $E_{Pt\,2}$ platinum electrode potentials are shown for both reactors. Reproduced by permission from J. Kosek and M. Marek, *J. Phys. Chem.*, submitted (1992).

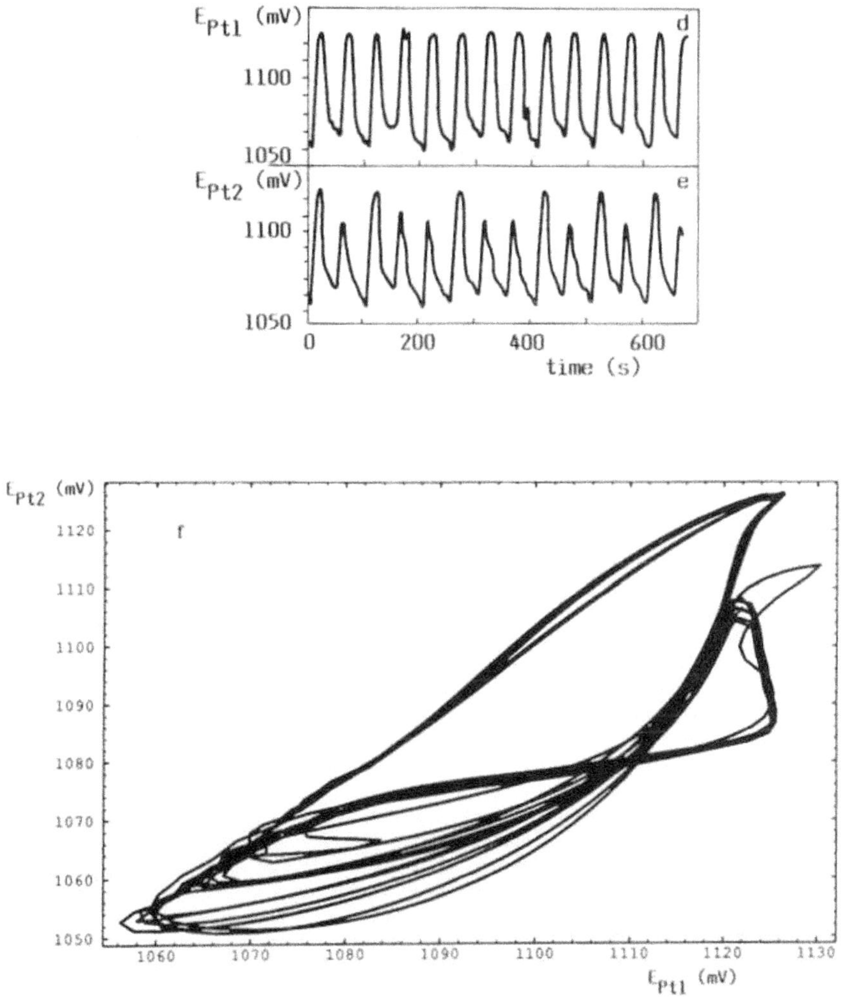

Figure 4.13. Continued (d - f).

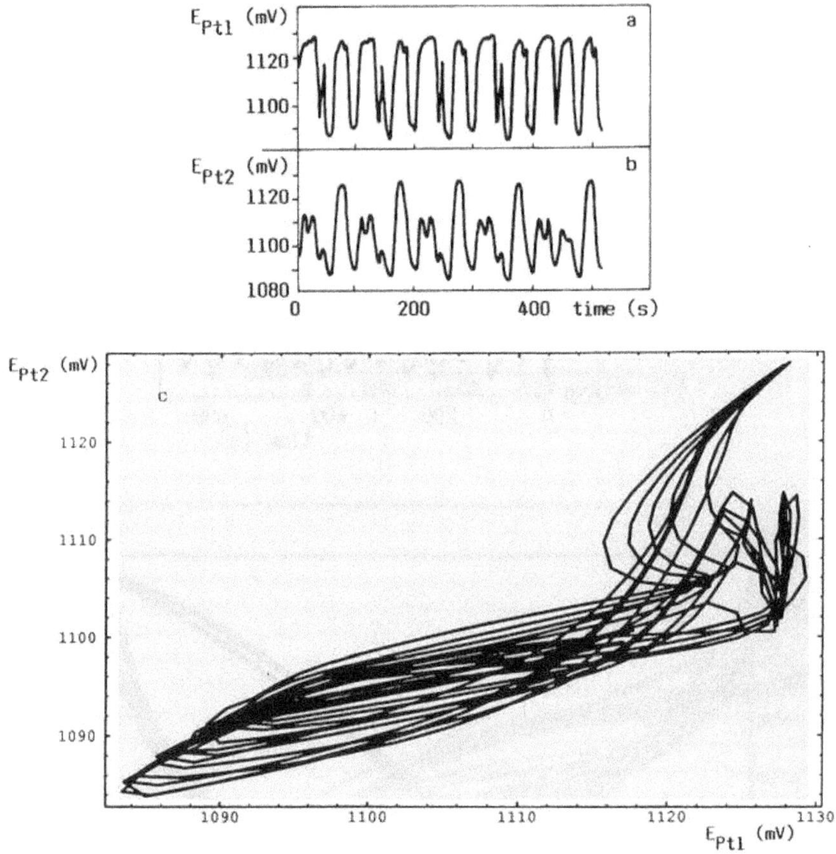

Figure 4.14. An aperiodic dynamical response of two coupled CSTRs subjected to periodic perturbations of the first cell: The relative coupling strength, z, is 10; T = 20 s; $[Ag^+]$ = 9.0 x 10^{-5} M. $E_{Pt\,1}$, $E_{Pt\,2}$ are the platinum electrode potentials in the two reactors. Reproduced by permission from J. Kosek and M. Marek, *J. Phys. Chem.*, submitted (1992).

Fig. 4.14. This regime is close to a phase locking with $\rho_{1,2}$ = (2/5,1/5), but in spite of that, the trajectory in the phase plane densely fills an enclosed area.

The phase shift between two coupled oscillators also has been a subject of experimental investigation. Both in-phase and out-of-phase regimes have been reported[72] for nearly identical mass-flow coupled oscillators. Numerical modelling of coupled Brusselators reveals similar regimes.[2]

Coupling and forcing of chemical oscillators and excitators also can occur in a single reactor if two chemically separate reacting subsystems are mutually coupled and/or forced via the interaction of intermediates common to both reactions. Complex dynamical behavior including aperiodic oscillations have been reported by Alamgir and Epstein[76] for the coupled chlorite-bromate-iodide and bromate-iodide-manganous systems.

4.4. Computer Modelling - A Comparative Study

We numerically examine in this section the effects of external sinusoidal perturbation on an oscillatory or excitable system, as well as the effects of coupling two oscillators via diffusion. Simple prototypical chemical systems are used as examples. One can see directly the relationship between forcing and coupling because they are applied to the same system. While some of the observed phenomena, in particular those occurring in the forced oscillatory systems, are well documented in the literature, others are much less familiar.

Our prototype of a chemical oscillator is the well-known Brusselator.[77]

$$[A] \rightarrow X$$

$$2X + Y \rightarrow 3X$$

$$X + [B] \rightarrow Y + \text{Products}$$

$$X \rightarrow \text{Products}$$

The [ ] denote pool components, while X and Y are the variable species. The prototype of excitable dynamics is a biochemical model due to Sel'kov[78] of the metabolism of cytoplasmic thiols. It has the chemical interpretation shown below.

While the Brusselator does not show any behavior other than a stable steady state or stable limit cycle oscillations, the Sel'kov model displays not only excitability but oscillations and various kinds of bistability as well. Both prototypes are described by the two dynamical variables, X and Y, but their chemistry is quite different. In both models X is involved in a higher-order direct autocatalysis, but the way in which Y controls the production of X is different in the two systems. In addition, there is

$$[A] \rightarrow Z$$

$$Z \rightarrow \nu_0 X$$

$$Z + \gamma X \rightarrow (\gamma + 1)X$$

$$X \rightarrow \beta Y$$

$$X + Y \rightarrow 2Y$$

$$Y \rightarrow \text{Products}$$

another species Z in the Sel'kov model analogous to Y in the Brusselator that is assumed to be in a quasi-steady-state and which accounts for the fractional exponent in the evolution equations.

Let us examine the oscillator first. The evolution equations for the Brusselator with X perturbed by an additive cosine term are,

$$\mathrm{d}x/\mathrm{d}t = a - (b + 1)x + x^2 y + A\,(1 + \cos(2\pi t/T)),$$

$$\mathrm{d}y/\mathrm{d}t = bx - x^2 y, \qquad (4.1)$$

where x and y are dimensionless concentrations of the species X and Y. This equation can be interpreted as perturbing the pool concentration, a, of the species A. The physically unrealistic situation of having the parameter a take on negative values is avoided by offsetting the cosine term by 1. This equation (with the offset equal to zero) has been extensively investigated by Kai and Tomita.[14] All the bifurcation diagrams presented here were calculated using a numerical continuation package[2] based on the work of Kubicek and coworkers.[79,80]

The resonance diagram is shown in Fig. 4.15. For the sake of clarity, only curves corresponding to bifurcations from the periodic solution with period equal to T are displayed. The bifurcation structure is well known and amply documented in the literature.[12-14, 17, 18, 23] There are basic *p/1* entrainment bands (1/1, 2/1 are shown in Fig. 4.15) and curves connecting them. The boundary of the entrainment band corresponds to a saddle-node (or a limit-point) bifurcation. The connection of the *p/1*-entrainment bands is realized by a pair of curves that correspond to the emergence of the torus and a closed curve corresponding to the emergence of the 3/2-periodic solution via period doubling.

We shall call this connecting system of curves a torus-period-doubling-torus (TPDT) structure and the area enclosed by the period-doubling curve alone a period-doubling bubble. The region enclosed by the T-axis, the 1/1 and 2/1 entrainment bands, and the TPDT structure is the region where all of the p/q-Arnold tongues ($1 < p/q < 2$) are located. A series of analogous structures exists to the right of the 2/1 band, and a similar region exists to the left of the 1/1 entrainment band as well, except for the absence of a 0/1 entrainment band. There is for our choice of

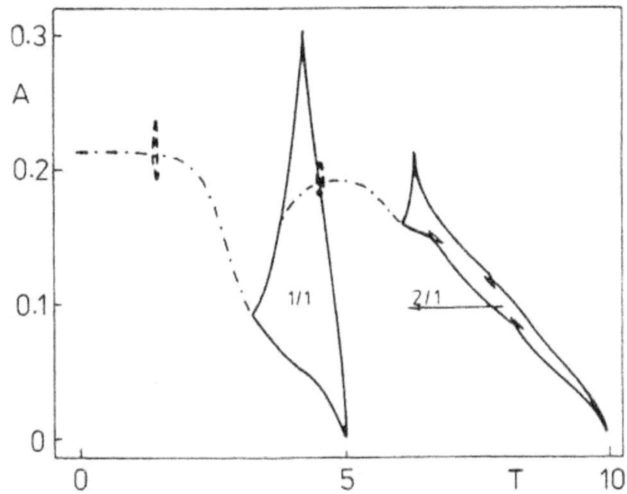

Figure 4.15. Resonance diagram of the periodically forced Brusselator: $a = 2$, $b = 5.9$; full line - saddle-node; dashed line - period doubling; dash and dotted lines - torus bifurcation.

parameters a large overlap of the Arnold tongues near the intersection of the 3/2-period doubling bubble with the 1/1 resonance band. This suggests the presence of chaos. A coexistence of periodic and chaotic regimes is expected since the 1/1-resonance is stable. The occurrence and abundance of chaos is strongly dependent on the form of perturbation; chaos is typically found near the period-doubling bubbles, and the infinite sequence of period doublings there is a prerequisite of chaos.[14]

Rather than doing numerical simulations to verify the presence of chaos and the dominant bifurcation structures accompanying its creation, we focus our attention on a system of two cells coupled by diffusion and use continuation methods. That is, we consider the equations,

$$\mathrm{d}x_1/\mathrm{d}t = a - (b_1 + 1)\,x_1 + x_1^2 y_1 + D_x\,(x_2 - x_1),$$

$$\mathrm{d}y_1/\mathrm{d}t = b_1 x_1 + x_1^2 y_1 + D_y\,(y_2 - y_1),$$

$$\mathrm{d}x_2/\mathrm{d}t = a - (b_2 + 1)\,x_2 + x_2^2 y_2 + D_x\,(x_1 - x_2),$$

$$\mathrm{d}y_2/\mathrm{d}t = b_2 x_2 + x_2^2 y_2 + D_y\,(y_1 - y_2), \tag{4.2}$$

where D_x and D_y are the exchange coefficients for X and Y. The special case of identical cells with $b_1 = b_2$ has been investigated.[2,74,75,81] Here we focus on the case of non-identical cells. We assign the values of $a = 2$, $b_2 = 5.9$ and $q = D_x/D_y = 0.1$. Thus we are left with two variable parameters, b_1 and D_x. The strength of the interaction is controlled by varying D_x. The cells are uncoupled when $D_x = 0$. The frequency and amplitude of both oscillators can be independently tuned in the uncoupled system by varying b_1 and b_2. If $b_i < (a_i^2 + 1)$, the i-th cell ceases to oscillate and approaches a stable steady state. Thus the choice $b_2 = 5.9$ defines the second cell as an oscillator, whereas the first cell is allowed to be either oscillatory or at the steady state. A non-zero coupling produces various nonlinear effects as suggested by Figs. 4.16 and 4.17.

Figure 4.16 presents the loci of bifurcations from stationary states. Hopf and saddle-node bifurcations occur. There can be up to five coexisting steady states, but two at most are stable. These are the simplest Turing structures and are found in the central region of the diagram. However, we are primarily interested in oscillatory regimes. The figure can be used to trace regions of oscillatory behavior because they are linked to stationary states via a Hopf bifurcation. A complete picture of the bifurcations in this system is very complex, as indicated by the number of Hopf bifurcation curves shown. We have marked three major oscillatory regions. The first one, O_1, is the most interesting because it contains a hierarchy of entrainment bands quite analogous to those found in the forced Brusellator. The second region, O_2, corresponds to the interaction of an oscillatory second cell with a non-oscillatory first cell. Not surprisingly, the oscillatory unit induces oscillations in the other one, but only for weak coupling. Strong coupling leads to a stable stationary state. An oscillator-oscillator interaction leads to oscillations for a weak coupling, steady state Turing structures for moderate coupling, and oscillations again for strong coupling in region O_3. While there is only a simple temporal pattern of oscillations in the

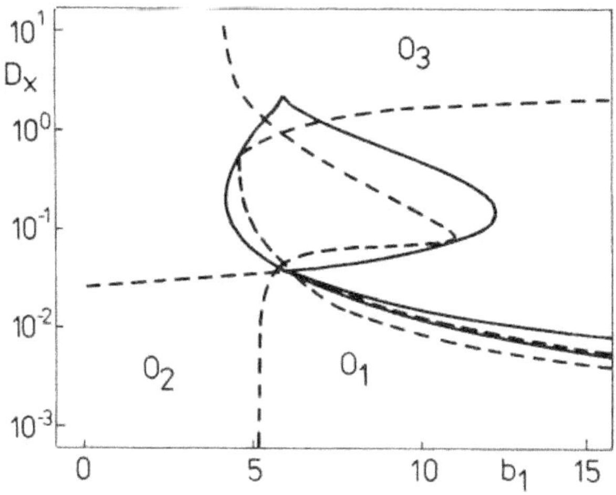

Figure 4.16. Bifurcations from stationary states for two coupled Brusselators in the b_1 - D_x plane: full line - saddle-node; dashed line- Hopf bifurcation; O_1, O_2, O_3 are the major oscillatory regions. See Fig. 4.17 for bifurcations of periodic solutions in the region O_1.

regions O_2 and O_3, the oscillatory behavior in the region O_1 is expected to be quite complex, as suggested by Fig. 4.17. The analogy of the bifurcation curves seen in Fig. 4.17 to the forced system (Fig. 4.15) discussed previously is obvious. Again, there are the main *p/1* entrainment bands connected by the TPDT structure. There are important differences in details, but the basic structure marked by a hierarchy of entrainment bands is retained. The period-doubling bubbles here are much larger, than in Fig. 4.15 and they overlap far more with the *p/1* entrainment bands, so that one can expect chaos to be abundant in the regions of the overlap. Indeed, simulations confirm a complex bifurcation pattern in which periodicity, quasiperiodicity and chaos intermingle in a highly nontrivial way.

As an illustration, we offer in Fig. 4.18 Poincaré plots of a quasiperiodic and a chaotic attractor. The Poincaré plots are constructed by plotting x and y in the second cell at the times when x reaches its maximum in the first cell. The closed curve in Fig. 4.18a is characteristic of a quasiperiodic response typical of small values of A. Fig. 4.18b displays a chaotic attractor from a region where the 3/2-period-doubling bubble overlaps with a 2/1-entrainment band. The topology of this attractor differs from the circle topology of the torus. It is simple in the sense that it is well

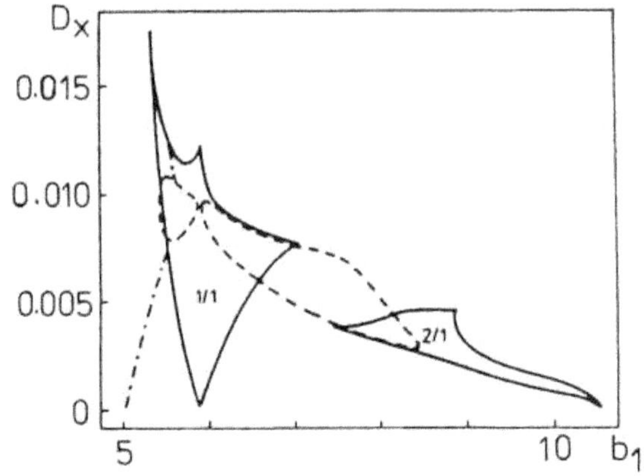

Figure 4.17. Resonance diagram of two coupled Brusselators in the b_1 - D_x parameter plane: full line - saddle-node; dashed line - period doubling; dash and dotted lines - torus bifurcation.

approximated by a one-dimensional curve. However, closer examination reveals a multiple-layered structure characteristic of chaotic attractors. An interesting feature of the oscillations is that the oscillator with the larger amplitude (which corresponds to a larger b_1) oscillates in a more regular fashion than does the other oscillator.

Let us now consider the Sel'kov model as a prototype excitator. The system is described by the following evolution equations when periodically forced in the same way as the Brusselator,

$$dx/dt = \alpha(\nu_0 + x^\gamma)/(1 + x^\gamma) - x(1 + y) + A(1 + \cos(2\pi t/T)),$$

$$dy/dt = x(\beta + y) - \delta y, \tag{4.3}$$

where x and y are the dimensionless concentrations of the species X and Y. The dimensionless parameters (stoichiometric factors) ν_0 and β, order γ of the autocatalytic reaction with respect to X, and the kinetic constant δ for the removal of Y, are assigned the values: $\nu_0 = 0.01$, $\beta = 1.5$, $\gamma = 2$, $\delta = 0.6$. The dimensionless

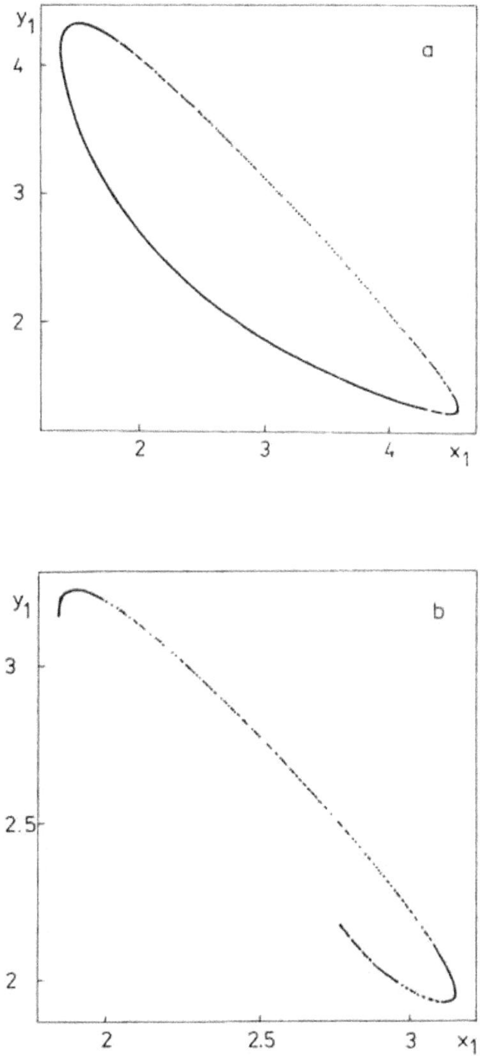

Figure 4.18. Poincaré plot of x_1 vs y_2 in two coupled Brusselators: a) quasiperiodic attractor, $b_1 = 8.0$, $D_x = 0.024$; b) chaotic attractor, $b_1 = 8.0$, $D_x = 0.038$.

concentration of A, α, is used to switch the dynamics of the autonomous system from an oscillator ($\alpha = 7.5$) to an excitator ($\alpha = 6.5$). The perturbation may be interpreted as a periodic variation of a new species producing X.

The unperturbed system possesses a unique stable but excitable steady state for $\alpha = 6.5$, which becomes unstable via a subcritical Hopf bifurcation as α is increased. The limit cycle is unique and stable for $\alpha = 7.5$. The perturbation does not have the effect of switching between the oscillator and the excitator, so that bursting is not expected to occur (as it would if α was slowly varying). The resonance diagram for the oscillator in Fig. 4.19a and the excitation diagram for the excitator in Fig. 4.19b allow direct comparison of the two dynamical modes. Only the 1/1 main entrainment band is shown in the resonance diagram because there is no $p/1$ resonance with $p > 1$ in the excitation diagram. The oscillator case is quite analogous to the forced Brusellator, Fig. 4.15, and many of these features are retained in the excitation diagram as well.

In particular, the TPDT structure is unchanged. The 1/1 resonance is analogous except that it is much narrower, and one cusp point is missing. The important difference appears in the bottom part of both diagrams. There is a second TPDT structure in the excitation diagram which is completely absent in the resonance diagram. Consequently, all the p/q-Arnold tongues that emanate from the T-axis in the forced oscillator case (not seen in the figure except for the 1/1 resonance) are shifted upwards in the excitation diagram and are localized between the two TPDT structures. In other words, phase locking occurs for arbitrarily small forcing amplitudes in the oscillator case, whereas there is a threshold below which phase locking does not occur in the excitator. However, unlike ordinary steady state behavior, the excitator is capable of producing phase-locked structures and toroidal oscillations when sufficiently strongly perturbed. In fact, the lower TPDT structure in the excitation diagram is generic, and only after switching from the excitator to the oscillator does it fall onto the T-axis. As mentioned in Sec. 4.2.1, the superharmonic resonances do not have any analogues in the forced excitator case discussed here.

It is also implied by the structure of the excitation diagram that chaos is more likely to occur here than in the resonance diagram as the lower period-doubling bubble overlaps with the 1/1 entrainment band. An overlap of many Arnold tongues nearby is expected and indeed occurs, as demonstrated by direct numerical simulation. Stroboscopic Poincaré plots of quasiperiodic, Fig. 4.20a and chaotic, Fig. 4.20b, behavior have been calculated as an example of the dynamics of the forced excitator. As with two coupled cells, quasiperiodicity manifests itself by a closed curve covered entirely by successive iterates of the stroboscopic map. On the

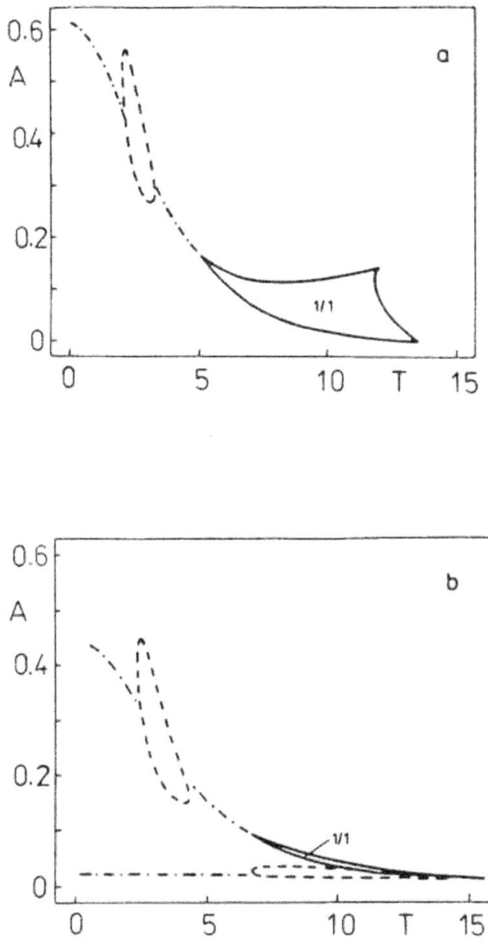

Figure 4.19. Periodically forced Sel'kov model: full line - saddle-node; dashed line - period doubling; dash and dotted lines - torus bifurcation; a) resonance diagram for α = 7.5; b) excitation diagram for α = 6.5.

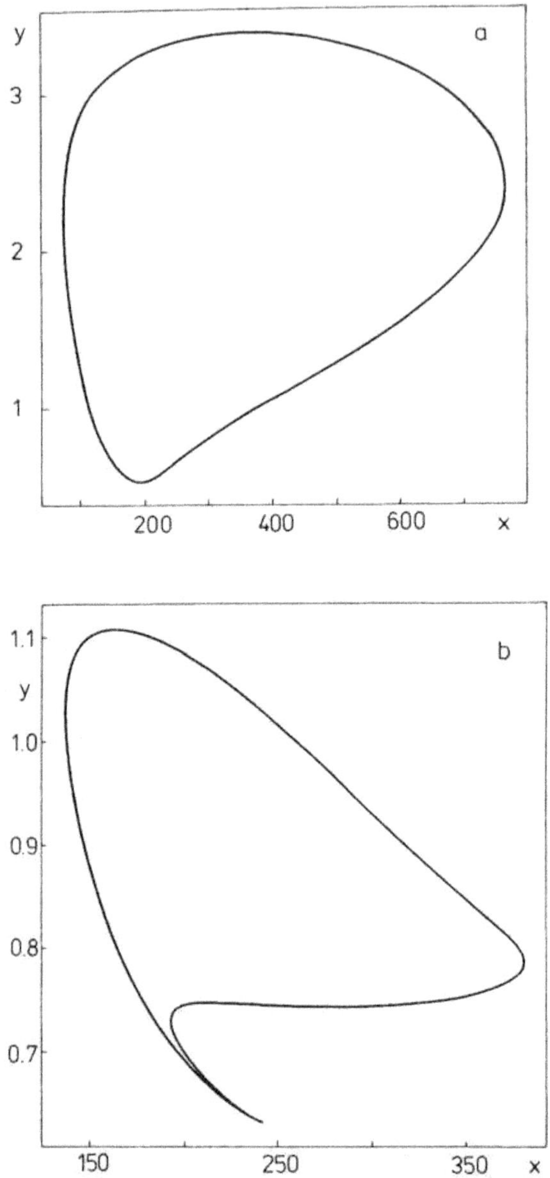

Figure 4.20. Poincaré plot of x vs y for the forced, excitable Sel'kov model: a) quasiperiodic attractor, $A = 0.0782$, $T = 4$; b) chaotic attractor, $A = 0.0782$, $T = 7$.

other hand, chaos is topologically closer to the torus in this case. However, the fold indicates that a break up of the torus has occurred and resulted in the chaos. The multi-layered structure observable with sufficient magnification again confirms the presence of chaos.

4.5. Conclusion

Comparison of experimental and modelling studies of chaotic dynamics in coupled and forced chemical systems reveals that we are still far away from a full understanding of the general features of mutually interacting periodic and chaotic dynamics. This is true even when the presence of chaos and in several cases also bifurcation routes to chaos are clearly demonstrated. Similar experimental results to those reported above need to be obtained for other chemical systems, including enzymatic reactions. The methodology of numerical studies presented here and based on the combination of studies of parametric dependence of bifurcations and direct numerical simulation needs to be explored in full using high-performance computers presently available to investigate more realistic and often more complicated models of experimental systems.

LITERATURE CITED

1. S.K. Scott, *Chemical Chaos* (Clarendon Press, Oxford, 1991) Chapters 5, 6.
2. M. Marek and I. Schreiber, *Chaotic Behaviour of Deterministic Dissipative Systems* (Cambridge Univ. Press, Cambridge and Academia Publ. House, Prague, 1991) Chapters 5, 6.
3. A.R. Bishop, G. Gruner and B. Nicolaenko, Eds., *Spatiotemporal Coherence and Chaos in Physical Systems* (*Physica* **23D**, 1986).
4. M.S. El Naschie, Ed., *Chaos, Solitons and Fractals*, Vol. 1 (Pergamon Press, Oxford, 1991).
5. D. Cambell, Ed., *Chaos* (*Int. J. on Bifurcations and Chaos*, 1 (4) 1991).
6. V.I Arnold, *Geometrical Methods in the Theory of Ordinary Differential Equations* (Springer, New York, 1982).
7. J.C. Alexander, E.J. Doedel and H.G. Othmer, *SIAM J. Appl. Math.* **50** (1990) 1373.
8. J. Rinzel, *A Formal Classification of Bursting Mechanisms in Excitable Systems* in *Proc. of the Int. Cong. of Mathematics, 1986* (AMS, Providence, RI, 1987) pp.1578-1593.
9. A.T. Winfree, *The Geometry of Biological Time* (Springer, Berlin, 1980); *When Time Breaks Down* (Princeton Univ. Press, Princeton, NJ, 1987).
10. M. Dolnik and M. Marek, *J. Phys. Chem.* **95** (1991) 7267.
11. M. Dolnik, I. Epstein and M. Marek, *J. Phys. Chem.* **96** (1992) 3218.

12. C. Hayashi, *Forced Oscillations in Nonlinear Systems* (Nippon Printing and Publ. Co., Osaka, 1953); *Nonlinear Oscillations in Physical Systems* (McGraw-Hill, New York, 1964).
13. W.N. Vance and J. Ross, *Chaos* **1(4)** (1991) 445.
14. T. Kai and K. Tomita, *Prog. Theor. Phys.* **61** (1979) 65, 54; K. Tomita, *Phys. Rep.* **86** (1982) 113.
15. I.G. Kevrekidis, R. Aris and L.D. Schmidt, *Chem. Eng. Sci.* **43** (1988) 1833.
16. M.A. Karnin, L.D. Schmidt and R. Aris, *Chem. Eng. Sci.* **43** (1988) 2833.
17. I. Schreiber, M. Dolnik, P. Choc and M. Marek, *Phys. Lett.* **128A** (1988) 66.
18. W.N. Vance and J. Ross, *J. Chem. Phys.* **91** (1989) 7654.
19. Y. Ueda, *Chaos, Solitons and Fractals* **1** (1991) 199.
20. L. Glass and M.C. Mackey, *From Clocks to Chaos* (Princeton Univ. Press, Princeton, NJ, 1988).
21. L. Glass, M.R. Guevara, A. Shrier and R.P. Perez, *Physica* **7D** (1983) 89.
22. M. Dolnik, I. Schreiber and M. Marek, *Phys. Lett.* **100A** (1984) 316.
23. J. Belair and L. Glass, *Physica* **16D** (1985) 143.
24. M. Dolnik, I. Schreiber and M. Marek, *Physica* **21D** (1986) 78.
25. P. Bak, T. Bohr and M.H. Jensen, *Circle Maps, Mode-Locking and Chaos*, in *Directions in Chaos*, Ed. Hao Bai-lin (World Scientific, Singapore, 1988). pp. 16-45.
26. L. Glass, *Simple Mathematical Models for Complex Dynamics in Physiological Systems*, in *Directions in Chaos*, Ed. Hao Bai-lin (World Scientific, Singapore, 1988). pp. 90-108.
27. M. Dolnik, J. Finkeova, I. Schreiber and M. Marek, *J. Phys. Chem.* **93** (1989) 2764.
28. J. Finkeova, M. Dolnik, B. Hrudka and M. Marek, *J. Phys. Chem.* **94** (1990) 4110.
29. J.R. Pasta, S.M. Ulam and E. Fermi, *Studies on Nonlinear Problems*, in *Collected Works of Enrico Fermi*, Vol.2 (Univ. of Chicago Press, Chicago, 1965). p. 978.
30. N.J. Zabusky and M.D. Kruskal, *Phys. Rev. Lett.* **15** (1965) 240.
31. J.P. Crutchfield and K. Kaneko, *Phenomenology of Spatio-Temporal Chaos*, in *Directions in Chaos*, Vol.1 (World Scientific, Singapore, 1987). p. 272.
32. S. Newhouse, D. Ruelle and F. Takens, *Commun. Math. Phys.* **64** (1978) 35.
33. C. Grebogi, E. Ott and J.A. Yorke, *Physica* **15D** (1985) 354.
34. Y. Kuramoto, *Chemical Oscillations, Waves and Turbulence* (Springer, New York, 1984).
35. Y. Kuramoto, *Physica* **50D** (1991) 15.
36. K. Kaneko, *Phys. Rev. Lett.* **63** (1989) 219; *Physica* **41D** (1990) 38.
37. K. Kaneko, *Collapse of Tori and Genesis of Chaos in Dissipative Systems* (World Sci. Pub., Singapore, 1986).
38. H. Sevcikova and M. Marek, *Physica* **D49** (1991) 114.
39. M. Markus and C. Schafer, in *Bifurcation and Chaos: Analysis, Algorithms, Applications*, R. Seydel, F.W. Schneider, T. Kupper, H. Troger, Eds. (Birkhauser, Basel, 1991).
40. V.Z. Vavilin, A.M. Zhabotinsky and A.N. Zaikin, *Russ. J. Phys. Chem.* **42** (1968) 1649.
41. A.N. Zaikin and A.M. Zhabotinsky, in *Biological and Biochemical Oscillators*, Eds. B. Chance, E.K. Pye, A.K. Ghosh and B. Hess (Academic Press, New York, 1973).
42. M. Marek, in *Synergetics. Far from Equilibrium*, Eds. A. Pacault and C. Vidal (Springer-Verlag, Heidelberg, 1979). p. 12.
43. E. Dulos and P. DeKepper, *Biophys. Chem.* **18** (1983) 211.
44. H. Sevcikova, D. Suchanova and M. Marek, *Sci. Papers of Prague Inst. Chem. Technol.* **K17** (1982) 137.
45. M. Marek, in *Temporal Order*, Eds. L. Rensing and N.I. Jaeger (Springer, Heidelberg, 1985). p 105

46. F. Buchholtz and F.W. Schneider, *J. Am. Chem. Soc.* **105** (1983) 7450.
47. F. Bucholtz, A. Freund and F.W. Schneider, in *Temporal Order*, Eds. L. Rensing and N.I. Jaeger (Springer, Heidelberg, 1985). p. 116.
48. F.W. Schneider, *Ann. Rev. Phys. Chem.* **36** (1985) 347.
49. J.L. Hudson, P. Lamba and J.C. Mankin, *J. Phys. Chem.* **90** (1986) 3430.
50. P. Lamba and J.L. Hudson, *Chem. Eng. Sci.* **42** (1987)1.
51. H. Sevcikova and M. Marek, *Physica* **D49** (1991) 114.
52. J. Lukesova, P. Pokorny and M.Marek, to be published (1992); J. Lukesova, MSc. Thesis, PICT, Prague, 1992.
53. A. Boiteux, A. Goldbetter and B. Hess, *Proc. Natl. Acad. Sci. USA* **72** (1975) 3829.
54. M. Markus, D. Kuschmitz and B. Hess, *FEBS Lett.* **172** (1984) 235.
55. H. Degn, A.V. Holden and L.F. Olsen, Eds., *Chaos in Biological Systems* (Plenum Press, New York, 1987).
56. N. Takanishi, Y. Hanyu, T. Musha, R. Kubo and G. Matsumoto, *Physica* **D43** (1990) 318.
57. M. Marek and I. Stuchl, *Biophys. Chem.* **3** (1975) 241.
58. M. Marek and E. Svobodova, *Biophys. Chem.* **3** (1975) 263.
59. M. Marek, M. Dolnik and I. Schreiber, in *Selforganization by Nonlinear Irreversible Processes*, Eds. W. Ebeling and H. Ulbricht (Springer, Berlin, 1986). p. 133.
60. M. Marek, in *Spatial Inhomogenities and Transient Behaviour in Chemical Kinetics*, Eds. P. Gray, G. Nicolis, F. Baras, P. Borckmans and S.K. Scott (Manchester Univ. Press, Manchester, 1990).
61. M. Marek, in *Lecture Notes in Biomathematics, Modelling of Patterns in Space and Time*, (Springer, Berlin, 1984). p. 214.
62. H. Fujii and Y. Sawada, *J. Phys. Chem.* **69** (1978) 3830.
63. K. Nakajima and Y. Sawada, *J. Chem. Phys.* **72** (1980) 2231.
64. M. Dolnik and M. Marek, *J. Phys. Chem.* **92** (1988) 2452.
65. M. Dolnik, E. Padusakova and M. Marek, *J. Phys. Chem.* **91** (1987) 4407.
66. J. Kosek and M. Marek, *J. Phys. Chem.*, submitted (1992).
67. M. Dolnik and M. Marek, in *Nonlinear Wave Processes in Excitable Media*, Eds A.V. Holden, M. Markus and H.G. Othmer (Manchester Univ. Press, Manchester, 1991).
68. I.S. Doumbouya, A. Muenster and F.W. Schneider, Universität Würzburg, private communication, 1992.
69. J.P. Laplante, Royal Military College of Canada, private communication, 1992.
70. M.F. Crowley and R.J. Field, in *Nonlinear Phenomena in Chemical Dynamics*, Eds. C. Vidal and A. Pacault (Springer, Berlin, 1981). p. 147.
71. M.F. Crowley and R.J. Field, *J. Phys. Chem.* **90** (1986) 1907.
72. M.F. Crowley and I. Epstein, *J. Phys. Chem.* **93** (1989) 2496.
73. I. Stuchl and M. Marek, *J. Phys. Chem.* **77** (1982) 1607.
74. I. Schreiber and M. Marek, *Physica* **5D** (1982) 258.
75. I. Schreiber and M. Marek, *Phys. Lett.* **253** (1982) 263.
76. M. Alamgir and I. Epstein, *J. Phys. Chem.* **88** (1984) 2848.
77. G. Nicolis and I. Prigogine, *Self-Organization in Nonequilibrium Systems* (Wiley, New York, 1977).
78. Ye. Sel'kov, *Biofizika* **15** (1970) 1065.
79. M. Kubicek, *ACM Trans. Math. Software* **2** (1976) 98; M. Holodniok and M. Kubicek, *J. Comp. Phys.* **55** (1984) 254.

80. M. Kubicek and M. Marek, *Computational Methods in Bifurcation Theory and Dissipative Structures* (Springer, New York, 1983).
81. I. Schreiber, M. Holodniok, M. Kubicek and M. Marek, *J. Stat. Phys.* **43** (1986) 489.

CHAPTER 5: CHAOS IN ELECTROCHEMICAL SYSTEMS

J. L. Hudson
Department of Chemical Engineering
School of Engineering and Applied Science
University of Virginia
Thornton Hall
Charlottesville, VA 22903-2442
U.S.A.

ABSTRACT

A review of chaotic behavior in electrochemical reactions is presented. The emphasis is on experimental studies and the analysis of data using methods of nonlinear dynamics. Three types of electrochemical systems are discussed: electrodissolution of metals, electrodeposition of metals, and electrocatalytic reactions.

5.1 Introduction

Nonsteady behavior of electrochemical systems was observed as early as 1828[1] and has been the subject of numerous investigations since that time. Both oscillations and surface patterns have been observed in a number of electrochemical experiments. The current can oscillate in potentiostatic (constant potential) systems, whereas in galvanostatic (constant current) systems the potential can oscillate. Both the current and potential can be nonsteady in less restricted systems.

Much of the earlier work on electrochemical oscillations has been reviewed by Wojtowicz.[2] Hachkar *et al.*[3] have presented more recently a survey of a subclass of oscillating electrocatalytic systems, those involving the oxidation of organics. Hudson and Bassett[4] have reviewed studies involving electrodissolution of metals with an emphasis on chaotic behavior. A comprehensive review of oscillations in all types of electrochemical reactions (electrodissolution, electrodeposition, and electrocatalysis) will be available;[5] papers concerning oscillations (including, but not restricted to, chaos), surface patterns, mechanisms and mathematical models are discussed.

In the present review the emphasis is on chaos. Electrochemistry has now furnished many clear cut examples of chaotic behavior. Furthermore, several one- and two-parameter studies have been carried out showing bifurcations into chaos and among different types of chaos. In the one-parameter studies, the potential or the current is an easily controlled quantity; for the two-parameter studies, a concentration or a disk-rotation rate is a second quantity that can be controlled. Only those papers are cited which deal with chaos and its characterization through the principles of nonlinear dynamics. Most of the studies in the literature deal with temporal chaos, although some work on waves and surface heterogeneities are discussed. Readers interested in more general aspects of dynamics in electrochemistry are urged to see earlier reviews.[5]

Electrochemical reactions have several features which make their use in dynamic studies interesting. The first, of course, is that many types of dynamic behavior can be observed. Secondly, the measurement of current or voltage is relatively straightforward. Furthermore, noise is often at a sufficiently low level that the data can be interpreted without ambiguity. Finally, the time scale of the oscillations under some conditions is short, both relative to the time scale of changes in the surface conditions, so that data are taken under stationary conditions, and absolutely so that large amounts of data can be obtained for analysis.

This review is organized in three sections: electrodissolution, electrodeposition, and electrocatalytic reactions. We begin with a short section in which some comments on experiments and data acquisition common to all three sections are discussed.

5.2 Electrochemical Experiments

Electrochemical experiments are often carried out in a three-electrode configuration employing a working electrode, a counter electrode, and a reference electrode. Occasionally, only two electrodes are employed. The working electrode is an anode in the case of the electrodissolution experiments, a cathode for the electrodeposition work, and either for electrocatalysis experiments.

The working electrode can be stationary, such as a metal foil or a wire immersed in the electrolyte. In many cases, however, a rotating-disk electrode is used. A rotating-disk electrode is simply a metal rod embedded in an insulating cylinder (such as Teflon) so that only the end of the metal rod is exposed. The potential is measured relative to a reference electrode. Current flows between the working and the counter electrodes. The arrangement of the three electrodes varies depending on the nature of the experiment. The reference electrode might be in a side chamber or might be close to the surface of the working electrode. The counter electrode can be either in a side chamber or located below the surface of the rotating disk. The makeup of the electrolyte, of course, depends on the reaction being investigated. The rotating disk is a standard electrochemical apparatus with which the mass-transfer resistance between the metal surface and the bulk fluid can be controlled by fixing the rate of disk rotation. More than one reference electrode has been used in some studies of spatial structure in order to measure potential distributions.

When the experiments are carried out potentiostatically, *i.e.*, at constant potential, the current is digitized as a function of time so that analysis of the oscillations can be performed. When the experiments are done galvanostatically (constant current), the potential is measured. Occasionally, a sweep is employed, *i.e.*, the potential is slowly ramped up and down.

5.3 Dynamics of the Anodic Dissolution of Metals

5.3.1 Overview

We discuss in this section some of the dynamics that have been observed during the electrodissolution of metals. This category of reaction has received more attention than the other types of electrochemical reactions[6,7] and has been the subject of a recent review.[4] Several different metals have been shown to produce oscillatory dissolution. Furthermore, many interesting types of temporal and spatial behavior have been observed.

Several of the tools of nonlinear dynamics have been applied in the past few years to time series obtained from electrodissolution of metals.[8,9,10-21] Copper electrodissolution in H_3PO_4 has been studied by Albahadily and Schell[8,9,22] and has been found to undergo Hopf bifurcations to oscillatory behavior and period-doubling bifurcations to simple chaos. This system also was found to give both periodic and chaotic mixed-mode oscillations. Lev *et al.* have studied the galvanostatic electrodissolution of nickel in H_2SO_4 and observed Hopf bifurcations, saddle loop bifurcations, period-doubling bifurcations, chaos, and quasiperiodicity.[18,19] In addition, Lev *et al.* have observed spatial structure during nickel dissolution.[20] Bassett and Hudson have shown that copper electrodissolution in NaCl and H_2SO_4 yields a large number of types of interesting behavior, including period doublings to simple chaos,[11] Shil'nikov chaos,[15] quasiperiodicity, chaos on a broken torus,[12] and period doubling of tori.[10] Diem and Hudson[16] and Wang and Hudson[23] have used methods such as the calculation of a correlation dimension[24,25] to show the existence of simple and higher-order chaotic structures during the electrodissolution of iron in H_2SO_4 at the limiting-current plateau. Gu and co-workers have analyzed time series from the electrodissolution of copper in sodium chloride/thiocyanate solutions with the aid of Hurst exponents.[26]

The first example to be discussed is the electrodissolution of copper in phosphoric acid and in acidic chloride solutions. These systems have produced a large set of fascinating dynamic behavior. We then turn to electrodissolution of iron in sulfuric acid. The oscillations in this system have probably received more attention than any other case. Finally, we discuss oscillations and spatial structures observed during nickel electrodissolution.

5.3.2 *Copper Electrodissolution in Phosphoric Acid*

Copper dissolution in phosphoric acid is known to give oscillatory current behavior at potentials where electropolishing occurs. This reaction system has produced a wealth of types of dynamic behavior.

Schell and coworkers have done a detailed study of copper electropolishing in phosphoric acid at reduced temperatures. They have observed both Hopf bifurcations and a period-doubling cascade to simple chaos.[8] In addition, they have found a wide variety of both periodic and chaotic mixed-mode oscillations.[9,22]

Period doubling of limit cycles is one of the several routes that a dynamic system can take on the way to chaos. This route is certainly the best known.[27] A few metal electrodissolution systems have exhibited this phenomenon. The period doubling can occur as a parameter such as potential or current is changed or as a film slowly grows on the electrode surface. Note below in later sections the work of Lev *et al.* on nickel dissolution and that of Bassett and Hudson on copper dissolution in acidic chloride.[13] Figure 5.1 shows a period-doubling sequence with copper in phosphoric acid due to Schell *et al.*[8] The parameter varied in this study is the applied potential. Mixed-mode oscillations are waveforms which consist of large- and small-amplitude oscillations and can be periodic or chaotic. An example is shown in Fig. 5.2.

Schell *et al.* also have observed two interesting sequences (a Farey sequence and a periodic-chaotic sequence) involving mixed-mode oscillations. The Farey sequence is a periodic sequence for which a one-to-one correspondence with ordered sequences of rational numbers can be established.[9] It was found that the ratio of the

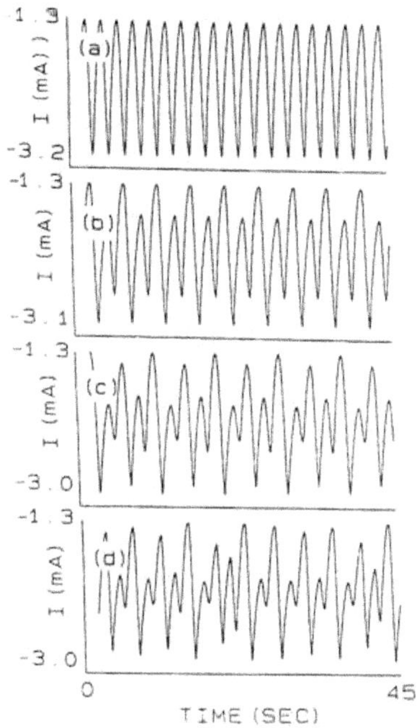

Figure 5.1. Period doubling during the electrodissolution of copper in phosphoric acid: (a) Period one; (b) Period two; (c) Period four; (d) Chaos. Reproduced by permission from F.N. Albahadily and M. Schell, *J. Chem. Phys.* **88** (1988) 4312.

number of large- to small-amplitude oscillations decreased as the potential was increased. An interesting feature of this Farey sequence is that it does not appear to be related to phase locking on a torus as has been the case with other systems exhibiting this sequence.[9] Schell *et al.* observed a periodic-chaotic sequence of mixed-mode oscillations in a different region of parameter space that is similar to that seen in the Belousov-Zhabotinskii reaction.[28] Chaotic mixed-mode oscillations were observed between each of the periodic states. These chaotic oscillations can be described in many cases as mixtures of waveforms from the adjacent periodic windows. In addition, a tangent bifurcation was observed at one edge of these periodic windows, while the other edge was found to be the limit of a sequence of period-doubling bifurcations.[22]

Time series from the electrodissolution of copper in phosphoric acid have been analyzed using nonlinear signal-processing techniques such as artificial neural networks.[29,30]

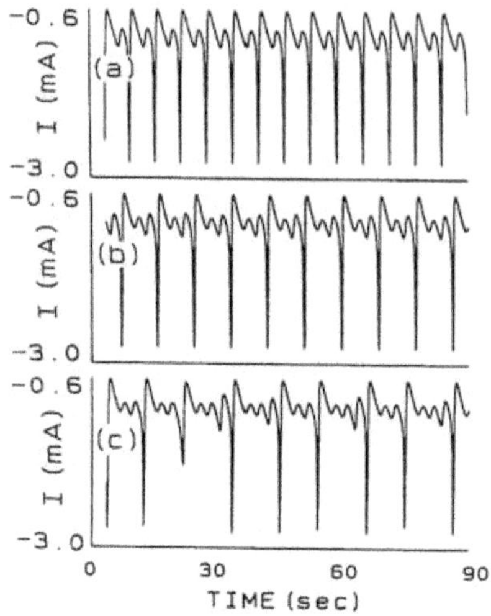

Figure 5.2. Mixed-mode oscillations during the electrodissolution of copper in phosphoric acid. Reproduced by permission from F.N. Albahadily and M. Schell, *J. Chem. Phys.* **88** (1988) 4312.

5.3.3 *Copper Electrodissolution in Acidic-Chloride Solution*

5.3.3.1 *Period Doubling and Chaos*

This reaction also can show a period-doubling sequence leading to chaos, but this doubling occurs as a film grows on the electrode surface, *i.e.*, as a parameter varies slowly.[13] We describe here the chaos which results from this period doubling. An example is shown in Fig. 5.3.

The time series is shown in Fig. 5.3a. The reconstructed attractor from Fig. 5.3a is shown in Fig. 5.3b. This attractor was constructed by the well-known method of time delays due to Packard *et al.*[31] and Takens.[32] It has a relatively simple and well-structured form. An example of a Poincaré section is given in Fig. 5.3c. A sequence of Poincaré sections is shown in Fig. 5.3e to explain the flow of trajectories around the attractor. In Fig. 5.3e only a few orbits around the attractor are shown so that the location and orientation of the sections can be seen. The arrow on each section is used to mark the points that started on the edge of the attractor in Section

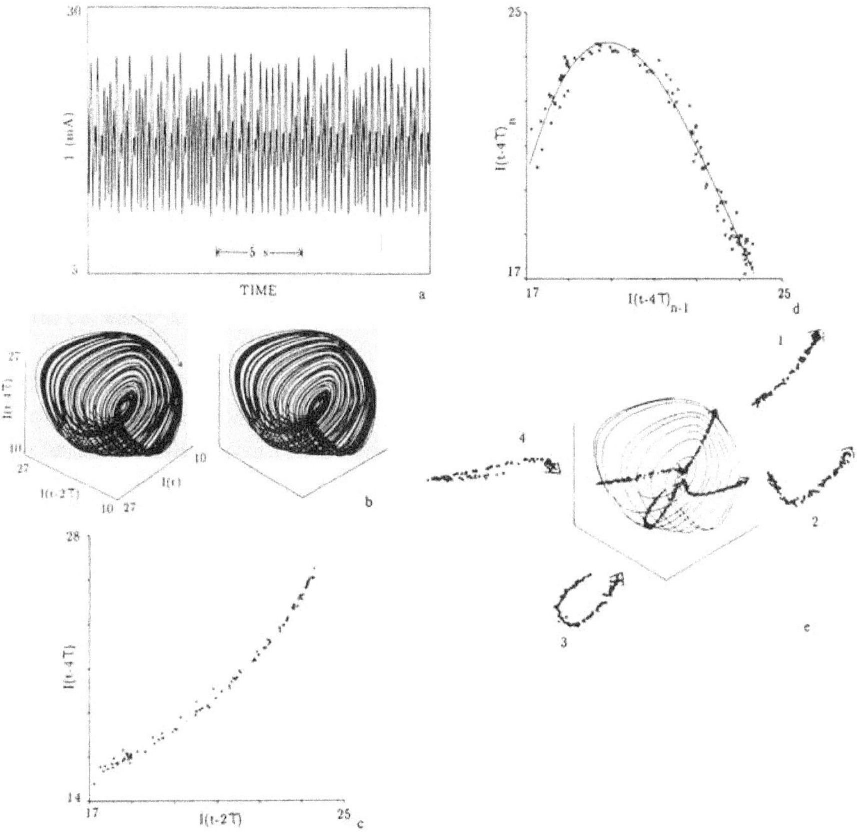

Figure 5.3. Simple-Chaos: a) Time series; b) Attractor. Two projections made at angles differing by 6°. The right-hand picture is for the left eye, and the left-hand picture is for the right eye. c) Poincaré section taken at I(t) = 17.5 mA; increasing I(t − 4τ). d) Return map constructed from the Poincaré section taken at I(t − 2τ) = 17.5 mA with increasing I(t − 4τ); using variable I(t − 4τ). e) Series of Poincaré sections showing the flow of trajectories around the attractor. Sections 1 and 3 were taken at I(t − 2τ) = 17.5 mA and sections 2 and 4 were taken at I(t − 4τ) = 17.5 mA. Reproduced by permission from M.R. Bassett and J.L. Hudson, *J. Phys. Chem.* **93** (1989) 2731.

1. This sequence shows the stretching, folding, mixing, and contraction processes common to these types of simple chaotic attractors.[33-35] A return map, made from the Poincaré sections, is shown in Fig. 5.3d.

5.3.3.2 *Shil'nikov Chaos*

Shil'nikov chaos[36] also has been observed in the copper/chloride ion system.[15] According to a theory developed by Shil'nikov, there are infinitely many nonperiodic trajectories in systems which display a homoclinic orbit bi-asymptotic to a saddle-focus at $t \rightarrow \infty$ and $t \rightarrow -\infty$, provided that $| \underline{P}/\mu | < 1$, where ($\underline{P} \pm i\Omega, \mu$) are the eigenvalues of the saddle-focus.[37]

A time series and attractor are shown in Figs. 5.4a and b, respectively. A portion of the trajectory making a single loop around the attractor is shown in Fig. 5.4c. There appears to be a saddle-focus at the point S. Trajectories spiral out from that point, make a large circuit around the attractor, and are then reinjected into the vicinity of S. The number of small-amplitude oscillations around the saddle focus also can vary because the point of reinjection varies slightly each time.

The facts that the trajectories are reinjected in a nonoscillatory manner near the supposed saddle-focus and leave the area of the supposed saddle-focus by oscillating away from it indicate that the saddle-focus has eigenvalues μ and $\underline{P} \pm i\Omega$ where μ is negative and $\underline{P}$ is positive. This then is a "spiral" type strange attractor as described by Rössler.[33-35] The return map of a Poincaré section taken transverse to the trajectories as they leave or return to the area of the saddle-focus is multibranched. It consists of a set of 1-D curves, each of which is associated with the number of small-amplitude oscillations a trajectory completes around the saddle-focus between two successive crossings of the Poincaré plane.[37] Such a return map can be found in reference 15.

5.3.3.3 *Period Doubling of Tori*

Period doubling of tori during the electrodissolution of copper in chloride solution also can occur.[10] The following sequence of transitions was observed: the current first oscillates in a simple periodic manner with frequency f_1. The attractor is a simple loop, and the Poincaré cross-section is a point. This state undergoes a transition to quasiperiodic behavior. There are now two frequencies f_1 and f_2 in the Fourier spectrum, the attractor has the shape of a torus, and the Poincaré cross

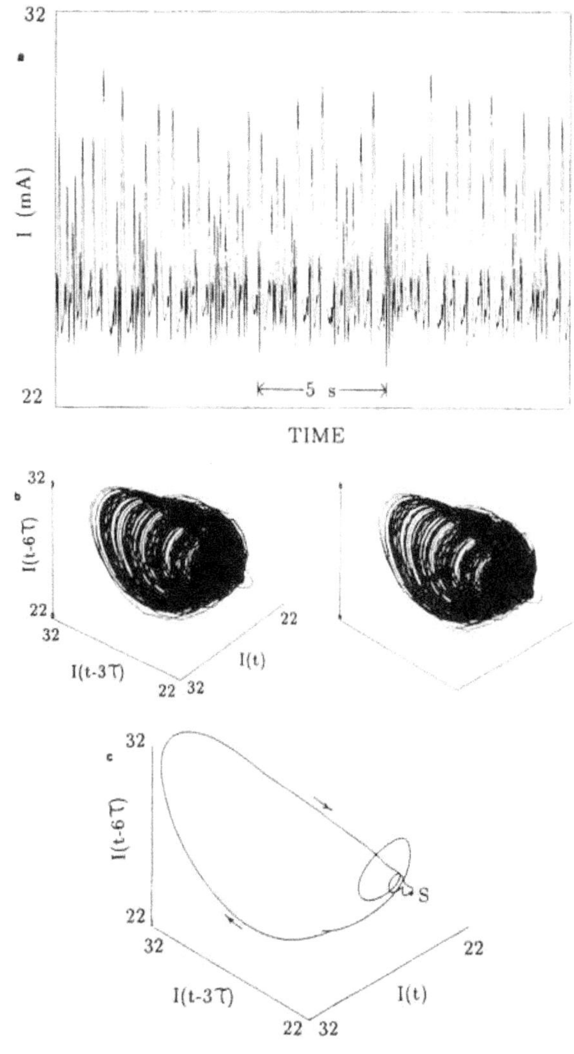

Figure 5.4. Shil'nikov chaos (E = 350 mV *vs* SCE); rotation rate = 200 rpm. a) Time series; b) Attractor; c) A single loop around the attractor. Reproduced by permission from M.R Bassett and J.L. Hudson, *J. Phys. Chem.* **93** (1989) 2731.

section is a closed curve or loop. A period-doubling bifurcation of the torus then occurs, *i.e.*, the frequency f_2 is halved to $f_2/2$. The attractor is now a torus with two loops (possible only in four dimensions), and the Poincaré cross section is a double loop. The period-doubling sequence was observed only to this stage. The system then underwent a transition to chaos on a deformed, period-doubled torus and then to chaos on a deformed torus.

5.3.4 Electrodissolution of Iron in Sulfuric Acid

One of the earliest studied and most important oscillating electrochemical reactions is the electrodissolution of iron in sulfuric acid solution.[16,38,39] Two quite different types of oscillations can occur in this system. The first is a slow oscillation near the active-passive transition, *i.e.*, where the current drops to nearly zero as the potential is raised. The second are oscillations on the mass-transfer limited plateau.

An important early study of the former type oscillation is the work of Franck and FitzHugh.[40] Oscillations can be obtained under some conditions in which the time signal swings between an active, high-current state and a passive state where the current is close to zero. These oscillations are not chaotic. The oscillations on the mass-transfer limited plateau are different in character. They normally have a much higher frequency (up to kilohertz), are more complicated (often chaotic), and have minima which are not zero, *i.e.*, the surface does not become completely passivated.[23] An excellent discussion of these latter oscillations can be found in the work of Russell and Newman.[38] These authors present time series showing the complicated nature of the oscillations. They also developed an explanation and a model for sustained oscillations in the iron/sulfuric acid system at the mass-transfer plateau based on a cycling of a portion of the electrode between active and passive states.[39]

The chaotic nature of these higher-frequency signals has been investigated.[16,23] The correlation dimension[24,25] and the information dimension (using the nearest-neighbor method[41,42]) have been calculated for the chaotic signals. It is shown[23] that the dimension generally increases with increasing electrode size. The signals progress as the electrode size is increased from periodic to low-order chaos with one apparent positive Lyapounov exponent, to more complicated chaos with more than one positive exponent. The increase in dimension is monotonic with increasing electrode size. This increase in the complexity of the dynamics with increasing electrode area may be due to increased interactions among reacting sites on the electrode surface. The dependence of the information dimension on electrode surface area is shown in Fig. 5.5.

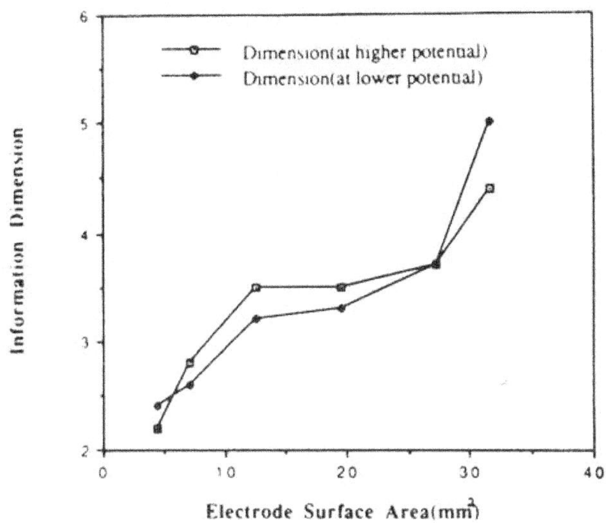

Figure 5.5. Attractor dimension dependence on electrode surface area. Reproduced by permission from Y. Wang and J.L. Hudson, *AIChE J.* 37 (1991) 1833.

5.3.5 Nickel Dissolution in Sulfuric Acid

The transpassive galvanostatic dissolution of a nickel wire in sulfuric acid has been extensively studied by Lev *et al.* At low sulfuric acid concentrations (<6 N), they observed both periodic and low-order chaotic structures, including the period-doubling route to chaos.[18] Quasiperiodic and chaotic behavior was observed at higher sulfuric acid concentrations, in addition to both simple and complex periodic behavior. The chaos in this case was found to have a dimension greater than three.[19] In addition to identifying various interesting types of dynamic behavior, Lev *et al.* experimentally traced a two-dimensional bifurcation map in the current density-sulfuric acid plane. Only simple oscillations exist below acid concentrations of 5.5 N. They are in a domain bounded by Hopf and saddle-loop bifurcations. The authors describe the bifurcation structure as follows: The saddle-loop bifurcation gives rise to an infinite-period transition and another steady-state branch. The saddle-loop bifurcation is contained in a domain of multiplicity which is bound by a saddle-node bifurcation. There exists another domain of birhythmicity between acid concentrations of 4.5 and 6.5 N which is bounded by a subcritical Hopf bifurcation. The transition from simple behavior to complexity is via quasiperiodicity (*i.e.*, a

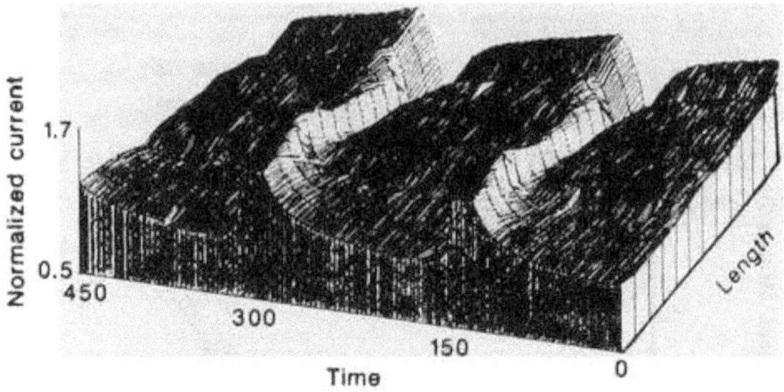

Figure 5.6. Local current distribution of antiphase oscillations during galvanostatic operation of a nickel wire. Reproduced by permission from O. Lev, M. Sheintuch, L. M. Pismen and Ch. Yarnitzky, *Nature* **336** (1988) 458.

secondary Hopf bifurcation) at the lower boundary of the complex domain. Chaos gives way at the upper boundary to usually simple, periodic behavior without any hysteresis in the transition. At higher concentrations, the sequence of behaviors observed at a given concentration while varying the current was: simple periodic, quasiperiodic, complex periodic, chaotic, complex periodic, chaotic, and periodic.[19]

In addition to studying the lumped behavior of the nickel wire, Lev *et al.* have also studied its distributed behavior.[20,21] A train of waves was observed under potentiostatic conditions. These waves start at one end of the electrode and travel to its other end at a speed of approximately 4.5 m/s.

The local current distribution obtained under galvanostatic conditions is shown in Fig. 5.6. Antiphase oscillations (a phase shift of π between the opposite oscillations at electrode edges) were observed under galvanostatic conditions. Coherent spatial organization was observed in the galvanostatic modes, even though the temporal behavior is aperiodic. In this mode, because the local current oscillates, the wire edges must lie out of phase so that their sum will eliminate the average current amplitude. The investigators believe that this symmetry-breaking causes much of the complex behavior discussed above.[20]

Haim and co-workers have developed in two recent papers mathematical models of both the temporal behavior[43] and the spatiotemporal patterns[44] of the nickel electrodissolution reaction. In the former paper they model behavior such as Hopf and saddle-loop bifurcations as well as period doubling to chaos. In the latter work they describe the potential and activity distribution along a dissolving nickel wire.

5.4 Electrodeposition of Metals

In contrast to electrodissolution of metals, there have been relatively few specific studies of oscillatory behavior during metal electrodeposition. The deposition process is certainly complicated and non-uniform (spatial) deposition is common. Dendritic growth on surfaces is an important practical concern. We limit this discussion to those studies in which autonomous non-steady deposition has been investigated.

Suter and Wong have considered oscillatory behavior observed during the electrochemical growth of zinc dendrites.[45] They note that a variety of patterns can form on the electrode surface (fractal, dendritic, *etc.*) and show that periodic oscillations can arise in the dendritic regime.

Argoul and Arneodo also investigated the electrodeposition of zinc, particularly in the intermediate regime between dendritic and diffusion-limited, aggregation-like clusters.[46] They present evidence for periodic oscillations, period doubling, and chaos.

In a recent study, Argoul and co-workers have observed homoclinic chaos during the electrochemical deposition of zinc.[47] They also obtained a transient period-doubling sequence to chaos. They analyzed the chaotic nature of the time signals through the use of attractor reconstructions, Poincaré sections, and return maps. By using video signals they were able to relate the temporal oscillations to oscillations of growth rate of the cathodic surface.

5.5 Electrocatalytic Reactions

5.5.1 Electrochemical Oxidation of Formic Acid, Formate, and Formaldehyde

Some recent studies have shown interesting dynamic behavior in such reactions. Schell *et al.* have investigated the galvanostatic oxidation of both formaldehyde and formate/formic acid solutions on a platinum rotating-disk electrode.[48,49] The time series consisted of combinations of small- and large-amplitude oscillations. They observed a sequence of states of alternating periodic and chaotic behavior. The principal states of this sequence are shown in Fig. 5.7.

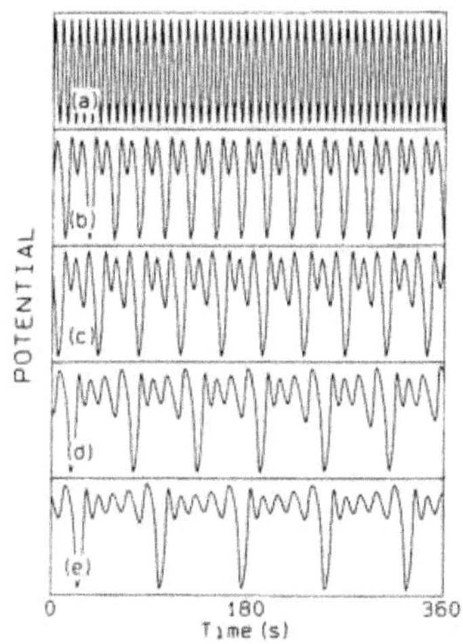

Figure 5.7. Periodic-chaotic sequence found in the oxidation of formaldehyde on platinum. Reproduced by permission from M. Schell, F.N. Albahadily, J. Safar and Y. Xu, *J. Phys. Chem.* **93** (1989) 4806.

They observed in the formate/formic acid system a sequence of Farey states, *i.e*, a sequence of periodic waveforms in which the numbers of small- and large-amplitude oscillations obey certain rules. Furthermore, they obtained measurements in the formaldehyde system that suggested that the periodic-chaotic sequence was terminated in the vicinity of where a torus is broken.[49]

5.5.2 *Oxidation of Hydrogen*

The oxidation of hydrogen, perhaps the electrocatalytic oscillatory system most ammenable to explanation, has received recent attention.[50-52] This reaction is typically studied under galvanostatic conditions in electrolytes through which hydrogen is bubbled and in which metal ions such as copper are present. The deposition and dissolution of the copper ions on the electrode (often platinum) plays a major role in the production of the oscillatory behavior.[53,54]

Interesting dynamics also have been observed with this system. An interior crisis has been reported by Krischer and co-workers.[50,51] In addition, a two-parameter study of the dynamics of this system has been carried out, where the two controlled variables were copper-ion concentration and current.[52] The experiments were done under galvanostatic conditions. A hierarchy of transitions into mixed-mode oscillations from a period-one harmonic oscillation, from a period-two oscillation, and via a period-doubling sequence to small-amplitude chaos, as well as a transition to large-amplitude chaos was observed. An overview of the dynamics of the system can be seen in Figure 5.8. Very recently the dynamics have been investigated on single-crystal platinum.[55]

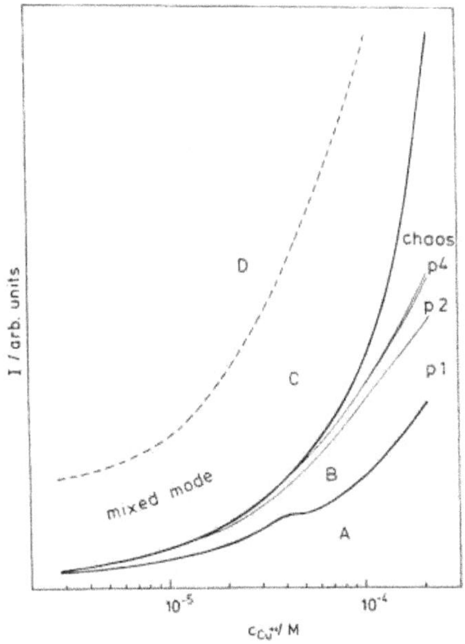

Figure 5.8. Overview of the dynamics seen in the oxidation of hydrogen on platinum: a) Steady state; b) P1--period-one harmonic oscillations, P--period two, *etc.* c) Mixed-mode ocillations. d) Period-one relaxation oscillations. Reproduced by permission from K. Krischer, M. Luebke, M. Eiswirth, W. Wolf, J.L. Hudson and G. Ertl, *Physica D* (1992), in press.

5.5.3 *Reduction of Hydrogen Peroxide*

Oscillations occur during the reduction of hydrogen peroxide on a platinum electrode under galvanostatic conditions even without the addition of metal ions, so that the concentration of metal ions is at the impurity level. Very complicated behavior, including cycles with a large number of oscillations, have been observed.[56]

5.5.4 *Other Catalytic Reactions*

Schell and Cai have investigated the oxidation of ethanol using voltammograms, that is, by sweeping the potential up and down.[57] The experiments were carried out on a rotating-disk electrode. They observed period-doubling bifurcations and chaotic voltammetric responses.

Koper and Gaspard have recently developed a mathematical model of an electrochemical oscillator and compared the results to experiments with the reduction of indium (III) on mercury in the presence of thiocyanate.[58] The three-variable model, which approximates the equations of reaction and diffusion near a mercury drop, produces both mixed-mode oscillations and chaotic behavior.

5.6 Concluding Remarks

Many examples of complex dynamic behavior have been observed during electrochemical reactions. Periodicity, quasiperiodicity, and low-order chaos have been obtained with the three main classes of electrochemical reactions, *viz.*, electrochemical dissolution, electrochemical deposition, and electrocatalytic reactions. Most of the studies have concentrated on temporal behavior and the analysis of time series using the methods of nonlinear dynamics. In a few cases, spatiotemporal behavior has been observed. This latter area, the observation of patterns on electrode surfaces and among sets of two or more electrodes, is one which is presently receiving increased attention and is the subject of Chapter 6.

Electrochemistry thus joins homogeneous reactions, such as the Belousov-Zhabotinsky reaction, and heterogeneous reactions, such as gas-solid reactions, as a source of interesting nonlinear dynamics in chemistry.

Acknowledgements. This work was supported in part by the National Science Foundation and by DARPA/ONR. I thank Theo Tsotsis, Katharina Krischer, Mark Schell, Mark Bassett, Moshe Sheintuch, Len Pismen, and Otto Rössler for their help.

Literature Cited

1. G.Th. Fechner, *Schweigg. J. für Chemie and Physik* **53** (1828) 129.
2. J. Wojtowicz, in *Modern Aspects of Electrochemistry 8*, J. Bockris and B. Conway, Eds. (Plenum Press, New York, 1972).
3. M. Hachkar, B. Beden and C. Lamy, *J. Electroanal. Chem. Interfacial Electrochem.* **287** (1990) 81.
4. J.L. Hudson and M.R. Bassett, *Oscillatory Electrodissolution of Metals*, in *Reviews in Chemical Engineering*, D. Luss and N.R. Amundson, Eds. (1991).
5. J.L Hudson and T.T. Tsotsis, *Chem. Eng. Sci.*, in press.
6. J.L. Hudson, *Z. Phys. Chem. (Leipzig)* **270** (1989) 497.
7. J.L. Hudson, in *IMA Volume in Mathematics and Its Applications*, H. Swinney, R. Aris and D. Aronson, Eds. (Springer, Berlin, 1991).
8. F.N. Albahadily and M. Schell, *J. Chem. Phys.* **88** (1988) 4312.
9. F.N Albahadily, J. Ringland and M. Schell, *J. Chem. Phys.* **90** (1989) 813.
10. M.R. Bassett and J.L. Hudson, *Physica* **35D** (1989) 289.
11. M.R. Bassett and J.L. Hudson, *Chem. Eng. Comm.* **60** (1987) 145.
12. M.R. Bassett and J.L. Hudson, *J. Phys. Chem.* **93** (1989) 2731.
13. M.R. Bassett and J.L. Hudson, *J. Electrochem.Soc.* **137** (1990) 922.
14. M.R. Bassett and J.L. Hudson, *J. Electrochem. Soc.* **137** (1990) 1815.
15. M.R. Bassett and J.L. Hudson, *J. Phys. Chem.* **92** (1988) 6963.
16. C.B. Diem and J.L. Hudson, *A.I.C.h.E. J.* **33** (1987) 218.
17. A.L. Kawczyński, W. Raczyński and B. Baranowski, *Z. Phys. Chemie (Leipzig)* **269** (1988) 596.
18. O. Lev, A. Wolffberg, M. Sheintuch and L.M. Pismen, *Chem. Eng. Sci.* **48** (1988) 1339.
19. O. Lev, A. Wolfberg, L.M. Pismen, and M. Sheintuch, *J. Phys. Chem.* **93** (1989) 1661.
20. O. Lev, M. Sheintuch, L.M. Pismen and Ch. Yarnitzky, *Nature* **336** (1988) 458.
21. O. Lev, M. Sheintuch, H. Yarnitsky, and L.M.Pismen, *Chem. Eng. Sci.* **45** (1990) 839.
22. M. Schell and F.N. Albahadily, *J. Chem. Phys.* **90** (1989) 822.
23. Y. Wang and J.L. Hudson, *AIChE J.* **37** (1991) 1833.
24. P. Grassberger and I. Procaccia, *Phys. Rev. Lett.* **50** (1983) 346.
25. P. Grassberger and I. Procaccia, *Physica* **9D** (1983) 189.
26. Z.H. Gu, T.Z. Fahidy and J.P. Chopart, *Electrochim. Acta* **37** (1992) 97.
27. M. Feigenbaum, *Physica* **7D** (1983) 16.
28. J.L. Hudson, M. Hart and D. Marinko, *J. Chem. Phys.* **71** (1979) 1601.
29. J.L. Hudson, M. Kube, R.A. Adomaitis, I.G. Kevrekidis, A.S. Lapedes and R.M. Farber, *Chem. Eng. Sci.* **45** (1990) 2075.
30. R. Rico-Martinez, K. Krischer, I. G. Kevrekidis, M. C. Kube, and J. L. Hudson, *Chem. Eng. Comm.* (1992), in press.
31. N.H. Packard, J.P. Crutchfield, J. D. Farmer and R.S. Shaw, *Phys. Rev. Lett.* **45** (1980) 712.
32. F. Takens, *Lect. Notes Math.* **898** (1981) 366.

33. O.E. Rössler, *Bull. Math. Biol.* **39** (1977) 275.
34. O.E. Rössler, *Z. Naturforsch.* **31A** (1976) 1664.
35. O.E. Rössler, *Z. Naturforsch.* **31A** (1976) 259.
36. L.P. Shil'nikov, *Soc. Math Dokl.* **6** (1965) 163.
37. F. Argoul, A. Armeodo and P. Richetti, *Phys. Lett.* **120** (1987) 269.
38. P. Russell and J. Newman, *J. Electrochem. Soc.* **133** (1986) 2093.
39. P. Russell and J. Newman, *J. Electrochem. Soc.* **134** (1987) 1051.
40. U.F. Franck and R. FitzHugh, *Z. Elektrochem.* **65** (1961) 156.
41. R. Badii and A. Politi, *J. Stat. Phys.* **40** (1985) 725.
42. E.J. Kostelich and H.L. Swinney, in *Chaos and Related Nonlinear Phenomena*, I. Procaccia and M. Shapiro, Eds. (Plenum, New York, 1987).
43. D. Haim, O. Lev, L.M. Pismen and M. Sheintuch, preprint.
44. D. Haim, O. Lev, L.M. Pismen and M. Sheintuch, preprint.
45. R.M. Suter, and P. Wong, *Physical Review* **39B** (1989) 4536.
46. F. Argoul and A. Arneodo, *J. Physics (Paris)* **51** (1990) 2477.
47. F. Argoul, J. Huth, P. Merzeau, A. Arneodo and H.L. Swinney, *Physica D,* (1992), in press.
48. M. Schell, F.N. Albahadily, J. Safar and Y. Xu, *J. Phys. Chem.* **93** (1989) 4806.
49. Y. Xu and M. Schell, *J. Phys. Chem.* **94** (1990) 7137.
50. K. Krischer, M. Luebke, W. Wolf, M. Eiswirth and G. Ertl, *Int. Series of Numerical Mathematics* **97** (1991) 211.
51. K. Krischer, M. Luebke, W. Wolf, M. Eiswirth and G. Ertl, *Ber. Bunsenges. Phys. Chem.* **95** (1991) 820.
52. K. Krischer, M. Luebke, M. Eiswirth, W. Wolf, J.L. Hudson and G. Ertl, *Physica D* (1992), in press.
53. T. Kodera, T. Yamazaki and N. Kubota, *Electrochim. Acta* **31** (1986) 1477.
54. T. Kodera, T. Yamazaki, M. Masuda and R. Ohnishi, *Electrochim. Acta* **33** (1988) 537.
55. M. Eiswirth, M. Luebke, K. Krischer, W. Wolf, J.L. Hudson, and G. Ertl, *Chem. Physics Letters*, in press.
56. N. Fetner and J.L. Hudson, *J. Phys. Chem.* **94** (1990) 6506.
57. M. Schell and X. Cai, *J. Chem. Soc. Faraday Trans.* **87** (1991) 2255.
58. M.T.M. Koper and P. Gaspard, J. Phys. Chem. **95** (1991) 4945.

CHAPTER 6: CHAOS IN SURFACE-CATALYZED REACTIONS

MARKUS EISWIRTH
Fritz-Haber-Institut der Max-Planck-Gesellschaft,
Faradayweg 4-6, W-1000 Berlin 33, Germany

ABSTRACT

Chaos has been obtained in heterogeneous reactions occurring on catalyst surfaces using a large variety of conditions, catalyst systems and reactor configurations. In most cases, the observed complexity seems to result from thermal or mass-transport coupling of oscillatory kinetics occurring on different parts of the catalyst surface. Spatially resolved measurements are therefore required for characterization of this dynamical behavior. The best understood system is the oxidation of CO on platinum, the mechanism of which has been elucidated with single-crystal studies at low pressures and was found to involve structural changes of the catalyst surface. Temporal chaos and hyperchaos as well as spatial turbulence have been observed.

6.1 Introduction

Heterogeneous catalysis in which a chemical reaction takes place at a phase boundary (usually a solid/gas interface) is of enormous technical importance. The world-wide production of catalysts reached a value of almost three billion dollars/yr in the mid-eighties.[1] The economic importance of these catalysts can be gauged by the fact that they usually represent only 1% of production costs. About 70% of all commercial chemicals are synthesized via heterogeneous catalysis.[2] Catalytic processes also are becoming ever more important in pollution control, *e.g.*, in automobile exhaust systems and in fossil-fuel burning power plants. Despite their technical relevance to large-scale chemical production,[3,4] relatively little was known for decades after the beginning of their use about the mechanisms of heterogeneous as compared to homogeneous systems, mainly because the critical steps occur on the catalyst surface and could not be directly monitored. Another complication is that catalysts mostly do not have well-defined surfaces. They instead usually consist of crystallites with different crystallographic orientations. Furthermore, the surface concentrations and the lateral distribution of different constituents of the catalyst as well as of impurities is often hard to determine. This led to the opinion that,

"Catalysis is still more an art than a science," attributed to D.D. Eley. A way out of this problem is the use of "idealized" model catalysts, namely pure, well-defined, single-crystal surfaces, and experiments under ultra-high vacuum (UHV) conditions. Modern surface-science techniques[5,6] then can be used to probe the surface directly and to clarify the elementary steps of a catalytic process.[7-10]

The kinetics of surface processes may become very complex, and phenomena associated with nonlinear dynamics were first investigated in the oxidation of carbon monoxide on a platinum catalyst.[11-13] Subsequently a large number of other reactions were studied,[14] including the oxidation of ammonia,[15-17] hydrocarbons,[18-21] alcohols,[22-26] ethers,[27] formic acid[28] and hydrogen,[29-34] as well as the reaction of nitrogen monoxide with carbon monoxide[35-40] or ammonia,[41,42] and the hydrogenation of carbon monoxide,[43] ethylene[44] and nitric oxide.[45,46] Besides the platinum metals, Pt, Pd, Ir and Rh, silver and nickel catalysts as well as metal-oxides also are used.[47] Nevertheless, the CO-oxidation is still the heterogeneous process studied in most detail. It exhibits a very rich variety of temporal and spatial self-organization that has been well reviewed.[48-55] Indeed, it plays a role in the study of nonlinear dynamics in heterogeneous catalysis systems similar to that of the celebrated Belousov-Zhabotinsky reaction in homogeneous systems. See Chapters 1, 2, 3 and 4.

Although aperiodic oscillations in CO oxidation had been reported before,[56-58] the first unambiguous identification of deterministic chaos in this system by Razón *et al.*[59] was obtained on a Pt ribbon under gradientless, isothermal conditions. A transition from periodic to chaotic behavior (via a sequence of period-doubling bifurcations) was first observed on a Pt (110) single-crystal surface.[60,61] Chaos subsequently was identified in several other processes using different catalysts and reactor configurations.

Many heterogeneous catalysts are best considered as consisting of individual (*i.e.*, spatially separated), but coupled oscillators, *e.g.*, catalytically active crystallites distributed on an inert support, different pellets in a packed-bed reactor, or even two single crystals.[62] Since the emergence of chaos in such systems is not fundamentally different from other coupled oscillators, we shall not go into details (See Chapter 4) but instead concentrate on a discussion of the specific coupling mechanisms between different catalyst particles and some representative examples with supported catalysts (Section 6.2). The influence of external perturbations has been well studied[63-71] but is not treated here. Rather we discuss the occurrence of chaos on unsupported, monolithic catalysts in the form of a polycrystalline wire, ribbon or foil (Section 6.3), as well as single-crystal surfaces (Section 6.4). Unlike homogeneous systems where efficient stirring can always suppress spatial inhomogeneities, surface reactions occur in an extended medium. Though uniform oscillatory media may actually synchronize, one nevertheless has to reckon with the formation of dynamic spatial structures even with uniform catalysts. It is therefore desirable to have not only data describing the (integral) temporal behavior, but also spatially resolved measurements. We shall consequently concentrate on studies for which such measurements are available.

Experimental techniques differ widely for different catalysts and reactor configurations and will be briefly described where the need arises.

6.2 Supported Catalysts

The catalysts relevant in the present context are Pd and Pt on ceramic supports, namely alumina, silica and zeolithes. The catalytically active metallic crystallites are typically 5-100 nm in size and distributed on the support. Essentially three different types of experimental setups have been used: packed-bed tubular reactors filled with pellets of typically several mm diameter, catalyst wafers and finely grained support particles ($<100\ \mu$m) distributed in a thin layer (shallow bed).

Very weak interaction among the catalyst particles clearly should result in random phase relationships without macroscopic oscillations, while very strong autocatalytic coupling would favor complete synchronization. Chaotic regimes can be expected between these extremes.

The individual catalyst particles in such systems basically can communicate in two ways: through heat transfer (thermal coupling) or via the concentration changes they cause in the gas phase (gas-phase coupling). The first type clearly cannot occur in isothermal systems, but it is practically impossible to suppress temperature changes with supported catalysts, even if the average reactor temperature can be kept fairly constant. The catalyst in the course of oscillations generally goes through a phase where it is reactive and hot ("ignited") and another phase where it is less reactive and cool ("passivated" or "extinguished"). Thermal coupling is for activated exothermic reactions (which include all cases studied) autocatalytic and consequently favors synchronization, since a hot particle will heat its cool neighbors, which subsequently ignite (and *vice-versa*). It is in theory conceivable for there to be a negative feedback through thermal coupling in endothermic reactions, but no such oscillator has yet been examined experimentally. The transfer of heat may occur along the support or through the gas phase, depending on mixing; radiative heat transfer can be neglected in most cases.

Coupling via pressure changes in the gas phase can *a priori* be either autocatalytic or inhibitory. The latter favors desynchronization. We use the oxidation of CO to illustrate gas-phase coupling.[64,72]

While pre-adsorbed CO blocks the adsorption of oxygen, an O-covered surface does not prevent CO from adsorbing (asymmetric inhibition). Therefore the O-covered state is reactive, and such a catalyst particle will lower the partial pressures of both CO and O_2 in its vicinity, while a passivated CO-covered particle does not significantly alter the pressures of CO and O_2. Since oxygen is always present under oscillatory conditions in (mostly large) excess, gas-phase coupling will mainly be mediated via the CO pressure. A reactive particle consequently will cause a much more significant drop in CO than in O_2 pressure. Another particle in the vicinity still in the passivated state cannot maintain its high CO coverage and also will ignite. This argument also works the other way round, *i.e.*, an unreactive particle will cause an increase of the CO/O_2 ratio in the gas phase, which will tend to extinguish nearby reactive states. Clearly the interaction caused by pressure changes of O_2 is inhibitory at the same time, but this effect does not become relevant as long as O_2 is present in large enough excess. However, one expects inhibitory gas-phase coupling when the reactant leading to a passivated surface is present in excess or for symmetric inhibition of adsorption, *e.g.*, as is the case in the NO/CO reaction.

There are numerous studies of spatio-temporal oscillations showing ignition phenomena, obtained with thermography[33] or the use of several thermocouples to record local temperature changes.[24,38,63,73-75] In most cases, however, both thermal and gas-phase coupling can occur, and it is hard to determine whether one coupling mechanism clearly dominates. Tsai *et al.*[76] showed the importance of gas-phase coupling in an elegant experiment on CO oxidation on a Pt-wire loop, which was cut into 2 pieces and separated by a glass plate, resulting in desynchronization. Jaeger, Plath and coworkers[24,77,78] varied the thermal conductivity between the catalyst particles (Pd/Al_2O_3) by putting finely powdered catalyst on either a silver or a teflon plate. This did not have a significant effect on CO oxidation, whereas synchronization occurred in the oxidation of methanol, ethanol and formic acid. This led to the conclusion that gas-phase coupling dominates with CO, whereas thermal effects are crucial for the other reactions.

Chaotic oscillations are found under various conditions. Onken and Wicke,[63,75,79] as well as Kapicka and Marek[80,81] studied CO oxidation on Pt/Al_2O_3 pellets in a tubular reactor and found both periodic and chaotic regimes. The catalyst was first reduced in one case[81] to increase its activity, which then decreased slowly during CO oxidation. This amounts to a slow, continuous change of a parameter (the catalyst activity), in the course of which a transition to chaos via two period-doublings was obtained (Fig. 6.1).

Kapicka and Marek[80] also used 5 thermocouples attached to different pellets in a tubular reactor. Interestingly, aperiodic oscillations were found both with the pellets in phase and when they were desynchronized. Desynchronization could also go so far that no further overall oscillations occurred; *i.e.*, constant total CO_2 production was monitored at the reactor exit.

Schüth and Wicke studied the oxidation of CO as well as the CO/NO reaction on catalyst wafers with both Pd and Pt catalysts.[38] Their observations could be successfully modeled using a 10 × 10 matrix of randomly placed thermokinetic oscillators (for CO/NO) coupled by heat transfer.[82]

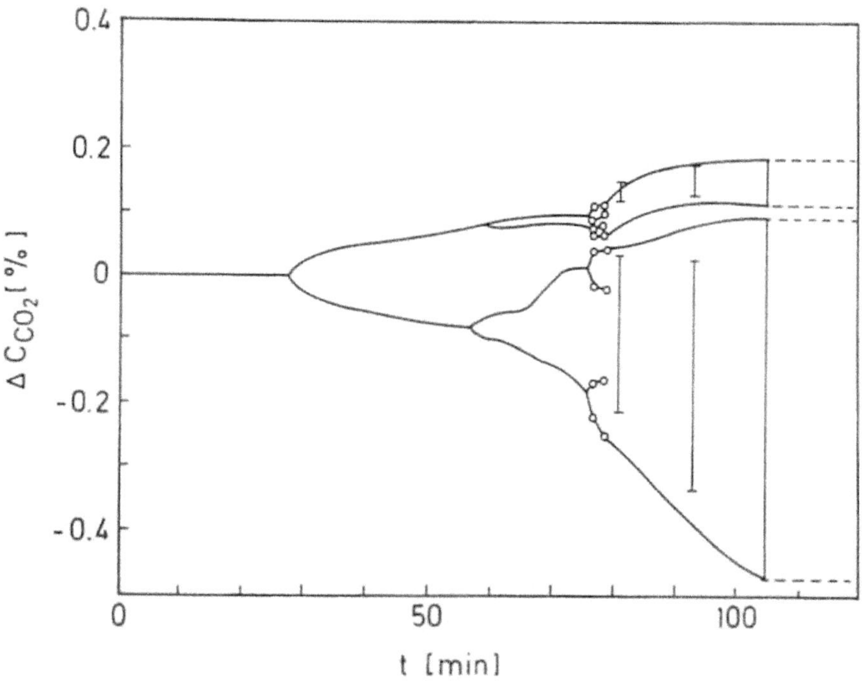

Figure 6.1. Bifurcation diagram of a period-doubling transition to chaos in the CO oxidation on a slowly changing Pt/Al_2O_3 catalyst. The maxima and minima of the outlet CO_2 concentration are shown *vs* time (which plays the role of the control parameter because of the slow change in the catalyst activity). Vertical bars indicate a continuous distribution of extrema. After J. Kapicka and M. Marek, *Surf. Sci.* **222** (1989) 885.

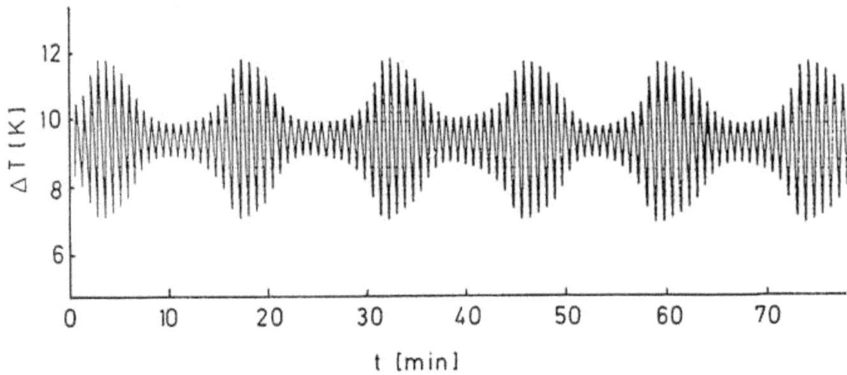

Figure 6.2. Quasiperiodicity in the oxidation of ethanol on zeolithe-supported Pd. ΔT denotes the difference between the catalyst and the reactor temperature, T_R, which was fixed. After H. Herzel, P. Plath and P. Svensson, *Physica*, **48D** (1991) 340.

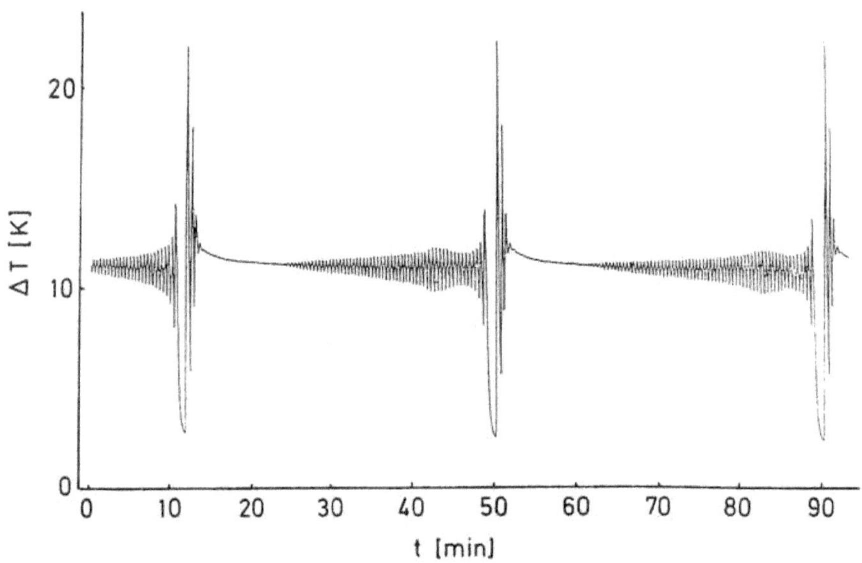

Figure 6.3. Near-homoclinic situation in the oxidation of ethanol. After H. Herzel, P. Plath and P. Svensson, *Physica*, **48D** (1991) 340.

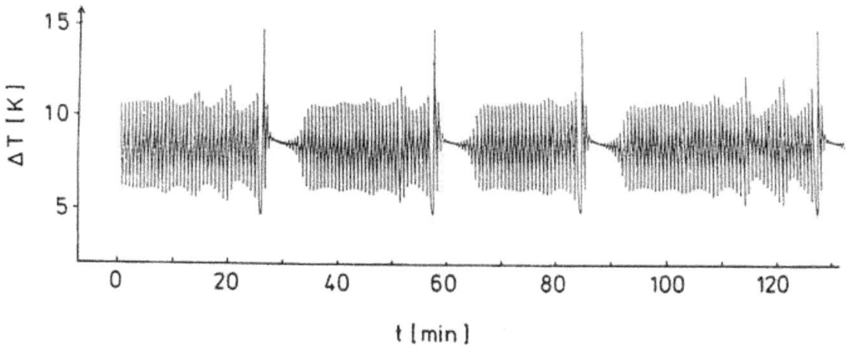

Figure 6.4. A time series indicating type-II intermittency. After H. Herzel, P. Plath and P. Svensson, *Physica*, **48D** (1991) 340.

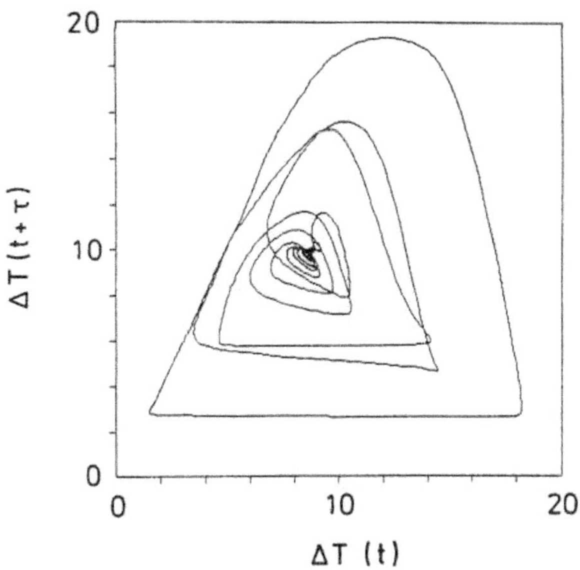

Figure 6.5. Part of a strongly inhomogeneous reconstructed attractor from the oxidation of ethanol on Pd/zeolithe. After H. Herzel, P. Plath and P. Svensson, *Physica*, **48D** (1991) 340.

Chaotic time series also were analyzed for the oxidation of CO on zeolithe-supported Pd distributed on a glass frit.[83,84] A variety of complex temporal oscillations were found by Herzel, Plath and Svensson[26] for shallow-bed conditions using a silver holder in the oxidation of ethanol. Besides harmonic oscillations of periodicity 1, 2 and 4, quasiperiodicity was observed (Fig. 6.2), indicating the occurrence of a Neimark-Sacker (secondary-Hopf) bifurcation. Moreover, the authors obtained bursting phenomena suggesting the presence of a homoclinic orbit (Fig. 6.3).

Careful analysis of data from the oxidation[26] of ethanol revealed that there is most likely no Shil'nikov-type chaos,[85] but Herzel *et al.* did observe a transition to chaos via type-II intermittency.[86,88] This route occurs due to the interaction of a subcritical Neimark-Sacker bifurcation and a homoclinic orbit. The shape of the time series (Fig. 6.4), which exhibits a beat-like modulation before the large burst, is clearly characteristic of type-II intermittency.

The correlation dimensions, D_2, were between 2 and 3 in all cases mentioned in this section. The value $D_2 = 2.1$ (without error limits) has been reported[79,84] using the Grassberger-Procaccia method,[89] and 2.4 ± 0.2 was found for the homoclinic chaos in reference 26. This very inhomogeneous attractor (Fig. 6.5) had to be analyzed using a nearest-neighbor algorithm.[90]

6.3 Monolithic Polycrystalline Catalysts

The catalysts treated in this section are metallic foils, ribbons or wires. These can be prepared to be much more uniform than the supported catalysts. They represent continuous two- or effectively one-dimensional active media. Diffusion of adsorbed particles on the surface in such systems has to be taken into account as another coupling mechanism, which also will play a role in the single-crystal studies described in Section 6.4. The thermal conductivity is homogeneous over such catalysts, in contrast to the systems described in the previous section. Oscillations under nonisothermal conditions seem usually to be connected with traveling waves, mediated mainly through thermal conduction, which is much faster than surface diffusion. These traveling waves can be observed as temperature waves[33,34] in real time using an infrared camera. No chaotic states were found in these studies, but such states were observed in others where the catalyst temperature was controlled.

As mentioned in the introduction, Razón, Chang and Schmitz[59] carried out experiments on the oxidation of CO on a Pt-ribbon in a well-stirred, gradientless system. Both the gas and the catalyst temperature were controlled. Temperature variations across the ribbon due to the intrinsic inhomogeneity of the catalyst itself were well below 3 K. Though no transition from periodic to chaotic oscillations was observed, data analysis with the usual methods[88,89] clearly revealed chaos with a correlation dimension between 3 and 4 (roughly 3.5). No *in-situ*, spatially resolved measurements with such a setup have been reported.

More recently Sheintuch and Schmidt[91] studied the oxidation of NH_3 on a Pt wire. The wire was heated electrically, and the experiments could be run either "isothermally" (constant wire resistance, hence constant average temperature) or nonisothermally (constant heating current). A bifurcation diagram in two parameters was mapped out[91] for isothermal conditions that revealed a complicated sequence of multiperiodic and aperiodic states. While some of the latter states were clearly quasiperiodic, chaotic oscillations also occurred. The time series and especially the return map suggest a type-I intermittency.[86,88]

Control of the average temperature[59,91] by no means guarantees that the process is actually isothermal. Rather there can be stationary or moving dynamical spatial structures, as was shown in a sequence of experiments by Luss and coworkers.

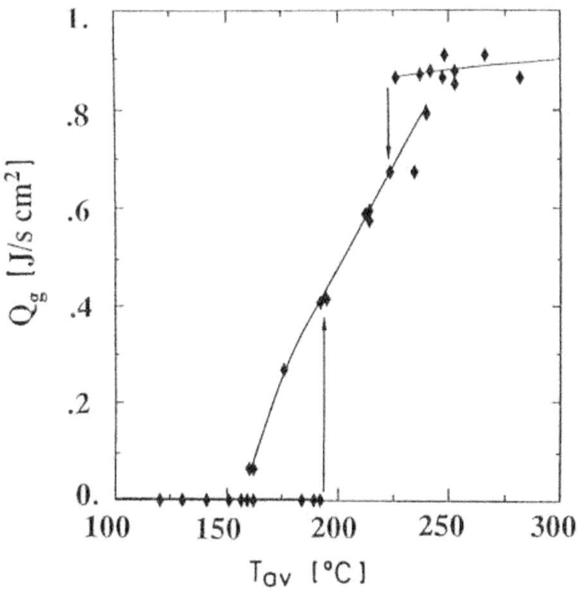

Figure 6.6. A bifurcation diagram showing the rate of heat generation by the oxidation of NH_3 on a Pt ribbon *vs* average catalyst temperature, T_{av}. The middle branch corresponds to a state with a standing temperature front (1% NH_3 in air). After L. Lobban, G. Philippou and D. Luss, *J. Phys. Chem.* **93** (1989) 733.

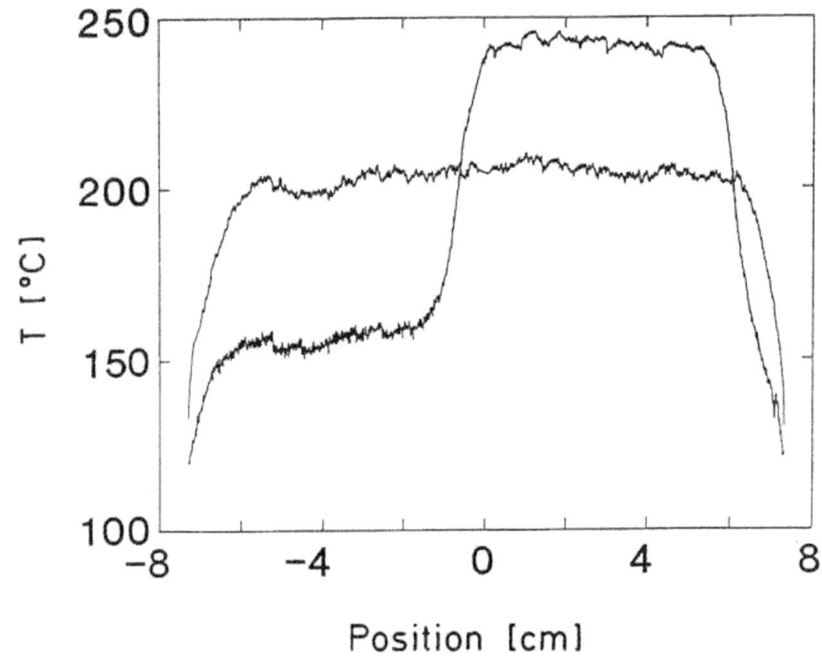

Figure 6.7. Coexistence of two different temperature profiles for T_{av} = 195° C. After L. Lobban, G. Philippou and D. Luss, *J. Phys. Chem.* **93** (1989) 733.

They studied NH_3 oxidation with a setup similar to that used by Sheintuch and Schmidt, but used a thin Pt ribbon as catalyst and a thermal imager to measure the temperature distribution along it.[17] Homogeneous extinguished and ignited states were found in constant-resistance mode at low and high T, respectively. A stationary front resulted at intermediate average temperatures, T_{av}, (Fig. 6.6). Hysteresis was observed in the transition from the uniform to the nonuniform states and back (Fig. 6.7). Note that the external control keeping the total resistance constant introduces a global feedback into this system. This causes the front between ignited and extinguished states to stand still.[17]

Philippou, Schultz and Luss[21] studied the oxidation of propylene in air with the same setup. The nonuniform state in this system exhibited an ignited section which oscillated in length and position. These oscillations were periodic (or quasiperiodic) at high temperatures and chaotic at low temperatures (Fig. 6.8). A splitting of the ignited section into two waves was observed in the transition range (region T). See

Fig. 6.9.

An integral time series of a typical chaotic state, the reconstructed attractor and a Poincaré section are shown in Fig. 6.10. The correlation dimension, D_2, with 0.3% propylene in the feed increased with decreasing average temperature from 2.8 to about 3.2. A D_2 even slightly above 5 was obtained for less propylene in the feed (0.2%). Interestingly, the correlation dimensions of attractors reconstructed from the local temperature at a fixed position were always smaller. Note that the transition to chaos in this system does not seem to fit any of the theoretically described scenarios.[86-88]

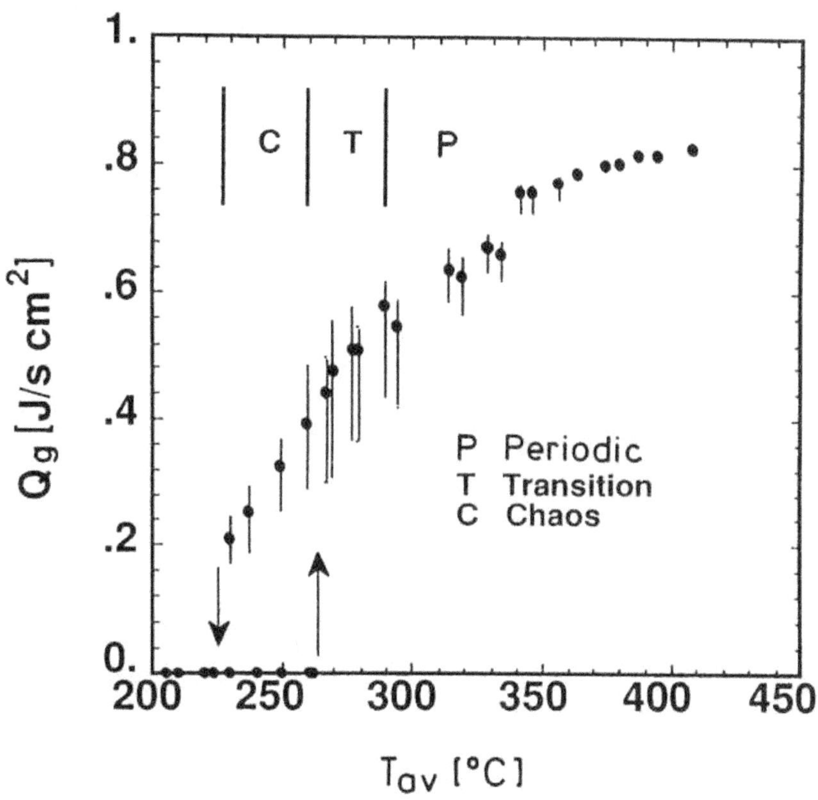

Figure 6.8. Overall heat generation in the oxidation of propene on Pt *vs* average temperature. Vertical bars indicate oscillation amplitudes (0.2% propene in air). After G. Philippou, F. Schultz and D. Luss, *J. Phys. Chem.* 95 (1991) 3224.

These authors also studied the profile and propagation of the waves more closely (Fig. 6.11) by keeping the camera stationary on a 1-cm segment of the wire

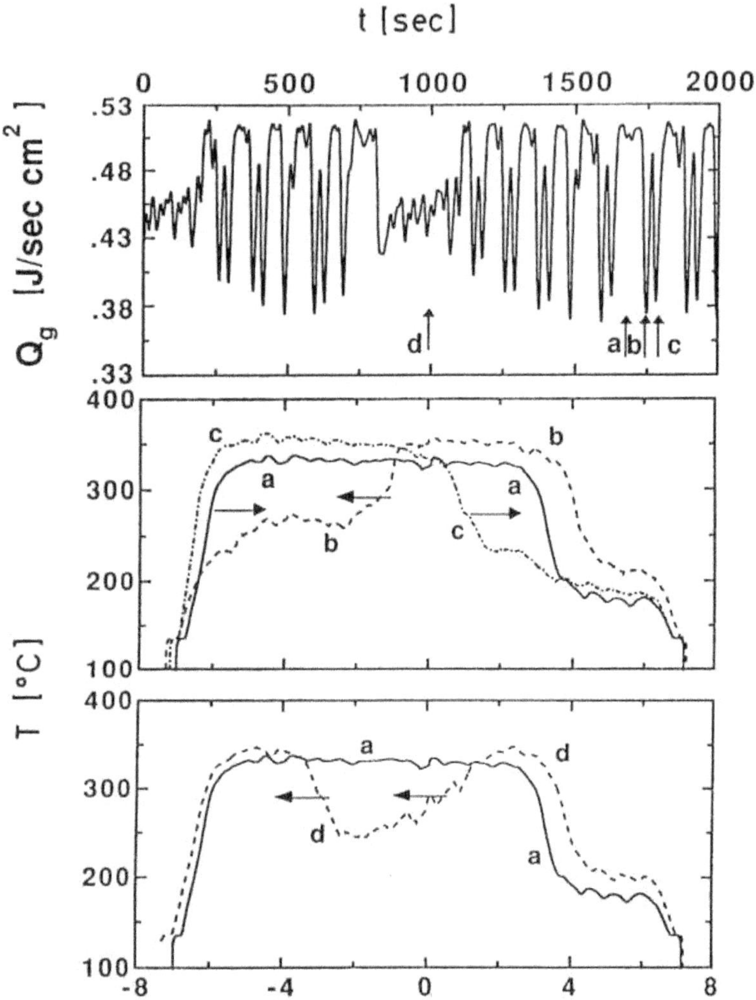

Figure 6.9. Time series of total heat generation at T_{av} = 269° C (corresponding to region *T* in Fig. 6.8) and related temperature profiles. After G. Philippou, F. Schultz and D. Luss, *J. Phys. Chem.* **95** (1991) 3224.

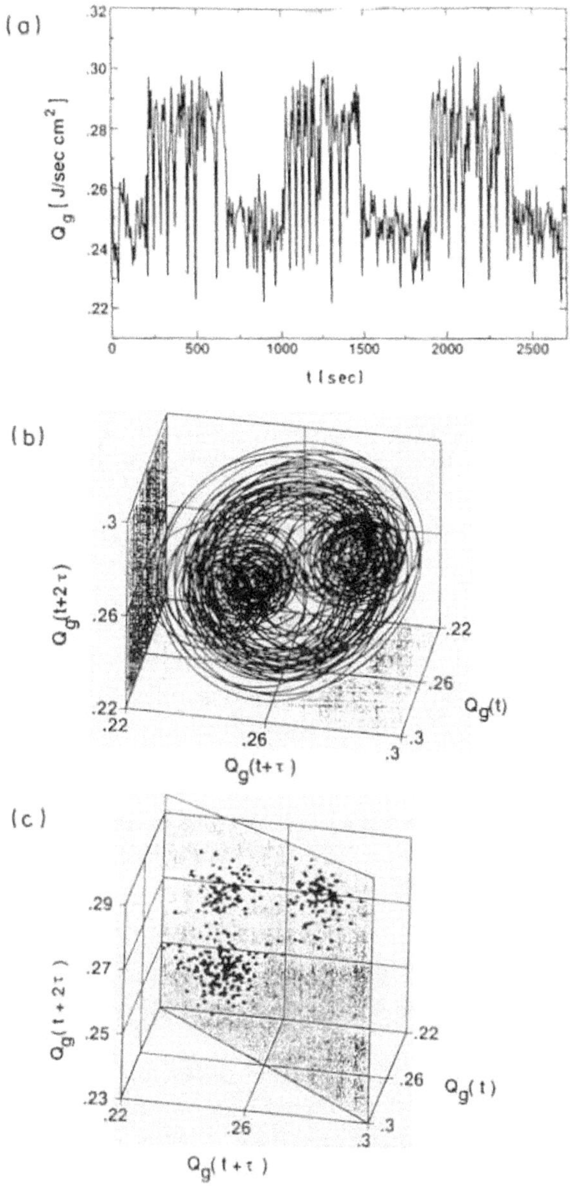

Figure 6.10. Time series a); reconstructed attractor b); and Poincaré section c) of a chaotic state in the oxidation of propene (0.3% propene, T_{av} = 225°C). After G. Philippou, F. Schultz and D. Luss, *J. Phys. Chem.* **95** (1991) 3224.

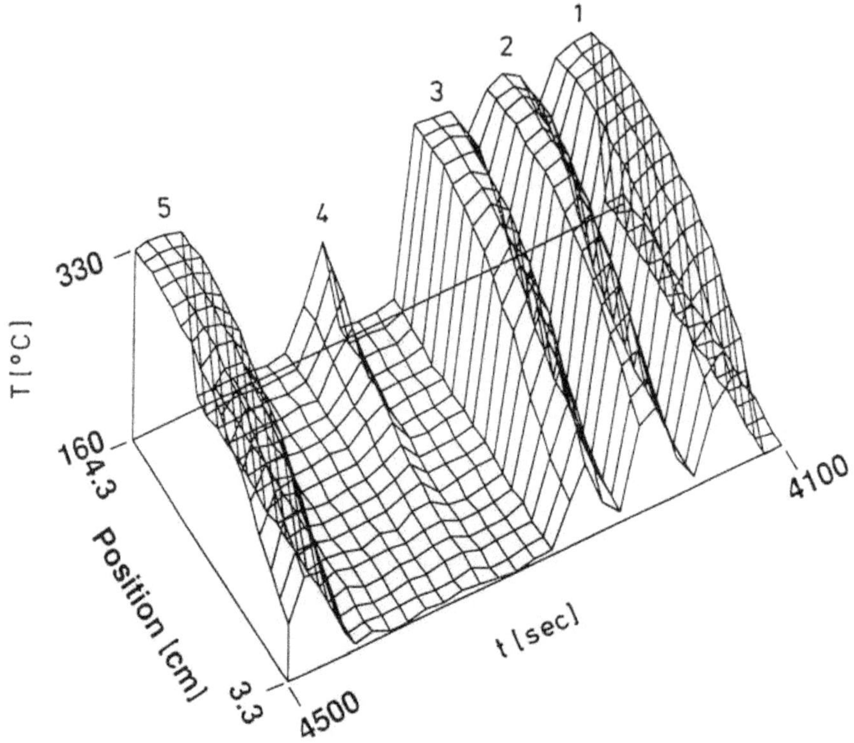

Figure 6.11. Detailed view of a temperature front moving back and forth on the Pt ribbon (0.3% propene, T_{av} = 243° C, chaotic state). After G. Philippou, F. Schultz and D. Luss, *J. Phys. Chem.* **95** (1991) 3224.

close to a position where the front changes its direction. Five waves reached the segment observed during the 400 s shown, although one of them (no. 4) barely entered it.

To summarize, failure to account for temperature nonuniformity can lead to pitfalls in the analysis of data[17] for (fairly) uniform (albeit polycrystalline) catalysts even under on-average isothermal conditions. Lobban and Luss[34] conjectured that chaotic oscillations in such systems may very well be due to dynamic spatial structures rather than to complicated kinetic mechanisms. The construction of intrinsically chaotic chemical reaction schemes for heterogeneous catalysis does not appear to be necessary unless spatially homogeneous chaotic oscillations are observed.

6.4. Single Crystals

The idea behind the use of single crystals as catalysts is to have a uniform surface in order to study nonlinear spatio-temporal phenomena which are genuinely due to the intrinsic reaction-diffusion dynamics. Single crystals suppress spatial nonuniformities that already exist in the surface composition and structure with other catalysts. Nevertheless, defects will always exist on real surfaces. Scanning tunneling microscope (STM) studies revealed that on carefully prepared surfaces typically about 1% of the sites are defect-related.[92,93] This means that large portions of the surface are uniform down to the atomic level.

These experiments are usually carried out under ultra-high vacuum (UHV) conditions, *i.e.*, at residual pressures below 10^{-9} mbar in order to maintain impurity concentrations low. Working pressures on the order of (or below) 10^{-4} mbar are used. This guarantees very good mixing of the gas phase because the mean free path of the gas molecules is of the same order as the size of the UHV chamber. The use of massive single crystals also allows strictly isothermal experiments. Thermal inhomogeneities are avoided because at these low pressures the heat generated by the (low) reaction rate quickly dissipates due to the high thermal conductivity of the metal. Theoretical estimates of overall temperature variations put them at < 2 mK[94] for pressures of $\approx 10^{-4}$ mbar. The experimental detection limit is much higher at $\approx$ 0.05 K.[71]

Oscillations in such systems were first reported for the oxidation of CO on Pt,[95] which has subsequently been studied in great detail.[51-55] More recently, Pd and Rh surfaces have been used, and other reactions, *e.g.*, CO/NO, NO/NH_3 and NO/H_2, found to undergo oscillation.[96]

The first such system for which a transition to chaos was observed (and still the system studied in most detail) is the oxidation of CO on Pt(110).[60] Besides the formation of a stationary dissipative structure,[97-100] this reaction was found to exhibit bistability, excitable regions with target patterns, spirals and solitary waves, as well as a rich variety of oscillatory phenomena.[61,71,94,101-115] We first concentrate on the integral temporal behavior in the following description of chaos and then point out the associated spatial effects.

The temporal oscillations in this system can be monitored by measuring the overall CO_2 production rate using mass spectrometry, or the laterally averaged work

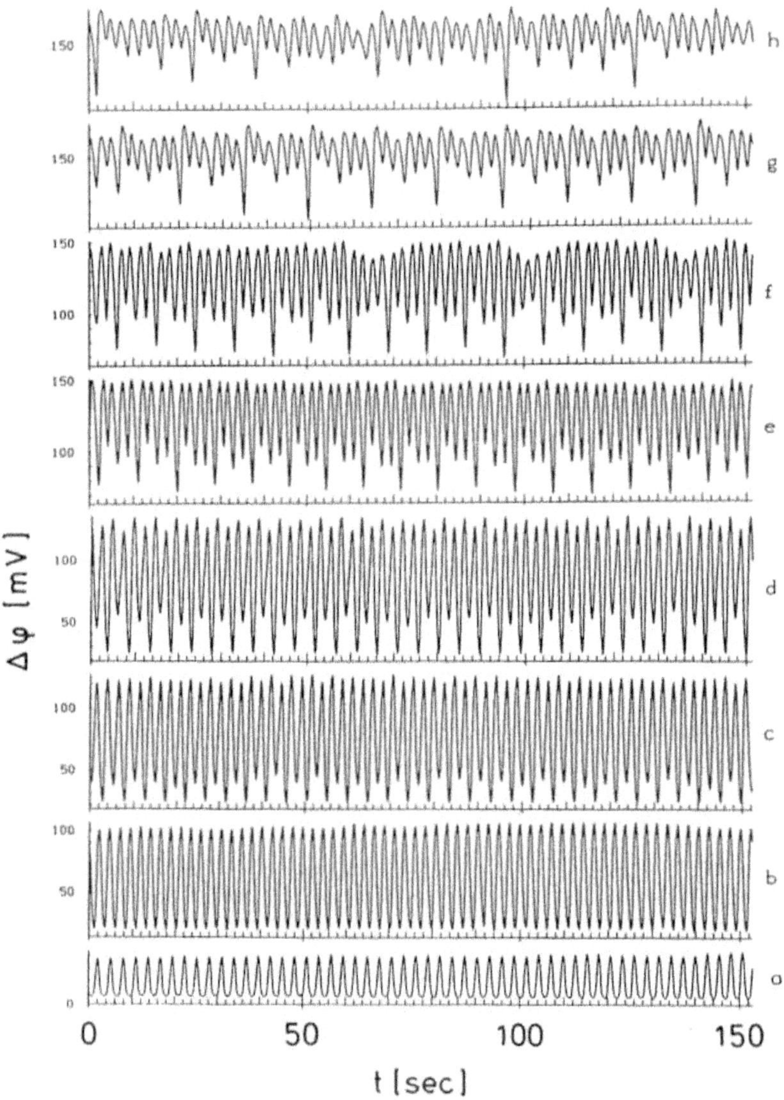

Figure 6.12. Sequence of oscillations in the oxidation of CO on Pt(110) obtained for T = 540 K and $p(O_2)$ = $1x10^{-4}$ mbar upon decreasing p_{CO} in small steps from a) 5.12 to h) 4.52 × 10^{-5} mbar.

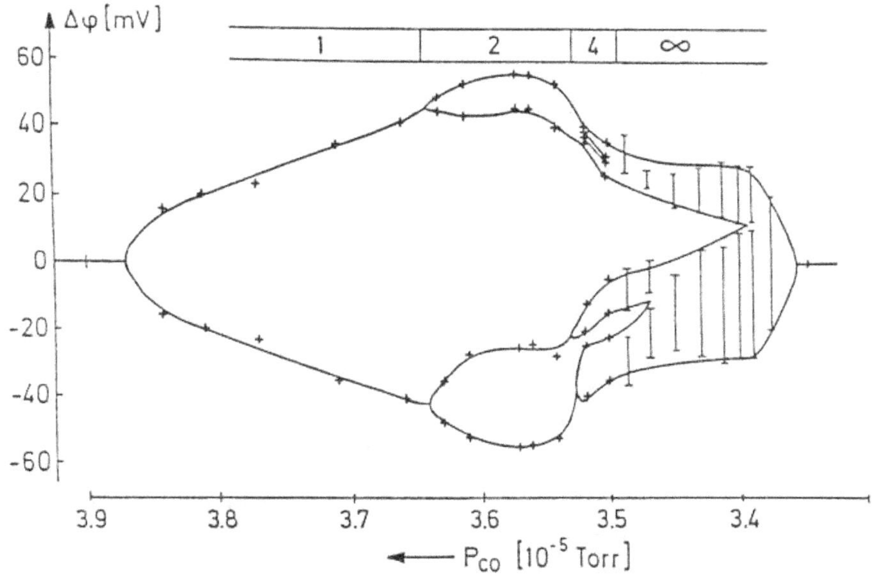

Figure 6.13. Bifurcation diagram constructed from the time series shown in Fig. 6.12. Vertical bars indicate continuous distributions of minima and maxima.

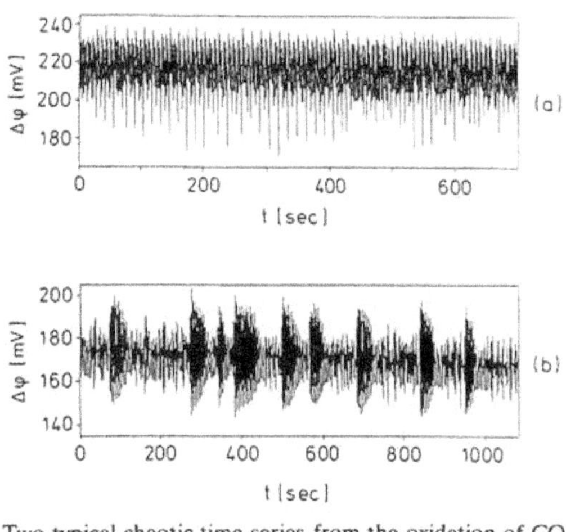

Figure 6.14. Two typical chaotic time series from the oxidation of CO on Pt(110). States a) close to, and b) further away from accumulation point. (See Table 6.1 for parameter values).

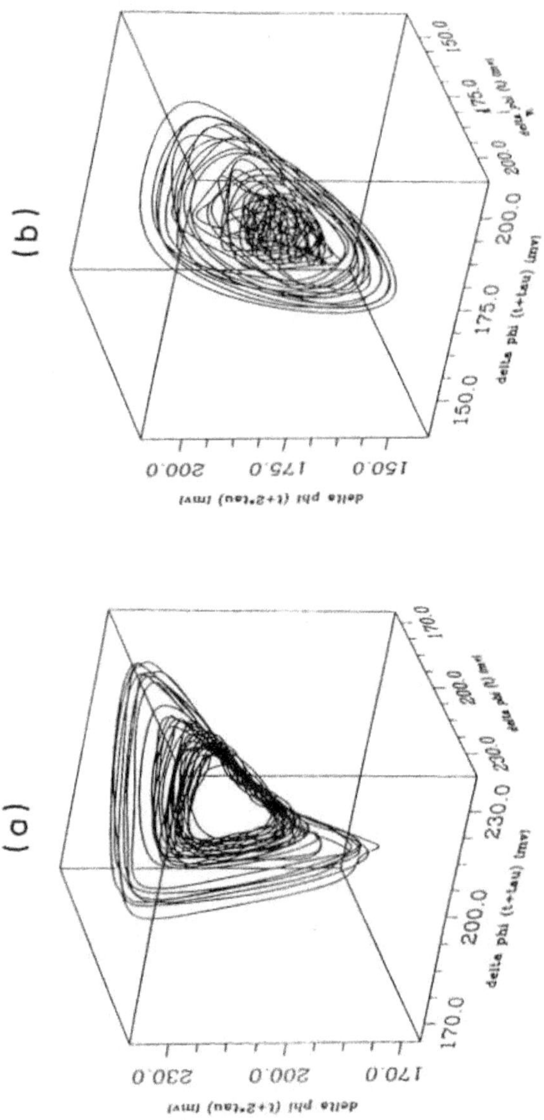

Figure 6.15. Attractors reconstructed from the time series in Fig. 6.14.

function $\Delta\Phi$ with a Kelvin probe.[71,94] The work function monitors (to a first approximation) the coverage of adsorbed O, which causes a large change[60,101] in $\Delta\Phi$. The two methods are equivalent because the reaction rate is proportional to the oxygen coverage as a consequence of the asymmetric inhibition discussed in Section 6.2. Oscillations set in as a result of a Hopf bifurcation when the CO pressure is lowered starting from the CO-covered stationary state. Two period-doublings could be observed before the onset of chaos, as is shown in Fig. 6.12.

The bifurcation diagram in Fig. 6.13 is compatible with the Feigenbaum route to chaos with respect to the separation of the period-doublings.[61] Furthermore, shortly after the accumulation point, shallower and deeper minima occur, alternating with separate distributions of size (2-band attractor). Farther away their distributions merge (bands no longer discernible) before also merging with the distribution of maxima. In addition, for slightly different conditions, a period-3 window in the chaotic regime could be clearly identified.[54,60,61]

Figure 6.14 shows two chaotic time series, one close to the accumulation point and one farther away from it under slightly different conditions. The corresponding reconstructed attractors are reproduced in Fig. 6.15. While the first attractor exhibits

Table 6.1. Parameters and characteristic measures of the chaotic time series from the oxidation of CO on Pt(110) at 540 K shown in Fig. 6.14. D_2 = correlation dimension; K_2 = lower bound for Kolmogorov entropy in s^{-1}; λ_{wssv} = largest Lyapounov exponent from the WSSV method in s^{-1}. Sampling time was 0.15 s.

Time series	Control Parameters (T = 540 K)	D_2	K_2	λ_{WSSV}
A Fig. 6.14a	$p(O_2) = 1.3 \times 10^{-4}$ mbar $p(CO) = 6.0 \times 10^{-5}$ mbar	2.4 ± 0.2	0.2 ± 0.05	0.12 ± 0.05
B Fig. 6.14b	$p(O_2) = 1.1 \times 10^{-4}$ mbar $p(CO) = 5.5 \times 10^{-5}$ mbar	4.5 ± 0.4	0.4 ± 0.05	0.12 ± 0.02

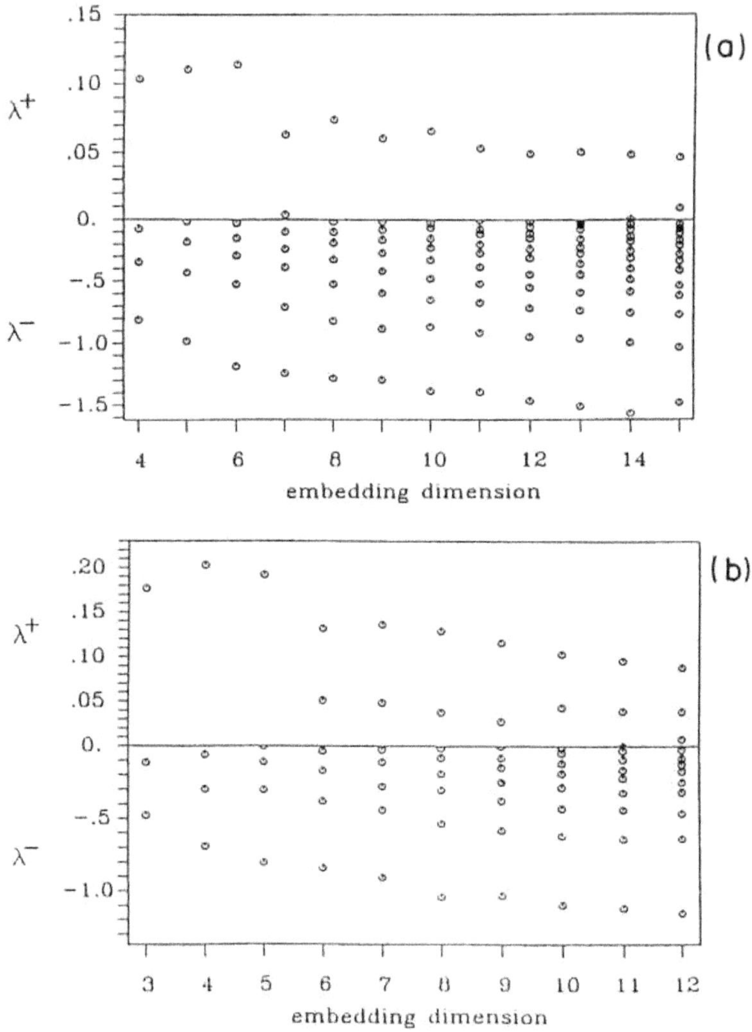

Figure 6.16. Lyapounov exponents as a function of the embedding dimension for the time series of Fig. 6.14 showing that the first data set, a), reflects simple chaos and the second, b), hyperchaos.

a 2-band structure, no such structure is discernible in the second one. Quantitative analysis revealed that the latter has significantly higher characteristic measures (such as the correlation dimension, D_2, lower bound, K_2, of the Kolmogorov entropy[89]), although the largest Lyapounov exponent computed with the WSSV method[116] is about the same for both attractors (Table 6.1).

Analysis with an algorithm that computes all Lyapounov exponents[94,117] (programmed after references 118-121) revealed that case A exhibited one and case B two positive Lyapounov exponents (Fig. 6.16), *i.e.*, the latter data are hyperchaotic. These measurements represent the first unambiguous identification of hyperchaos in a chemical oscillator.[94,114,115]

More recently, multiple periodicities and chaos (correlation dimension 2.6) also have been detected in the NO/H_2 reaction on Pt(100), but no bifurcation diagram could be obtained.[122]

Spatially resolved measurements with oscillations on single-crystal surfaces were first carried out using scanning LEED (low-energy electron diffraction), which gives information about the substrate and/or adsorbate structure. The resolution of this technique is not very good because about 10 s is required for a 2-dimensional scan with a spatial resolution of about 1 mm. Nevertheless, traveling waves could be observed in the oscillatory CO oxidation on Pt(100).[123] While the lateral resolution could be significantly improved to $\approx 3\mu m$ using a scanning photoemission

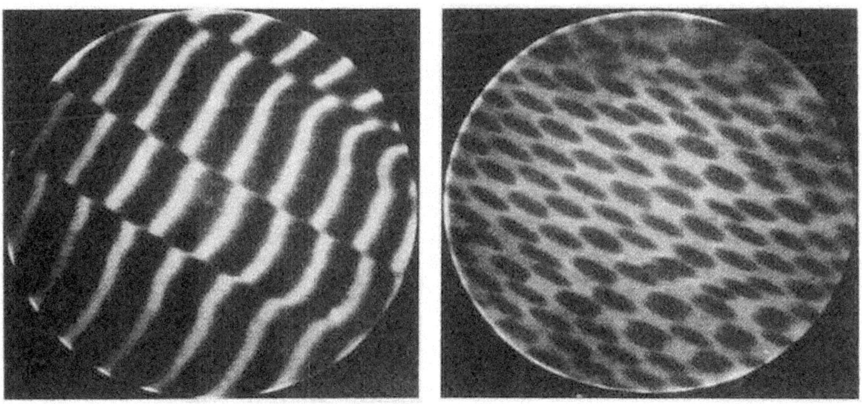

Figure 6.17. Standing waves associated with periodic oscillations in the oxidation of CO on Pt(110); stripes (left) and cells (right) are seen. The diameter of the PEEM images is 400 μm in this and following figures. Courtesy of H.H. Rotermund.

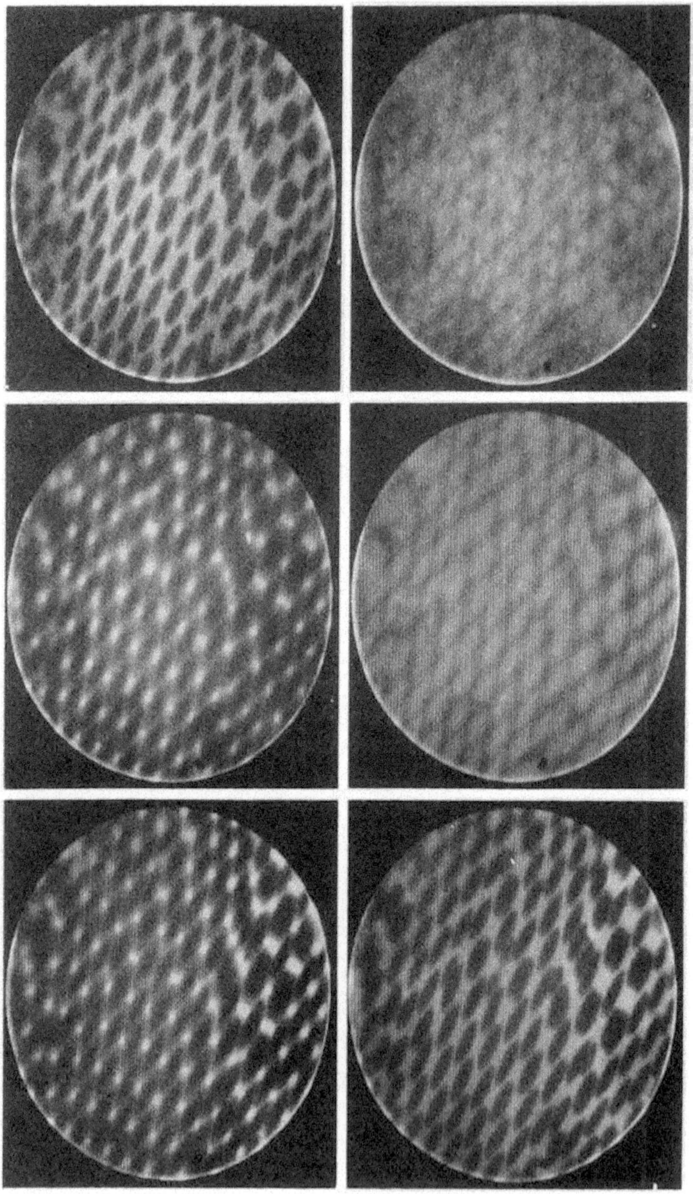

Figure 6.18. Time evolution of a standing wave pattern over half an oscillation period on Pt(110); The time between pictures is 0.2 sec. Courtesy of H.H. Rotermund.

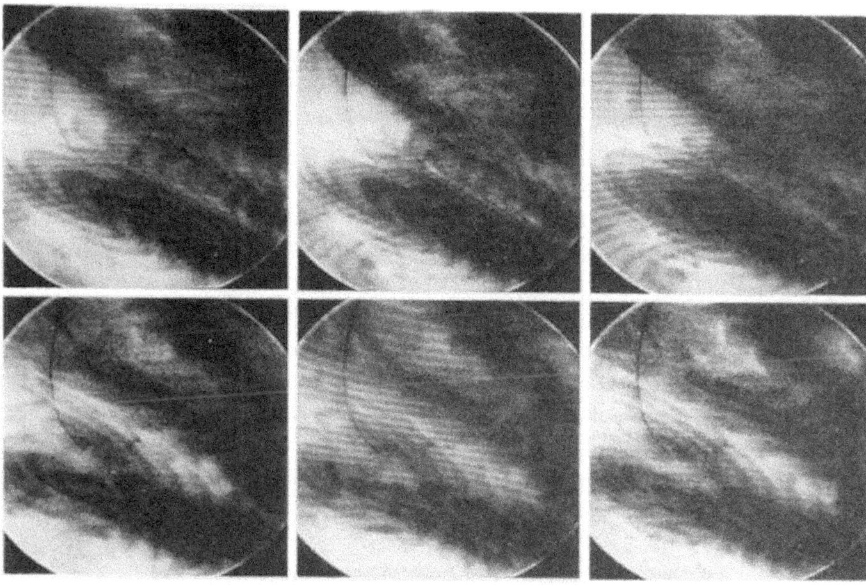

Figure 6.19. Time evolution of an irregular wave pattern on Pt(110). The time between pictures is 0.5 sec. The second row was taken 5 s after the first in order to illustrate the long term changes. Courtesy of H.H. Rotermund.

technique,[109,124-126] it still took about 10 s to record a picture. This is too slow to image the fast oscillations at high temperature on Pt(110). Subsequently, the photoemission electron microscope (PEEM) was developed,[104,127] which allows a spatial resolution of $\approx 0.2\ \mu m$ in real time. It essentially probes the local work function. Although it is possible in principle to detect work function differences of ≈ 1 mV with PEEM, the accuracy becomes poorer (typically 5-10 mV) when high spatio-temporal resolution is required. PEEM measurements of the oscillatory CO oxidation on Pt(110) have been carried out by Rotermund and coworkers.[104-107,109] They reproduced the above-described transition to chaos via two period doublings. No spatial nonuniformities were discernible in these experiments, *i.e.*, the whole surface oscillated strictly in phase. However, structures below the PEEM limits, in this case smaller than ≈ 0.5 μm, or differing by less than ≈ 0.01 monolayers of oxygen, cannot be excluded. The synchronization is attributed to very fast coupling through the gas phase.[71] Only simple chaos was obtained in these experiments.

For very similar control parameters near the CO-covered stationary state, periodic oscillations were found which did exhibit spatial structures in the form of oscillating standing waves in the form of either stripes or cells (Fig. 6.17).

The temporal evolution of the cells is reproduced in Fig. 6.18. These waves dissolved upon reduction of the CO pressure, resulting in fast-moving turbulent patterns such as are shown in Fig. 6.19. The integral time series for these did not exhibit oscillations, but remained constant except for small fluctuations.

Irregular (turbulent) spatial patterns with constant integral behavior also have

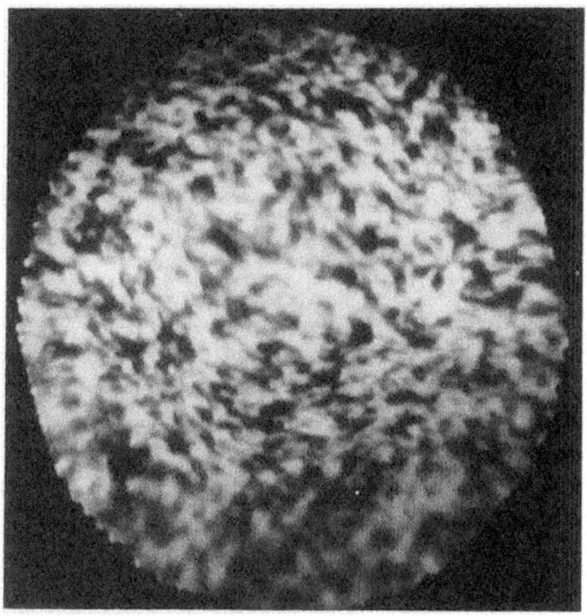

Figure 6.20. Irregular spatial pattern in the NO/NH_3 reaction on Pt(100). After G. Vesar, F. Esch and R Imbihl, *Catal. Lett.* **13** (1992) 371.

been obtained more recently by Imbihl and coworkers in the CO/NO and NO/NH_3 reactions on Pt(100).[128,129] These look quite different from the ones shown above (Fig. 6.20). They also moved more slowly by nearly two orders of magnitude.

Spatial coupling on isothermal single-crystal surfaces only can occur through surface diffusion, which is a slow, local process, or through pressure changes in the gas phase, which almost instantaneously affect the whole surface (global coupling). Generally one would not expect complete synchronization of the oscillations from diffusion alone in typical experimental situations, although it is theoretically conceivable. However, sufficiently strong, autocatalytic gas-phase coupling can easily synchronize a surface, and this is almost certainly the case in the CO oxidation on Pt(110). Note, however, that the excess of O_2 is not very large at the temperatures in question. Thus coupling may become too weak (or even effectively inhibitory, *cf.* Section 6.2) giving rise to the standing waves as well as the breakdown of spatial order, presumably due to phase turbulence.[130,131]

Coupling through the gas phase is inhibitory in the case of the CO/NO reaction on Pt(100), mainly because of symmetric inhibition of adsorption. Actually, only damped macroscopic oscillations have been observed with this system for moderate temperatures,[30] and these die out after a couple of periods because of desynchronization. However, it has been shown that periodic variation of the CO pressure (introducing "artificial" positive gas-phase coupling) leads to sustained oscillations of substantial amplitude.[132] The same effect has been obtained by

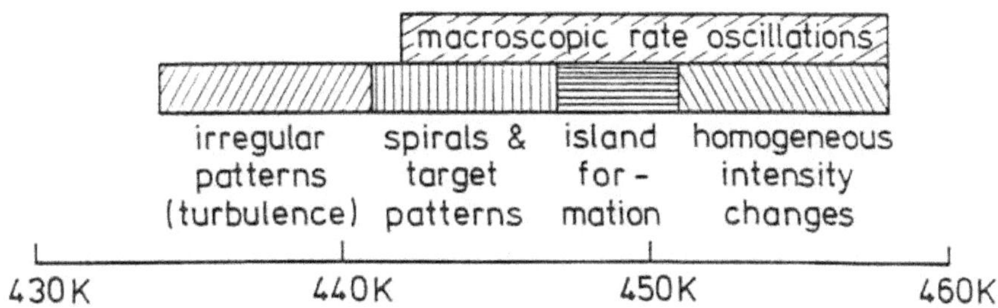

Figure 6.21. Existence regions of various types of spatial patterns in the NO/NH_3 reaction on Pt(100). Turbulent patterns are not associated with macroscopic oscillations. After G. Veser, F. Esch and R. Imbihl, *Catal. Lett.* **13** (1992) 371.

modulation of the catalyst temperature. Sustained macroscopic oscillations were found in the case of NO/NH_3 on Pt(100) at temperatures above ≈ 442K. Irregular patterns with constant overall behavior resulted below this temperature (Fig. 6.21). The interplay of the different coupling processes is not yet understood for this system because of its complicated mechanism, although very likely the gas-phase coupling changes sign with increasing temperature. This effect is presumably connected with the 1×1-hex structural phase transition of the Pt(100) surface.

In summary, single-crystal studies reveal a Feigenbaum transition to chaos in the temporal behavior without detectable spatial structures in the CO oxidation on Pt(110), as well as "chemical turbulence"[133] in several systems. Spatially resolved experiments are not yet available for any system exhibiting temporal hyperchaos.

6.5. Discussion and Mechanistic Considerations

Chaos in surface reactions has been found in a considerable variety of experimental setups. All systems are unavoidably spatially distributed. It is therefore crucial to elucidate the nature of the spatial coupling in order to understand the emergence of chaos. It appears in most cases that because of the spatial effects a relatively simple kinetic scheme would suffice to explain complex dynamical behavior.

Supported catalysts can serve to study the behavior of a very large number of coupled oscillators. In a typical reactor there may be 10^{15} or more individual crystallites. There are, however, complications arising from the fact that these particles are not identical (there is usually a distribution of sizes) and not evenly spaced.

Those studies on monolithic polycrystalline catalysts for which spatially resolved data are available clearly show the importance of spatial structures for the emergence of chaos.

It does not appear very useful to construct complicated chemical mechanisms to model chaotic behavior using ordinary differential equations unless chaos is found to occur in the absence of spatial inhomogeneities. At present, the best candidate for such a system is the CO oxidation on Pt(110), which exhibits (among other behaviors) a transition to chaos spatially homogeneous down to the PEEM detection limit. The mechanism of this reaction therefore will be qualitatively discussed here. For numerical studies see references 102, 108, 134.

The isothermal oxidation of CO proceeds via the classical Langmuir-Hinshelwood (LH) mechanism.

$$CO_{(g)} \rightleftarrows CO \tag{6.1}$$

$$2^* + O_{2\,(g)} \rightarrow 2O \tag{6.2}$$

$$CO + O \rightarrow CO_{2\,(g)} \tag{6.3}$$

$$O + CO_{(g)} \rightarrow [O+CO] \rightarrow {}^* + CO_{2\,(g)} \tag{6.4}$$

Here * denotes an empty site. The gaseous species, which are nonessential,[135] have the subscript "*g*"; all other species are adsorbed. Only process 6.1 is significantly reversible. Process 6.4 takes asymmetric inhibition into account, but note that CO spends a finite (albeit very small) time in the adsorbed state (indicated by brackets) before reacting. The short-lived coexistence of adsorbed O and CO can be neglected in the kinetic scheme.

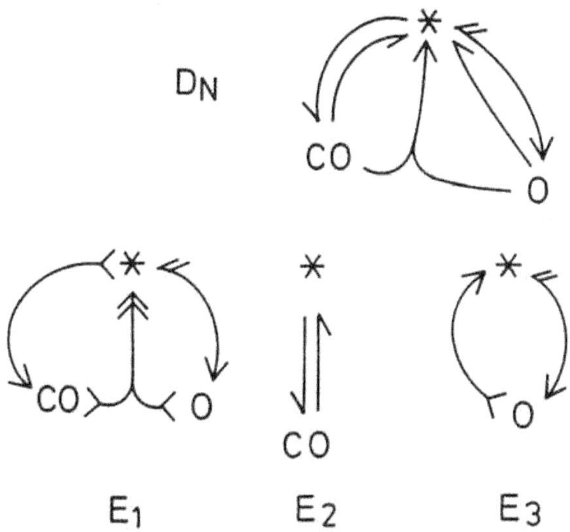

Figure 6.22. Network diagram, D_N, of the Langmuir-Hinshelwood mechanism with asymmetric inhibition and decomposition into the extreme currents E_1, E_2 and E_3.

The network diagram[136] of the above scheme is depicted in Fig. 6.22. One can readily see that it decomposes into 3 extreme currents.[136] E_2 and E_3 are stable and correspond to the CO-covered (passivated) and O-covered (reactive) states, respectively. E_1 is unstable and dominates the saddle point between the stable states.

Thus, this system is bistable, but cannot oscillate for lack of a negative feedback loop with sufficient time delay.[135] The chemical scheme shown in Fig. 6.22 contains three species. However, there is a conservation constraint (as can readily be seen from the left null space of the stoichiometric matrix) that requires the sum over *, CO and O to be a constant reflecting the fixed surface area of the catalyst. The two remaining variables (say CO and O) are not independent, but are highly anticorrelated because

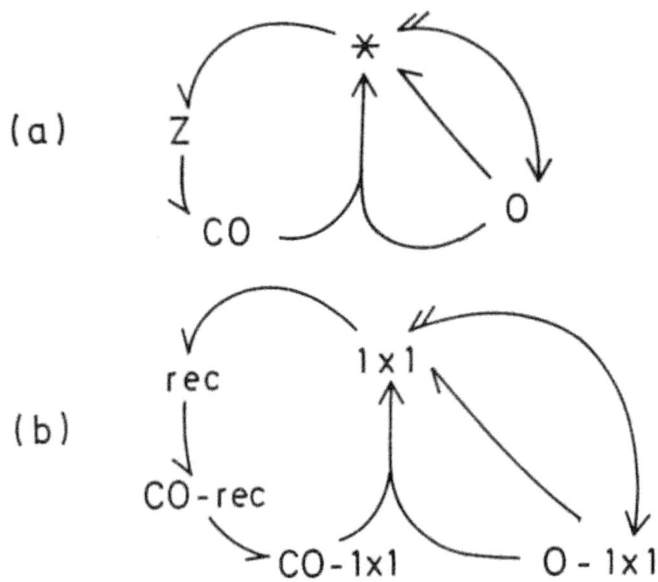

Figure 6.23. a) An extension of the LH-mechanism leading to oscillations. b) An oscillatory subnetwork for the reconstruction model of surface oscillations.[108]

reaction 6.3 is fast. The kinetic scheme is thus effectively a 1-variable system, which can clearly not exhibit oscillations; it is restricted to bistable behavior.

One (but not the only) very simple way to obtain oscillations in a LH-scheme is to introduce (at least) one extra species, Z, in the way shown in Fig. 6.23a.[137] Such an extension comes in on Pt(110) because of the adsorbate-driven structural change of the surface. The clean uncovered Pt(110) surface is reconstructed, more precisely, the topmost atomic layer exhibits a 1×2 ("missing row") structure. This reconstruction is lifted (*i.e.*, the bulk-like 1×1 forms) upon adsorption of CO, but is not affected by oxygen. The crucial point here is that the sticking coefficient of oxygen is higher on the 1×1 than on the reconstructed surface. Consequently, in the course of oscillation, the reconstructed surface will become CO covered because oxygen adsorption is slow. Beyond a critical CO coverage, the 1×1-phase forms, which exhibits faster oxygen adsorption, leading to reactive removal of CO. The resulting surface with low CO coverage will then reconstruct, and the cycle starts again. Note that this scenario leads to a time-delayed negative feedback loop because high CO coverage causes a slow structural change, which leads to enhanced oxygen adsorption, in turn lowering the CO coverage. Such a loop can in general

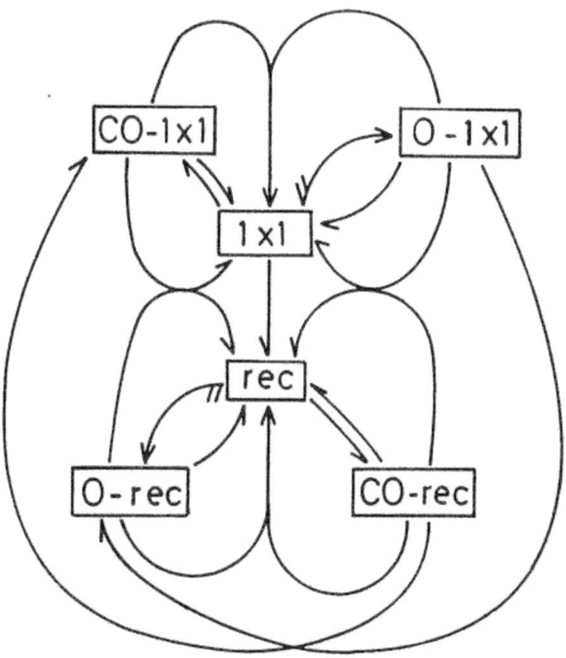

Figure 6.24. Complete network diagram including all conceivable processes for the oxidation of CO on Pt(110). Note that CO lifts the reconstruction (CO-rec → CO-1×1), while empty or O-covered states reconstruct (1×1 → rec, O-1×1 → O-rec).

give rise to oscillation in an otherwise bistable system. An oscillating subnetwork for the oxidation of CO on Pt(110) is shown in Fig. 6.23b. This mechanism exhibits only simple periodic oscillations (as does the one shown in Fig. 6.23a). It consists of just a stable and an unstable extreme current and is still too simple to produce chaos. However, the full mechanism is considerably more complex, since both gases can adsorb on both surface structures, and cross-reaction terms (such as O-1×1 + CO-rec) become possible. The complete, heavily branched network (shown in Fig. 6.24) contains no less than 4 unstable extreme currents (besides 12 stable ones). Complex dynamical behavior would therefore not be surprising.

There is, however, still no reliable way to predict chaos simply on the basis of mechanistic features. Therefore a set of 5 ordinary differential equations (the 6-th variable can be dropped because of the conservation constraint) was integrated numerically for various sets of rate constants. A period-doubling transition to chaos indeed was obtained[134] for certain sets of parameter values. No evidence for hyperchaos was found. This is attributed to the fact that there are always two very fast-relaxing directions in phase space (roughly corresponding to the O species), so that the model effectively has only a 3-dimensional phase space.

Experimentally, not only temporal chaos but also spatial turbulence has been detected in the oxidation of CO on Pt(110). It is therefore tempting to speculate that hyperchaos may originate from a combination of temporal and spatial instabilities. However, there are not yet any experimental results to support such an assumption.

6.6 Conclusions

Chaos can occur on all levels in surface-catalyzed reactions, *i.e.*, whole reactors, individual pellets and single-crystal surfaces. Spatially resolved measurements are crucial to elucidate the origin of complex dynamical behavior. So are studies of the various coupling processes between different areas of the catalyst surfaces, which (depending on their sign) may favor either synchronization or desynchronization. The system exhibiting the richest variety of nonlinear phenomena is the oxidation of CO on a Pt(110) surface. The spatio-temporal self-organization in this system can be understood on the basis of the reconstruction model of surface oscillations and the interplay of local surface diffusion and global gas-phase coupling.

LITERATURE CITED

1. *Chemical Week*, June 26th (1985).
2. E.G. Schlosser, *Heterogene Katalyse*, (VCH, Weinheim 1972).
3. A. Mittasch, *Geschichte der Ammoniaksynthese* (VCH, Weinheim 1951).
4. B. Timm, "The Ammonia Synthesis and Heterogeneous Catalysis-A Historical Review." in *Proceedings of the 18th International Congress on Catalysis (Berlin 2-6, July 1984)* Vol. 1 (1984) 7.
5. G. Ertl and J. Küppers, *Low Energy Electrons and Surface Chemistry*, 2nd ed. (VCH, Weinheim-Deerfield Beach, FL 1985).
6. A. Zangwill, *Physics at Surfaces* (Cambridge University Press, Cambridge, 1988).
7. W.D. Mroβ and J. Kronenbitter, *Chem. Ing. Techn.* **54** (1982) 33.
8. G.A. Somorjai and S.M. Davis, *Plat. Met. Rev.* **27** (1983) 54.
9. R. Schwankner and M. Eiswirth, *Umschau* **85** (1985) 471.
10. G. Ertl, *J. Mol. Catal.* **54** (1989) 343.
11. P. Hugo, *Ber. Bunsenges. Phys. Chem.* **74** (1970) 121.
12. H. Beusch, P. Fieguth and E. Wicke, "Chemical Reaction Engineering," *Advances in Chemistry Series* **109** (1972) 615.
13. H. Beusch, P. Fieguth and E. Wicke, *Chem. Ing. Tech.* **44** (1972) 445.
14. R.A. Schmitz, G.A. D'Netto, L.F. Razón and J.R. Brown in *Chemical Instabilities*, G. Nicolis and F. Baras, Eds., NATO ASI Series C, *Mathematical and Physical Sciences*, Vol 120 (D. Reidel & Co., Dordrecht, 1984).
15. M. Flytzani-Stephanopoulos, L.D. Schmidt and R. Caretta, *J. Catalysis* **64** (1980) 346.
16. J. Schmidt and M. Sheintuch, *Chem. Eng. Commun.* **46** (1986) 289.
17. L. Lobban, G. Philippou and D. Luss, *J. Phys. Chem.* **93** (1989) 733.
18. M. Sheintuch and D. Luss, *J. Catalysis* **68** (1981) 245.
19. C.G. Vayenas, B. Lee and J. Michaels, *J. Catalysis* **66** (1980) 36.
20. M.M. Slinko and A.A. Unharskii, in *Proceedings of the 18th International Congress on Catalysis (Berlin 2-6 July, 1984)* Vol. III (1984) 243.
21. G. Philippou, F. Schultz and D. Luss, *J. Phys. Chem.* **95** (1991) 3224.
22. N.I. Jaeger, P.J. Plath and E. Van Raaij, *Z. Naturforsch*, **36a** (1981) 395.
23. M. Gerhardt, H. Schuster and P.J. Plath, *Ber. Bunsenges, Phys. Chem.* **90** (1986) 1040.
24. N.I. Jaeger, R. Ottensmeyer and P.J. Plath, *Ber. Bunsenges. Phys. Chem.* **90** (1986) 1075.
25. N.I. Jaeger, R. Ottensmeyer, P.J. Plath and H. Engel-Herbert, *Chem. Eng. Sci.* **45** (1990) 947.
26. H. Herzel, P. Plath and P. Svensson, *Physica*, **48D** (1991) 340.
27. M. Stoukides and C.G. Vayenas, *J. Catalysis*, **74** (1982) 266.
28. N.I. Jaeger, K. Koblitz, M. Liauw and P.J. Plath, to be published.
29. J.E. Zuniga and D. Luss, *J. Catalysis*, **53** (1978) 312.
30. K. Rajagopalan and D. Luss, *J. Catalysis*, **61** (1980) 289.
31. E. Wicke, P. Kummann, W. Keil and J. Schiefler, *Ber. Bunsenges. Phys. Chem.* **84** (1980) 315.
32. Z. Kurtanjek, M. Sheintuch and D. Luss, *J. Catalysis*, **66** (1980) 11.
33. J.R. Brown, G.A. D'Netto and R.A. Schmitz, in *Temporal Order*, L. Rensing and N.I. Jaeger, Eds. (Springer, Berlin-Heidelberg, 1985). p.86.
34. L. Lobban and D. Luss, *J. Phys. Chem.* **93** (1989) 6530.

35. W. Adhoch, H.G. Lintz and T. Weisker, *Surf. Sci.* **103** (1981) 576.
36. R.L. Klein, S. Schwartz and L.D. Schmidt, *J. Phys. Chem.* **89** (1985) 4908.
37. S.B. Schwartz and L.D. Schmidt, *Surf. Sci.* **185** (1987) 268.
38. F. Schuth and E. Wicke, *Ber. Bunsenges. Phys. Chem.* **93** (1989) 191; ibid. **93** (1989) 491.
39. Th. Fink. J.-P. Dath, R. Imbihl and G. Ertl, *J. Chem. Phys.* **95** (1991) 2109.
40. V. Schmatloch and N. Kruse, *Surf. Sci.* **269/270** (1991) 488.
41. C. Takoudis and L. Schmidt, *J. Phys. Chem.* **87** (1983) 964.
42. S.J. Lombardo, F. Esch and R. Imbihl, *Surf. Sci.* **271** (1992) L367.
43. T.T. Tsotsis, V.U.S. Rao and L.M. Polinski, *AICHE J.* **28** (1982) 346.
44. H. Niyama and Y. Suzuki, *Chem. Eng. Commun.* **14** (1980) 145.
45. J. Siera, P. Cobden, K. Tanaka and B.E. Nieuwenhuys, *Catal. Lett.* **10** (1991) 335.
46. M. Slinko, T. Fink, T. Löher, H.H. Madden, S.J. Lombardo, R. Imbihl and G. Ertl, *Surf. Sci.* **264** (1992) 157.
47. H.H. Kung, *Transition Metal Oxides: Surface Chemistry and Catalysis* (Elsevier, Amsterdam, 1988).
48. M. Sheintuch and R.A. Schmitz, *Catal. Rev. - Sci. Eng.* **15** (1977) 107.
49. M.G. Slin'ko and M.M. Slin'ko, *Catal. Rev. - Sci. Eng.* **17** (1978) 119.
50. L.F. Razön and R.A. Schmitz, *Catal. Rev. - Sci. Eng.* **28** (1986) 89.
51. R. Imbihl, *Optimal Structures in Heterogeneous Reaction Systems*, P. Plath, Ed. Springer Series in Synergetics, Vol. 44 (Springer, Berlin, 1989).
52. G. Ertl, in *Spatial Inhomogeneities and Transient Behavior in Chemical Kinetics*, P. Gray, G. Nicolis, F. Baras and S.K. Scott, Eds. (Manchester University Press, Manchester, 1990). p. 565.
53. G. Ertl, *Phys. Bl.* **46** (1990) 339.
54. M. Eiswirth, K. Krischer and G. Ertl, *Appl. Phys.* A. **51** (1990) 79.
55. G. Ertl, *Adv. Catalysis,* **37** (1990) 79.
56. J. Rathousky, J. Puszynski and V. Hlavacek, *Z. Naturforsch.* **35a** (1980) 1238.
57. N.I. Jaeger, K. Möller and P. Plath, *Z. Naturforsch.* **36a** (1981) 1012.
58. M. Eiswirth, R. Schwankner and G. Ertl, *Z. Phys. Chem. NF* **144** (1985) 59.
59. L.F. Razón, S.-M Chang and R.A. Schmitz, *Chem. Eng. Sci.* **41** (1986) 1561.
60. M. Eiswirth and G. Ertl, *Surf. Sci.* **177** (1986) 90.
61. M. Eiswirth, K. Krischer and G. Ertl, *Surf. Sci.* **202** (1988) 565.
62. M. Ehsasi, O. Frank, J.H. Block and K. Christmann, *Chem. Phys. Lett.* **165** (1990) 115.
63. H.U. Onken and E. Wicke, *Ber. Bunsenges. Phys. Chem.* **90** (1986) 976.
64. S.C. Capsaskis and C.N. Kenney, *J. Phys. Chem.* **90** (1986) 4631.
65. M. Marek, in *Temporal Order*, L. Rensing and N.I. Jaeger, Eds. (Springer, Berlin, 1985).
66. R.J. Schwankner, M. Eiswirth, P. Möller, K. Wetzl and G. Ertl, *J. Chem Phys.* **87** (1987) 742.
67. P. Svensson, N.I. Jaeger and P.J. Plath, *J. Phys. Chem.* **92** (1988) 1882.
68. M. Eiswirth and G. Ertl, *Phys. Rev. Lett.* **60** (1988) 1526.
69. M. Eiswirth and G. Ertl, *Appl. Phys.* A **47** (1988) 91.
70. M. Eiswirth, P. Möller and G. Ertl, *Surf. Sci.* **208** (1989) 13; *J. Vac. Sci. Technol.* **7A** (1989) 1882.
71. K. Krischer, M. Eiswirth and G. Ertl, *J. Chem. Phys.* (in press).
72. M. Eiswirth, P. Möller, K. Wetzl, R. Imbihl and G. Ertl, *J. Chem. Phys.* **93** (1985) 321.
73. D.J. Kaul and E.E. Wolf, *J. Catalysis,* **89** (1984) 348; **91** (1985) 216; **93** (1985) 321.

74. R. Sant and E.E. Wolf, *J. Catalysis,* **110** (2988) 249.
75. H.U. Onken and E. Wicke, *Z. Phys. Chem. NF* **165** (1989) 23.
76. P.K. Tsai, M.B. Maple and R.K. Herz, *J. Catalysis,* **313** (1988) 453.
77. N.I. Jaeger, K. Möller and P.J. Plath, *J. Chem. Soc., Faraday Trans. I* **82** (1986) 3315.
78. R. Ottensmeyer, Ph.D. Thesis (Universität Bremen 1987).
79. H.U. Onken, Ph.D. Thesis (Universität Münster 1987).
80. J. Kapicka and M. Marek, *J. Catalysis,* **229** (1989) 508.
81. J. Kapicka and M. Marek, *Surf. Sci.* **222** (1989) L 885.
82. F. Schüth, X. Song, L.D. Schmidt and E. Wicke, *J. Chem. Phys.* **92** (1990) 745.
83. M.M. Slin'ko, N.I. Jaeger and P. Svensson, *J. Catalysis,* **118** (1989) 349.
84. M.M. Slin'ko, N.I. Jaeger and P. Svensson, in *Unsteady State Processes in Catalysis*, Yu. Sh. Matros, Ed. (VSP, Utrecht 1990).
85. L.P. Shil'nikov, *Soc. Math. Doc.* **6** (1965) 163.
86. H.G. Schuster, *Deterministic Chaos* (VCH, Weinheim 1988).
87. J.M.T. Thompson and H.B. Stewart, *Nonlinear Dynamics and Chaos* (Wiley, Chichester, 1986).
88. M. Marek and I. Schreiber, *Chaotic Behaviour of Deterministic Dissipative Systems* (Cambridge University Press, Cambridge, 1991).
89. P. Grassberger and I. Procaccia, *Phys. Rev. Lett.* **50** (1983) 346; *Phys. Rev.* A **28** (1983) 2591; *Physica* **9D** (1983) 189.
90. W.v.d. Water and P. Schram, *Phys. Rev.* **37A** (1988) 3118.
91. M. Sheintuch and J. Schmidt, *J. Phys. Chem.* **92** (1988) 3404.
92. T. Gritsch, D. Coulman, R.J. Behm and G. Ertl, *Phys. Rev. Lett.* **63** (1989) 1086; *Appl. Phys.* **49A** (1989) 403.
93. W. Hösler, E. Ritter and R.J. Behm, *Ber. Bunsenges. Phys. Chem.* **90** (1986) 205.
94. Th.-M. Kruel, M. Eiswirth and F.W. Schneider, *Physica D* (in press).
95. G. Ertl, P.R. Norton and J. Rüstig, *Phys. Rev. Lett.* **49** (1982) 177.
96. R. Imbihl, *Prog. Surf. Sci.* (in preparation).
97. S. Ladas, R. Imbihl and G. Ertl, *Surf. Sci.* **197** (1988) 153; ibid. **198** (1988) 42.
98. R. Imbihl, J. Falta, D. Kaletta and M. Henzler, *J. Vac. Sci. Technol.* **9A** (1990) 1749.
99. R. Imbihl, A.E. Reynolds and D. Kaletta, *Phys. Rev. Lett.* **67** (1991) 275.
100. M. Sander and R. Imbihl, *Surf. Sci.* **255** (1991) 61.
101. M. Eiswirth, Ph.D. Thesis (Ludwig-Maximilians-Universität, München, 1987).
102. K. Krischer, Ph.D. Thesis (Freie Universität, Berlin, 1990).
103. A.L. Vishnevskii and V.I. Savchenko, *React. Kinet. Catal. Lett.* **38** (1989) 167; *Kinetia i Kataliz* **30** (1989) 1444.
104. H.H. Rotermund, W. Engel, M. Kordesch and G. Ertl, *Nature,* **343** (1990) 355.
105. S. Jakubith, H.H. Rotermund, W. Engel, A. von Oertzen and G. Ertl, *Phys. Rev. Lett.* **65** (1990) 3013.
106. H.H. Rotermund, S. Jakubith, A. von Oertzen and G. Ertl, *Phys. Rev. Lett.* **66** (1991) 3083.
107. H.H. Rotermund, W. Engel, S. Jakubith, A. von Oertzen and G. Ertl, *Ultramicroscopy,* **36** (1991) 164.
108. K. Krischer, M. Eiswirth and G. Ertl, *Surf. Sci.,* **251/252** (1991) 900; *J. Chem. Phys.* **96** (1992) 9161.
109. S. Jakubith, Ph.D. Thesis (Freie Universität, Berlin, 1991).
110. M. Bär, C. Zülicke, M. Eiswirth and C. Ertl, *J. Chem. Phys.* **96** (1992) 8535.

111. M. Bär, M. Falcke, C. Zülicke, H. Engel, M. Eiswirth and G. Ertl, *Surf. Sci.* **269/270** (1992) 471.
112. M. Bär, M. Falcke and M. Eiswirth, *Physica A* (in press).
113. M. Bär, M. Eiswirth, H.H. Rotermund and G. Ertl, *Phys, Rev. Lett.* (in press).
114. M. Eiswirth, Th.-M Kruel, G. Ertl and F.W. Schneider, *Chem. Phys. Lett.* **193** (1992) 305.
115. G. Ertl, *Science,* **254** (1991) 1750.
116. A. Wolf, J.B. Swift, H.L. Swinney and J.A. Vastano, *Physica* **26D** (1985) 285.
117. Th.-M. Kruel, Ph.D. Thesis (Universität Würzburg, Würzburg, 1992).
118. M. Sano and Y. Savada, *Phys. Rev. Lett.* **55** (1985) 1082.
119. M. Sano, S. Sato and Y. Savada, *Prog. Theor. Phys.* **76** (1986) 945.
120. R. Stoop and P.F. Meier, *J. Opt. Soc. Am. 5B* (1988) 1037.
121. R. Stoop, J. Peinke, J. Parisi, B. Röhricht and R.P. Heubener, *Physica* **35D** (1989) 425.
122. P.D. Cobden, J. Siera and B.E. Niewenhuys, *J. Vac. Sci. Technol.* (in press).
123. M.P. Cox, G. Ertl and R. Imbihl, *Phys. Rev. Lett.* **54** (1985) 1725.
124. H.H. Rotermund, S. Jakubith, A. von Oertzen and G. Ertl, *J. Chem. Phys.* **91** (1989) 4942.
125. H.H. Rotermund, G. Ertl and W. Sesselmann, *Surf. Sci.* **217** (1989) L383.
126. H.H. Rotermund, S. Jakubith, S. Kubala, A. von Oertzen and G. Ertl, *J. El. Spectroscopy,* **52** (1990) 811.
127. W. Engel, M.E. Kordesch, H.H. Rotermund, S. Kubala and A. von Oertzen, *Ultramicroscopy,* **36** (1991) 148.
128. G. Veser and R. Imbihl, *J. Chem. Phys.* **96** (1992) 7155; *Surf. Sci.* **271** (1992) L367.
129. G. Veser, F. Esch and R. Imbihl, *Catal. Lett.* **13** (1992) 371.
130. Y. Kuramoto, *Chemical Oscillations, Waves and Turbulence* (Springer, Berlin, 1984).
131. M. Bär, Ph.D. Thesis (Freie Universität Berlin, in preparation).
132. J.-P. Dath, Th. Fink, R. Imbihl and G. Ertl, *J. Chem. Phys.* **96** (1992) 1582.
133. Q. Ouyang and H.L. Swinney, *Chaos J.* **1** (1991) 411.
134. K. Krischer, M. Eiswirth and G. Ertl (unpublished).
135. M. Eiswirth, A. Freund and J. Ross, *J. Phys. Chem.* **95** (1991) 1294; *Adv. Chem. Phys.* **80** (1991) 127.
136. B.L. Clarke, *Adv. Chem. Phys.* **43** (1980) 1.
137. J. Bürger and M. Eiswirth (to be published).

CHAPTER 7: CHAOS IN BIOCHEMICAL SYSTEMS: THE PEROXIDASE REACTION AS A CASE STUDY

RAIMA LARTER[+], LARS FOLKE OLSEN[++], CURTIS G. STEINMETZ[+,++]

and TORBEN GEEST[+,++]

[+] Department of Chemistry,
Indiana University-Purdue University at Indianapolis
1125 E. 38th St., Indianapolis, IN 46205

[++]Institute of Biochemistry
Odense University, Campusvej 55
Odense M, 5230 Denmark

ABSTRACT

The peroxidase-oxidase reaction is the oxidation of organic electron donors by molecular oxygen catalyzed by the enzyme horseradish peroxidase. The reaction is carried out *in vitro* in a flow reactor with no overflow and shows bistabilty, simple and complex oscillations, and chaos. It is the best understood example of an enzyme-catalyzed system showing behaviors associated with nonlinear dynamic laws. Its experimental behavior is described here, as well as the results from a number of examples of reaction mechanisms of varying complexity that reproduce much of its experimentally observed behavior.

7.1 Introduction

Chaos was discovered in 1977 in an innocuous enzyme reaction[1] only a few months before it was first observed in the Belousov-Zhabotinsky (BZ) reaction.[2] The reaction in question is the peroxidase-oxidase (PO) reaction, which is the oxidation of organic electron donors by molecular oxygen, catalyzed by the enzyme horseradish peroxidase. When this reaction takes place in a flow system with reduced nicotinamide adenine dinucleotide (NADH) as the reductant, the concentrations of reactants (oxygen and NADH) as well as some enzyme intermediates can be seen to oscillate with periods ranging from several minutes to about an hour, depending on the experimental conditions. Aperiodic oscillations resembling random fluctuations are also observed for some conditions; these irregular oscillations were shown to be chaotic in a study by Olsen and Degn[1] published nearly simultaneously with the observation of chaos in the BZ reaction by Schmitz, Graziani and Hudson.[2]

The PO reaction occurs as the first step in a sequence of reactions in plants which eventually culminates in the production of lignin;[3,4] it also is involved in the important processes of the photosynthetic dark reactions.[5] It is not known at this point whether the oscillations observed in the flow system have any bearing on behavior *in vivo*. However, oscillatory behavior is known to occur in other biochemical settings and seems to be a ubiquitous phenomenon at many levels in biological systems. The existence of chaotic behavior in the PO reaction brings to mind many intriguing questions. For example, if chaos can occur in a single-enzyme reaction (such as the PO reaction), then does the existence of multiple enzyme networks of reactions make chaos inevitable *in vivo*? Is chaos a sign of health or disease?

We will review in this chapter the experimental and theoretical studies of the peroxidase oscillations and related behavior which have been carried out since their discovery in the mid 1960s. This reaction provides an interesting case study of the types of questions that can be answered with the tools now available for the investigation of nonlinear phenomena. It is hoped that the approach taken in the study of the PO reaction will be successfully applied to other biochemical reactions so that we can begin to find answers to the more general questions posed above.

7.2 Experimental Studies

7.2.1 Early Work

The overall stoichiometry which we refer to as the PO reaction is:

$$2YH_2 + O_2 \rightarrow 2Y + 2H_2O,$$

where YH_2 is an organic electron donor. If YH_2 is NADH the reaction becomes:

$$2NADH + 2H^+ + O_2 \rightarrow 2NAD^+ + 2H_2O \qquad (7.1)$$

The reaction is catalyzed by horseradish peroxidase, which exists in one of five different forms:[6] ferrous peroxidase (Per^{2+}), ferric peroxidase (Per^{3+}), and three other forms known as compounds I, II and III (coI, coII and coIII). The native form of the enzyme is Per^{3+}. CoII, coI and coIII have oxidizing equivalents of +4, +5 and +6, respectively, and are critical intermediates in the mechanism leading to oscillations. The kinetics of reactions of coIII, in particular, is of special interest in understanding the oscillations, as will be described later.

Another critical species in the oscillatory reaction mechanism (both regular and chaotic) is the free radical NAD•. This species is thought to be produced autocatalytically by a sequence of reactions involving NADH, O_2, hydrogen peroxide, superoxide radical (O_2^-•) and several of the enzyme species. Autocatalysis in a reaction mechanism which also involves a termination pathway for the autocatalytic species (here NAD•) is known to be a sufficient explanation for the existence of oscillatory behavior.[7] Whether or not this structure exists in the PO reaction will be examined further in the following sections.

Yamazaki and co-workers discovered[8] in 1965 that the peroxidase-catalyzed oxidation of NADH occurs via damped oscillations when oxygen is supplied continuously by bubbling a mixture of oxygen and nitrogen through the reaction mixture. A very similar system, where the bubbling of oxygen through the solution is replaced by diffusion through the gas/liquid interface from a gas head space, was shown to exhibit bistability, *i.e.*, the existence of two simultaneously stable steady states for the same oxygen concentration in the gas phase.[9] Temporary perturbations in the oxygen concentration in the gas phase could induce reversible switches from one steady state to the other. The bistability phenomenon is thought to be due to inhibition of the enzyme by O_2, and evidence was presented that the inactive enzyme intermediate is identical with coIII.[9]

Degn showed[10] somewhat later (1969) that damped oscillations also could be obtained using the substrates dihydroxyfumaric acid and indoleacetic acid instead of NADH. The oscillations were accompanied by measurable chemiluminescence. The chemiluminescence was ascribed to free-radical intermediates and taken as evidence of the presence of autocatalysis in the reaction mechanism.

Sustained oscillations in the PO reaction were first obtained by Nakamura *et al.*[11] using NADPH as the substrate; NADPH was regenerated from the oxidized form ($NADP^+$) by glucose-6-phosphate and glucose-6-phosphate dehydrogenase. Sustained oscillations were found only when the modifiers 2,4-dichlorophenol (DCP) and methylene blue (MB) were present. Olsen and Degn later reported sustained oscillations[12] with a constant infusion of NADH, thus demonstrating that glucose-6-phosphate dehydrogenase was unnecessary to sustain the oscillations (See Fig. 7.1). However, these authors found that the presence of DCP (and, perhaps,

MB) were critical. Olsen and Degn[12] (1978) provided further evidence that the oscillatory behavior is more likely due to autocatalysis than to substrate inhibition by oxygen.

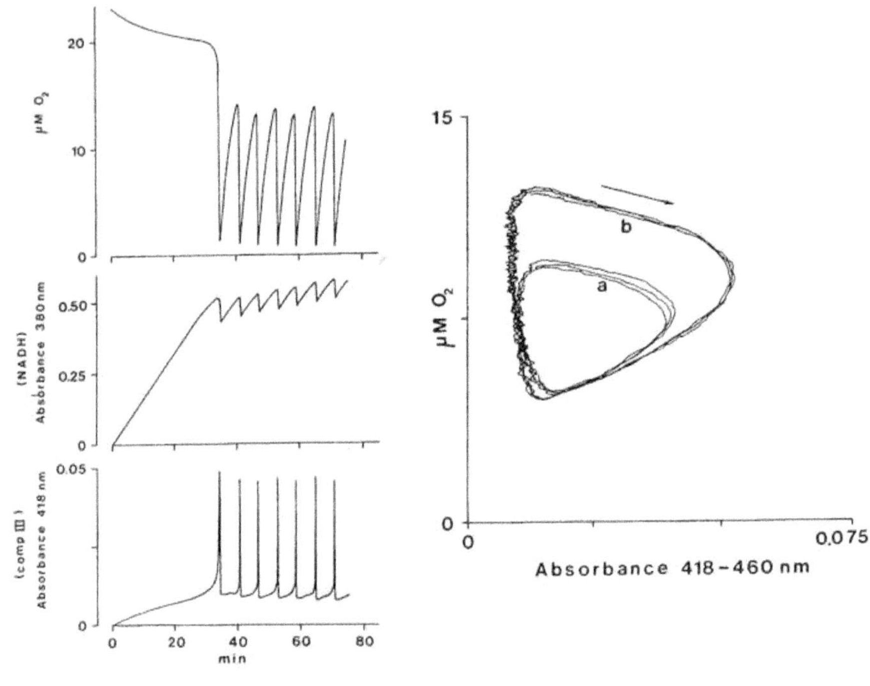

Fig. 7.1 a-left) Oscillatory behavior in the peroxidase-NADH-O_2 reaction. Simultaneous measurements of concentrations of O_2, NADH and coIII. Experimental conditions: 0.9 μM horseradish peroxidase (HRP); 0.2 μM methylene blue (MB) and 10 μM 2,4-dichlorophenol (DCP) in 0.1 M Na-acetate buffer, pH = 5.1. The O_2 content in the gas was 1.9% by volume. Temperature = 28 °C. The experiment was started by infusion of 0.2 M NADH at a rate of 11 μl/h. b) Phase-plane plot of O_2 concentration against concentration of coIII at two different infusion rates of NADH. Experimental conditions: 1.8 μM HRP, 0.6 μM MB and 20 μM DCP in 0.1 M Na-acetate buffer, pH = 5.1. Temperature = 25 °C. The O_2 content in the gas was 1.8% by volume; 0.2M NADH was infused at rates of a: 11 μl/h; and b: 7 μl/h. Reproduced by permission from L.F. Olsen and H. Degn, *Biochim. Biophys. Acta* **523** (1978) 321.

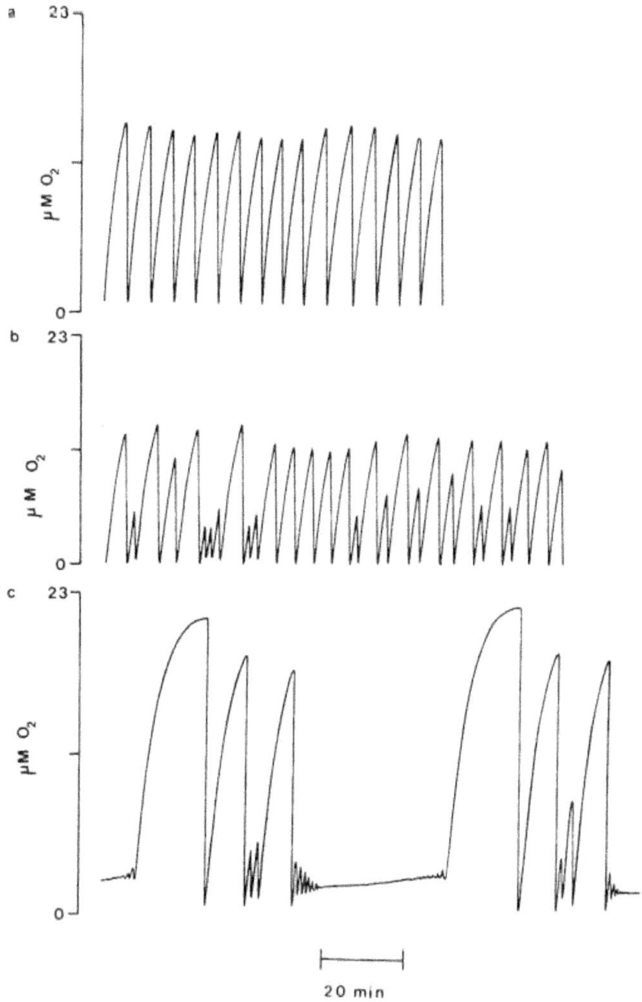

Figure 7.2. Chaos in the PO reaction. Experimental conditions: 0.2 μM MB and 10 μM DCP in 0.1M acetate, pH = 5.1. The O_2 content in the gas was 1.9% by volume. Temperature = 28 °C. The infusion rate of 0.2 M NADH was 11 μl/h. HRP concentration: a) 0.90 μM; b) 0.55 μM; c) 0.45 μM. Reproduced by permission from L.F. Olsen and H. Degn, *Nature* **267** (1977) 177.

7.2.2 *Observations of chaos*

Olsen and Degn observed in 1977 that the waveform of the PO oscillations depends strongly on the concentration of peroxidase.[1] Simple periodic oscillations with a period of about 5 min were obtained at enzyme concentrations of about 1 μM, whereas bursting oscillations with periods of up to 60 min were seen at enzyme concentrations below 0.5 μM. The oscillations were aperiodic and irregular at enzyme concentrations slightly above 0.5 μM (Fig. 7.2). A smooth curve fitted to the points in a next-amplitude map of the data was used to carry out symbolic dynamics; a period-three cycle was found for certain initial conditions. The theorem of Li and Yorke[13] that the existence of a period-three oscillation implies chaos was then used to argue that the irregular oscillations were in fact chaotic. This observation was made less than a year after publication of the pioneering paper by Rössler[14] suggesting that chaos might be found in chemical reactions. Schmitz, Graziani and Hudson reported observations of chaos in the BZ reaction[2] only a few months after its observation in the PO reaction.

The PO reaction remains as of this writing the only enzyme reaction shown to behave chaotically without the imposition of periodic forcing. Markus, Kuschmitz and Hess[15] have shown that the glycolytic reaction exhibits a chaotic response when the supply of glucose is periodic. However, although the existence of autonomous chaos in glycolysis has been predicted theoretically,[16] there is still no experimental verification.

Application of the theorem of Li and Yorke[13] to the next-amplitude maps of the aperiodic oscillations in the PO reaction remained for 15 years the only experimental evidence for chaos in this reaction; this experimental evidence was supported by numerical simulations which yielded next-amplitude maps that were very similar to the experimental maps. Recent experiments by Geest *et al.*[17] have demonstrated that chaos in the PO reaction arises by the well-known period-doubling route as the concentration of DCP is varied over a critical range (Fig. 7.3). A similar period-doubling route to chaos has been predicted from simulations with a detailed model of the PO reaction for variations of the enzyme concentration.[18] This prediction has yet to be verified experimentally, however.

The Lyapounov exponent and fractal dimension were computed in a further study by Geest *et al.*[19] for the experimental data and compared with those obtained from models of the reaction; these values are listed in Table 7.1. The Lyapounov exponent measures the average rate at which two initially close trajectories converge or diverge. A positive exponent implies chaos, and hence the Lyapounov exponents listed in Table 7.1 for the experimental time series suggest chaotic motion. The

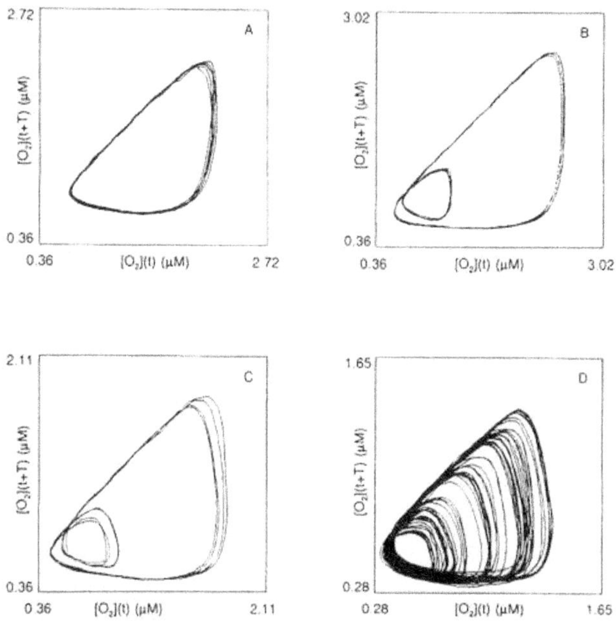

Figure 7.3. Two-dimensional projections of phase plots of oscillations in the oxygen concentration observed as DCP is varied showing: a) period-1; b) period-2; c) period 4 and d) chaos. The method of time delays was used to generate these plots, with a delay of $T = 6$ sec. Experimental conditions: 0.7 μM HRP (Boehringer Mannheim, immunoassay purity), 0.1 μM MB and DCP in 0.1 M Na-acetate at pH = 5.1. Temperature = 28°C. The oxygen concentration in the gas mixture was 1.42%. A 0.08 M NADH solution was infused at an initial rate of 39 μl/h, which was later decreased to 36 μl/h to stabilize the oscillations. The stirring rate was 1420 ± 5 rpm, corresponding to an oxygen-transfer rate constant of 0.224 min^{-1}. DCP concentration: a) 20 μM; b) 25 μM; c) 30.8 μM and d) 32.2 μM. Reproduced by permission from T. Geest, C.G. Steinmetz, R. Larter and L.F. Olsen, *J. Phys. Chem.* **96** (1992) 5678.

fractal dimension measures the information needed to specify the position of a point on an attractor to within a given accuracy and hence, in some sense, expresses the complexity of the motion. Simple types of motion have low-integer dimensions. For example, the dimension of a steady state is zero whereas a limit cycle oscillation has a dimension of one. Chaotic motions usually have finite but non-integer dimensions, also known as fractal dimensions. The dimensions of 2.45 - 2.7 obtained for the experimental data are indicative of chaotic motion and are also strikingly in accord with theoretical values previously predicted from a simple model of the PO reaction. These recent experimental results thus confirm that the irregular oscillations observed by Olsen and Degn in 1977 are indeed chaotic.

Table 7.1. Maximum Lyapounov exponent (λ_1) and fractal dimension (D_F) of chaotic oscillations in simulated and experimental data from the PO reaction. Taken from T. Geest, L.F. Olsen and W.M. Schaffer, *Lyapounov Exponents and the Fractal Dimension of Chaos in the Peroxidase-Oxidase reaction*, (preprint, 1992).

Species Measured	No. of Oscillations	λ_1 (bit/oscillation)	D_F
coIII	48	0.68	2.45
O_2	48	0.63	2.64
O_2	55	0.36	2.70
Simulation[a]	>2000	0.29-0.48[b]	2.09-2.26[b]

a) Model from reference 35. b) Depends on a rate constant.

Recent experiments at Stanford University[20] have revealed the existence of quasiperiodic oscillations in the PO reaction using the same experimental configuration as Olsen and Degn; whether or not this quasiperiodicity is associated with a different route to chaos is presently unknown. Both period doubling and quasiperiodicity are well-known routes to chaos, and both have been found to be associated with chaos in models of the PO reaction.

Motivated by the earlier studies, Baier *et al.*[21-23] have studied peroxidase-catalyzed reactions of the type:

$$BQH_2 + H_2O_2 \rightarrow BQ + 2H_2O,$$

where BQ and BQH_2 are the oxidized and reduced states of benzoquinone, respectively. Only steady state behavior is found if this reaction is carried out in a flow system. However, if it is carried out in a unstirred solution and benzohydroquinone (BQH_2) is regenerated electrochemically from benzoquinone (BQ) at the surface of a glassy carbon-disc electrode inserted into the reaction mixture, then damped oscillations, sustained periodic oscillations and even aperiodic oscillations are observed. The authors persuasively argue that the explanation for these observations must include diffusive and perhaps even convective effects in addition to the usual chemical kinetic nonlinearities. The system is, hence, similar in nature to the catalytic solid surfaces at which oxidation reactions have been observed to lead to a whole gamut of nonlinear phenomena.[24] See Chapters 5 and 6.

7.2.3 *Other Experimental Investigations*

7.2.3.1 *Bistability between Oscillations and Steady State*

Computational studies of a detailed model by Aguda and Larter[25] led them to predict that, under certain conditions, the PO reaction might possess coexisting stable states, one oscillatory and the other stationary. This type of bistability between a limit cycle and a steady state also is known as a "hard excitation." Aguda, Hofmann Frisch and Olsen[26] confirmed this prediction experimentally (See Fig. 7.4). Specifically, it had been predicted that a perturbation of the steady state by a sudden cut-off of the oxygen flow would lead to a transition from the steady state to the coexisting oscillatory state. Furthermore, a transition from the oscillatory state to the steady state could be induced by perturbing the reaction with a spike of H_2O_2. The transition from one state to the other was found to be reversible, as predicted by the simulations.

Some quantitative disagreement between experimental observations and the computational predictions were noted, however. Experimentally, the steady state was found to correspond to an oxygen concentration midway between the minimum and maximum O_2 concentrations of the oscillatory state, as opposed to the simulations which predicted the steady state O_2 concentration to be lower than the minimum O_2 level of the oscillatory state. The computational study further predicted that the observed bistability was due to an S-shaped steady-state curve in which the

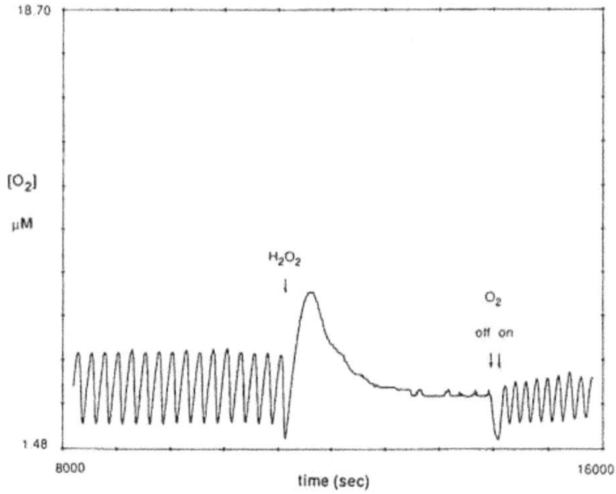

Figure 7.4. An experiment showing bistability between oscillations and a steady state. The oscillatory state was perturbed at the first arrow by adding 20 μl of a 1.8 mM H_2O_2 solution. The steady state which resulted was stable for at least 0.5 h. The second perturbation, indicated by the second set of arrows, was a temporary cut-off of the oxygen flow. Other experimental conditions: 0.53 μM HRP; 0.2 μM MB; 21.6 μM DCP in 0.1 M Na-acetate at pH = 5.1. Temperature = 28.3°C, The oxygen concentration = 1.3 %, and the stirring speed corresponded to an oxygen-transfer rate constant of 0.14 min^{-1}. NADH added at rate of 4.3 μM/min. Reproduced by permission from B.D. Aguda, L-L. Hofmann Frisch and L.F. Olsen, *J. Am. Chem. Soc.* **112** (1990) 6652.

upper branch undergoes a Hopf bifurcation. The experimental results seem to suggest that if an S-shaped steady-state curve underlies the observed dynamics, then $[O_2]$ is not the variable which is multiple valued. Rather, the S-shaped curve might correspond to multiple steady state values in one of the other species, each of which yields the same, or nearly the same, O_2 concentration. An alternative and equally valid explanation of the experimental results would involve a stable limit cycle surrounding an unstable limit cycle, which itself surrounds a stable steady state, *i.e.*, bistability would exist between a locally stable steady state and a surrounding stable limit cycle. No evidence for this latter scenario was found in the computational studies, but it cannot be ruled out experimentally.

7.2.3.2 *Periodic Perturbations of the Flow*

Lazar and Ross have carried out studies[27,28] of the effect of periodic perturbations of the oxygen inlet flow on the PO reaction and have found that these perturbations affect the overall reaction rate and the free-energy dissipation. The system studied was similar to that used by Nakamura *et al.*[11] in that a second enzyme, glucose-6-phosphate dehydrogenase, was used to regenerate NADH from NAD^+. (The glucose-6-phosphate dehydrogenase used by Lazar and Ross[27,28] was different from the enzyme used by Nakamura *et al.*[11] in that it reduces NAD^+ and $NADP^+$ equally well). The rate of reaction in these experiments was determined from the slope of the [NADH] time series while the free energy of reaction was calculated from the defining equation and the measured concentrations of substrates. The perturbation frequency was varied, and it was found that while all frequencies lower the dissipation (and hence increase the efficiency) the smallest effect occurred for frequencies near the autonomous frequency of the reaction. This latter observation was attributed to the fact that a perturbation at the autonomous frequency increases the free energy but lowers the reaction rate, and that these two effects compensate for each other, thus causing the dissipation to remain nearly unchanged.

7.2.3.3 *The Role of Methylene Blue*

One aspect of the experimental investigations of oscillations and chaos in the PO reaction that is still not well understood is the role which the additives 2,4-dichlorophenol (DCP) and methylene blue (MB) play in the oscillatory mechanism. Some studies[29] have indicated that DCP accelerates the rate of decomposition of coIII, possibly leading to a second autocatalytic loop in the mechanism. The role of MB is less well understood. Olsen and Degn[12] found that sustained oscillations in the PO reaction can be obtained with only DCP present. However, these oscillations were not stable over long times unless MB also was present.

Recently Sevcík and Dunford[30] published kinetic studies of the catalysis of NADH oxidation by MB itself. They pointed out that a related reaction, the MB catalyzed oxidation of sodium sulfide, was found to be oscillatory by Burger and Field.[31] In this latter reaction the MB catalysis was thought to occur autocatalytically. Sevcík and Dunford also pointed out that the flow rate of O_2 into the reaction mixture is a critical parameter in both the Burger and Field experiment and in the experimental configuration leading to oscillations in the PO reaction. Their studies of MB catalysis of NADH oxidation were carried out at oxygen flow rates outside the range of values where oscillations are to be expected.

The Sevcík-Dunford results raise the issue of whether it is possible for MB to play a role in the PO mechanism as a second catalyst for the oxidation of NADH (peroxidase is the first catalyst). Evidence against this possibility is given by the experiments of Degn,[10] which showed that nearly sustained oscillations may be obtained without MB when substrates other than NADH, such as dihydroxyfumaric acid and indoleacetic acid, are used. However, chaotic oscillations were never observed with these latter substrates; hence MB may play a role in the more complex behavior observed in the PO reaction. If the role of MB in the PO reaction is to provide an additional autocatalytic route to oxidation of the substrate (as it does in the Burger-Field reaction), this could explain the existence of chaos in the presence of MB. More experiments are needed to verify these speculations.

7.3 Theoretical Studies

7.3.1 Introduction

Theoretical understanding of the kinetics of the PO reaction is quite detailed. Much of this study has been based upon the early work of Olsen and Degn[12] who proposed a four-variable model composed of two coupled autocatalytic cycles. Further computer simulations and theoretical explanations of the behavior of this model were given by Degn, Olsen and Perram,[32] so this model has come to be known as the DOP model. Degn *et al.* reported in their 1979 paper that the DOP model seemed to be incapable of supporting chaotic behavior. Later work by Larter *et al.*[33,34] showed that chaos could, in fact, be found within narrow ranges of parameter values. Olsen[35] suggested a slightly modified model prior to this discovery to explain the existence of chaos in the PO reaction.

Several groups simultaneously were carrying out computer simulation studies of detailed models based on the twenty or so possible reaction steps which may occur in the PO reaction. Yokota and Yamazaki[36] proposed a detailed mechanism and found some agreement between simulations based on it and the induction kinetics, but never reported observing oscillatory behavior. Fed'kina, Bronnikova and Ataullakhanov[37] studied a similar mechanism by reducing it to a subset of approximate rate equations for the two species H_2O_2 and NAD•. Oscillatory behavior was found in this (greatly reduced) subsystem. Using a formalism known as stoichiometric network analysis, Aguda and Clarke[38] extracted a subset of ten crucial steps from the twenty possible reactions and showed that these ten steps are sufficient to explain the bistability and damped oscillations. Later studies[25] showed that the ten-step mechanism also can explain the existence of sustained oscillations. More recently, the addition of two more steps[18] has led to the discovery of chaotic behavior in the latest detailed model. It is interesting that certain details of the experimental observations which were not reproduced by the four-variable models are, in fact, seen in simulations with the detailed mechanisms.

The rest of this section will be divided between a discussion of the simple, four-variable models originally proposed by Degn *et al.*[32] and by Olsen[35] as well as a recent model proposed by Dunford *et al.*,[39] and a discussion of the state of understanding of the detailed models described above, which is considerably less advanced than understanding of the small models.

7.3.2 *Four-Variable Models*

7.3.2.1 *The Models*

The first model found to explain the oscillatory properties of the PO reaction postulates the existence of two critical free-radical intermediates, X and Y; the reactants oxygen (A) and NADH (B) comprise the remainder of the four-variable system. The enzyme in its various forms and additional species such as DCP and MB do not appear explicitly in this model. The enzyme, however, plays an implicit role through the values of the first and third rate constants in the scheme shown below:

$$
\begin{aligned}
A + B + X &\xrightarrow{k_1} 2X & \qquad Y &\xrightarrow{k_5} Q \\
2X &\xrightarrow{k_2} 2Y & \qquad X_0 &\xrightarrow{k_6'} X \\
A + B + Y &\xrightarrow{k_3} 2X & \qquad A_0 &\underset{k_{-7}'}{\overset{k_7'}{\rightleftarrows}} A \\
X &\xrightarrow{k_4} P & \qquad B_0 &\xrightarrow{k_8} B
\end{aligned}
\tag{7.2}
$$

The first three reactions in this model comprise the two autocatalytic cycles. The termination reactions, steps 4 and 5, are necessary to complete the mechanism by which oscillations arise from autocatalysis. The remaining steps in the mechanism involve an initiation reaction (step 6) for one of the free-radical species, a reversible reaction (step 7) that models the equilibration of gaseous oxygen with dissolved oxygen, and the slow perfusion of NADH into the reactor vessel (step 8). The PO reaction has typically been carried out experimentally in reactors which are not true CSTRs but, rather, open systems involving only inflow (not outflow) of the reactants. Model 7.2 is, therefore, consistent with the typical experimental configuration.

Model 7.2 as described above was initially thought to be incapable of supporting chaos, so a slightly modified version of it was proposed by Olsen[35] in 1983. Most of the steps in the two models are identical, but the first and third steps in the 1983 model are taken to be:

$$\mathrm{B} + \mathrm{X} \xrightarrow{k_1} 2\mathrm{X}$$

$$\mathrm{A} + \mathrm{B} + \mathrm{Y} \xrightarrow{k_3} 3\mathrm{X} \qquad (7.3)$$

This modification changes the relative stoichiometries of the two autocatalytic cycles and, in addition, eliminates A as a reactant from the first step.

Simulations based on the four-variable models described above have been carried out by numerically integrating (using either a Runge-Kutta or a Gear-type algorithm) the associated system of ordinary differential equations (ODEs). The ODEs for the DOP model are:

$$\mathrm{d}A/\mathrm{d}t = -k_1ABX - k_3ABY + k_7 - k_{-7}A$$

$$\mathrm{d}B/\mathrm{d}t = -k_1ABX - k_3ABY + k_8$$

$$\mathrm{d}X/\mathrm{d}t = k_1ABX - 2k_2X^2 + k_3ABY - k_4X + k_6$$

$$\mathrm{d}Y/\mathrm{d}t = -k_3ABY + 2k_2X^2 + k_5Y, \qquad (7.4)$$

where $k_6 = k_6'X_0$, $k_7 = k_7'A_0$ and $k_8 = k_8'B_0$. The ODEs for the modified model proposed by Olsen in 1983 are similar:

$$dA/dt = -k_3ABY + k_7 - k_{-7}A$$

$$dB/dt = -k_1ABX - k_3ABY + k_8$$

$$dX/dt = k_1ABX - 2k_2X^2 + 3k_3ABY - k_4X + k_6$$

$$dY/dt = -k_3ABY + 2k_2X^2 - k_5Y \qquad (7.5)$$

Numerical solution of either of these models reproduce the simple as well as the complex and chaotic behavior seen experimentally (See Fig. 7.5). The chaotic oscillations computed from Eqs. 7.5 using the parameter values listed in the legends of Figs. 7.5 and 7.6 seem to be qualitatively and quantitatively more similar to the chaos seen experimentally. Figure 7.6 shows a next-amplitude map made by overlaying the experimental data with a map taken from the chaotic simulations of Eqs. 7.5. As can be seen, the agreement is quite good.

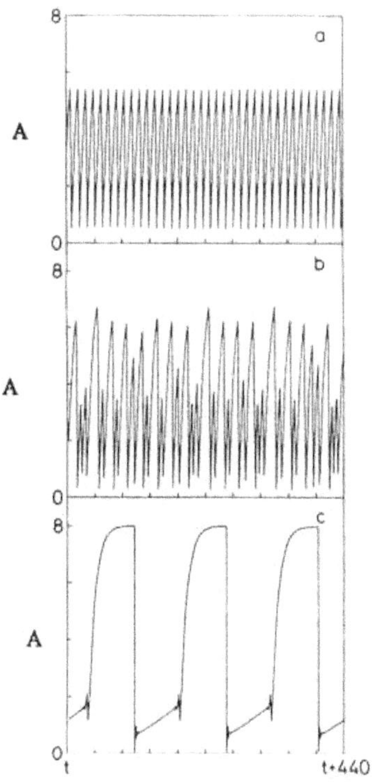

Figure 7.5. (a-c). See next page for legend.

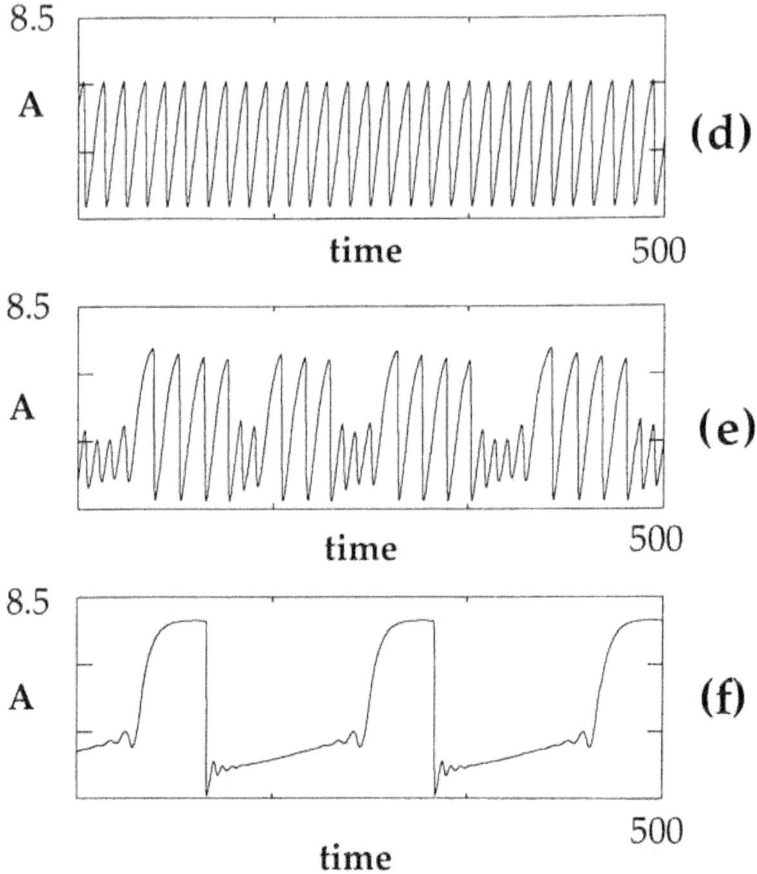

Figure 7.5. (Continued, d-f) Oscillations in species A computed from the 83 Model, Eqs. 7.5, for different values of k_1: a) $k_1 = 0.41$; b) $k_1 = 0.35$; c) $k_1 = 0.16$. Other parameter values are: $k_2 = 2.5 \times 10^2$; $k_3 = 3.5 \times 10^{-2}$; $k_4 = 20$; $k_5 = 5.35$; $k_6 = 1 \times 10^{-5}$; $k_7 = 0.8$; $k_{-7} = 0.1$; $k_8 = 0.825$. Reproduced by permission from L.F. Olsen, *Phys. Lett.* **94A** (1983) 454. In parts d) through f) results for the DOP model are shown for three different values of k_1: d) 0.12; e) 0.09; and f) 0.048. The values of the other parameters are: $k_2 = 1.25 \times 10^3$; $k_3 = 0.046\ 875$; $k_4 = 20$; $k_5 = 1.104$; $k_6 = 10^{-3}$; $k_7 = 0.89$; $k_{-7} = 0.1175$; $k_8 = 0.5$. Reproduced by permission from C.G. Steinmetz and R. Larter, *J. Chem. Phys.* **94** (1991) 1388. Compare these theoretical results to the experimental results in Fig. 7.2.

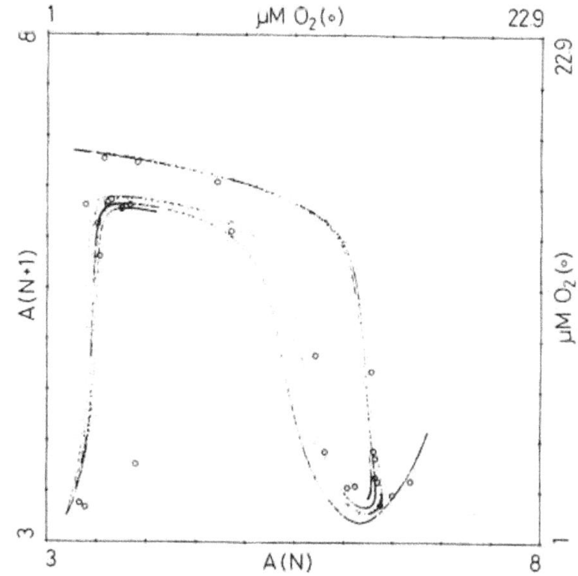

Figure 7.6. Next-amplitude map constructed from experimental (o) and theoretical (·) data. The theoretical results are computed from a chaotic solution of the 83 Model, Eqs. 7.5, with the following parameter values: k_1 = 0.35; k_2 = 2.5 x 10^2; k_3 = 3.5 x 10^{-2}; k_4 = 20; k_5 = 5.35; k_6 = 1 x 10^{-5}; k_7 = 0.8; k_{-7} = 0.1; k_8 = 0.825. Experimental data corresponds to the following conditions: 0.5 μM HRP; 0.2 μM MB and 20 μM DCP in 0.1 M Na-acetate buffer at pH = 5.1. The O_2 concentration in the gas phase was 2.15%. A 0.25 M NADH solution was fed in at a rate of 15 μl/h. Temperature = 28°C. Reproduced by permission from L.F. Olsen, in *Stochastic Phenomena and Chaotic Behavior in Complex Systems,* P. Schuster, Ed. (Springer-Verlag, Berlin, 1984). p. 116.

The chaotic states calculated by solving Eqs. 7.4 are accompanied by a rich and elaborate sequence of periodic states arranged in a grouping of Devil's staircases.[40] This type of sequence has been seen in other chaotic chemical reactions, the Belousov-Zhabotinsky reaction and the electrodissolution of copper. The chaotic states for Eqs. 7.4 are found in narrow intervals of parameter space, nestled between the steps of the staircases. A detailed study of these staircases and the associated transitions between the periodic states and the chaotic states has been worked out by

Steinmetz and Larter.[40,41] A summary of these findings is given in the next section.

7.3.2.2 The Quasiperiodic Route to Chaos in the DOP Model

Several routes to chaos have been observed in experimental and theoretical studies of nonlinear systems. Among these routes are period doubling, quasiperiodicity associated with a torus attractor, and intermittency. All of these routes to chaos have been found to occur in the DOP model, Eqs. 7.4, so it provides an especially interesting case study of the phenomenon of chaos in chemical oscillators.

Oscillations arise in the DOP model by way of a Hopf bifurcation. Figure 7.7 shows a curve in the k_1 - k_7 parameter plane corresponding to a set of Hopf bifurcations; this curve is the locus of points at which Hopf bifurcations give rise to simple oscillations from a steady state. The region of the plane above and to the left of the bifurcation curve contains a second curve; this second curve corresponds to a second Hopf bifurcation at which a second frequency converts the simple oscillations into quasiperiodic and, then, complex oscillations. The region of the parameter plane enclosed by this second curve contains solutions whose dynamics are associated with a torus attractor. In addition to quasiperiodic and complex oscillations, chaos is also found within this region of the k_1 - k_7 parameter plane.

Quasiperiodicity is found very close to the second Hopf bifurcation curve and just inside it. Quasiperiodicity corresponds to the introduction of a second frequency which is not a multiple of the frequency of the simple oscillations. Quasiperiodicity is associated with motion over the surface of a torus attractor; this attractor is shown in Figure 7.8. A rich variety of bifurcations of the torus takes place within the region enclosed by the second Hopf bifurcation. These bifurcations give rise to complex periodic oscillations known as mixed-mode oscillations and to irregular, or chaotic, oscillations. Both period-doubling bifurcations as well as tangent bifurcations associated with intermittency have been found to occur within this region.

The Poincaré sections of the torus attractor shown in Fig. 7.9 mirror the interesting changes in dynamical state which occur as the dashed line pathway in Fig. 7.7 is followed. Here, k_1' is defined as $k_1' = k_{SHB} - k_1$, where k_{SHB} is the value of k_1 at which the secondary Hopf bifurcation occurs. The Poincaré sections shown in Fig. 7.9 correspond to half of the cross section of the full torus; the full cross section would contain two closed, elliptical curves for the state corresponding to Fig. 7.9a. Here, the torus attractor just past the second Hopf bifurcation is smooth, with a fairly

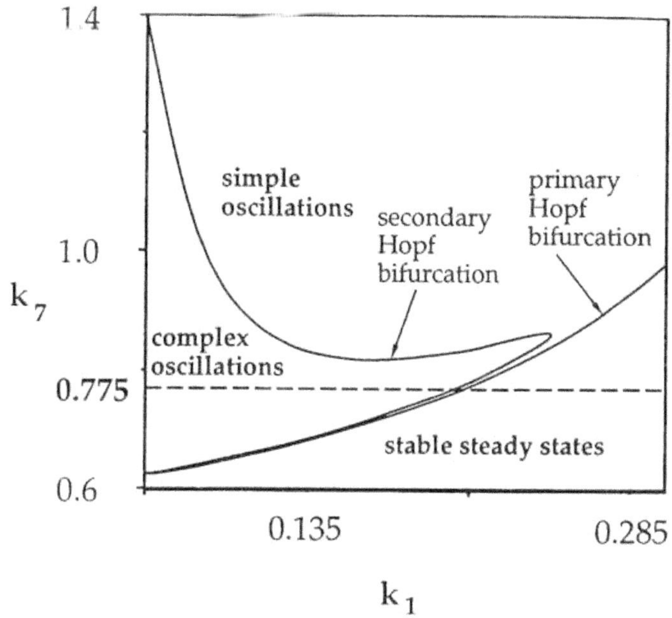

Figure 7.7. Phase diagram for the DOP model. All parameters except k_1 and k_7 are as in Fig. 7.5. The primary Hopf bifurcation separates the region of oscillatory solutions from the region of stable steady states. The secondary Hopf bifurcation separates the region of simple, single-frequency oscillations from the region of complex oscillations and chaos. The dashed line shows the one parameter path taken in studying the transition to chaos. Reproduced by permission from C.G. Steinmetz and R. Larter, *J. Chem. Phys.* **94** (1991) 1388.

symmetric, elliptical cross-section. This is typical of quasiperiodic motion. As the region bounded by the second Hopf bifurcation curve is penetrated further, *i.e.*, the dotted line pathway in Fig. 7.7 is followed further to the left, the Poincaré section becomes more irregular in appearance. Fig. 7.9b shows the section for a state corresponding to a wrinkled torus; the wrinkles appear on the surface of the torus near its inner hole. These wrinkles become more pronounced (although only slightly) in Fig. 7.9c, which actually corresponds to a fractal (as opposed to merely wrinkled)

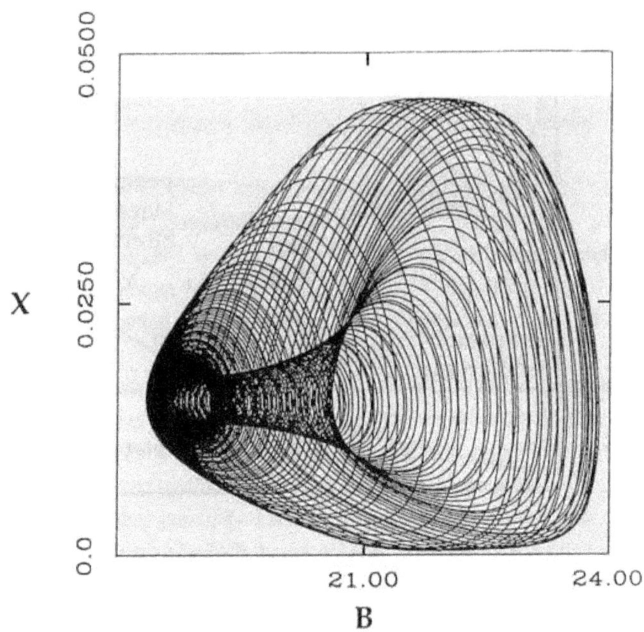

Figure 7.8. Quasiperiodicity on a 2-torus in the DOP model. All parameter values used in the creation of this phase portrait are the same as those in Fig. 7.5 except k_1 = 0.2 and k_7 = 0.775.

torus. The nature of the bifurcation corresponding to the transition from the torus of Fig. 7.9b to the torus of Fig. 7.9c will be explained in more detail below. The torus of Fig. 7.9d is more properly called a broken torus; it can be seen that the breaks occur near the places where wrinkling was seen in the wrinkled and fractal tori.

The states corresponding to the sections shown in Fig. 7.9 are either quasiperiodic or chaotic in nature. Periodic states also are found within this region, alternating with the quasiperiodic and chaotic states. These periodic states are always complex, *i.e.*, they cannot be characterized by a single frequency as a simple

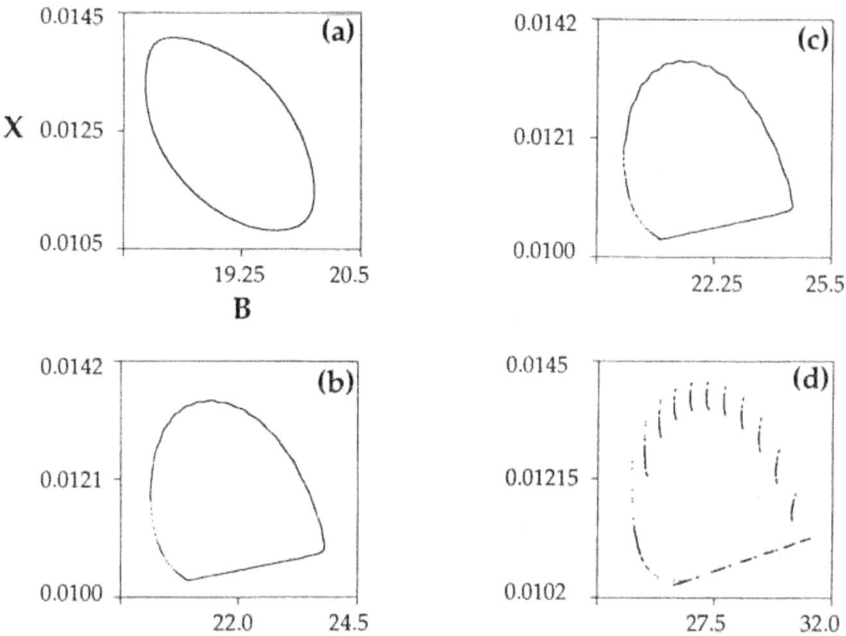

Figure 7.9. The four stages of the torus in the DOP model. The Poincaré sections are shown for: a) the smooth, undistorted torus; b) the wrinkled torus; c) the fractal torus; and d) the broken torus. Here, $k_1' = k_{SHB} - k_1$ with $k_{SHB} = 0.206$, the value of k_1 at which the secondary Hopf bifurcation occurs. The values of k_1' are a) 0.001; (b) 0.036; (c) 0.0426; and (d) 0.0882. All other parameter values are as given in Fig. 7.5. Reproduced by permission from C.G. Steinmetz and R. Larter, *J. Chem. Phys.* **94** (1991) 1388.

oscillation can. Poincaré sections for typical periodic states in these regions are shown in Fig. 7.10. The periodic states in Figs. 7.10a and 7.10b correspond to phase-locking on the smooth, undistorted and wrinkled tori, respectively; the phase-locking which occurs on the broken torus (Fig. 7.10c) gives rise to complex oscillatory states referred to as mixed-mode oscillations. The significance of this particular type of complex oscillation will be described in more detail later.

Another way in which the dynamics in this region can be described relies on the concept of the circle map. A map can be formed from the Poincaré sections by assigning an angle to each point which makes up the section. Ordering the angles as $\theta(1)$, $\theta(2)$, . . . ,$\theta(n)$, . . . , and plotting $\theta(n+1)$ vs $\theta(n)$ yields a map which

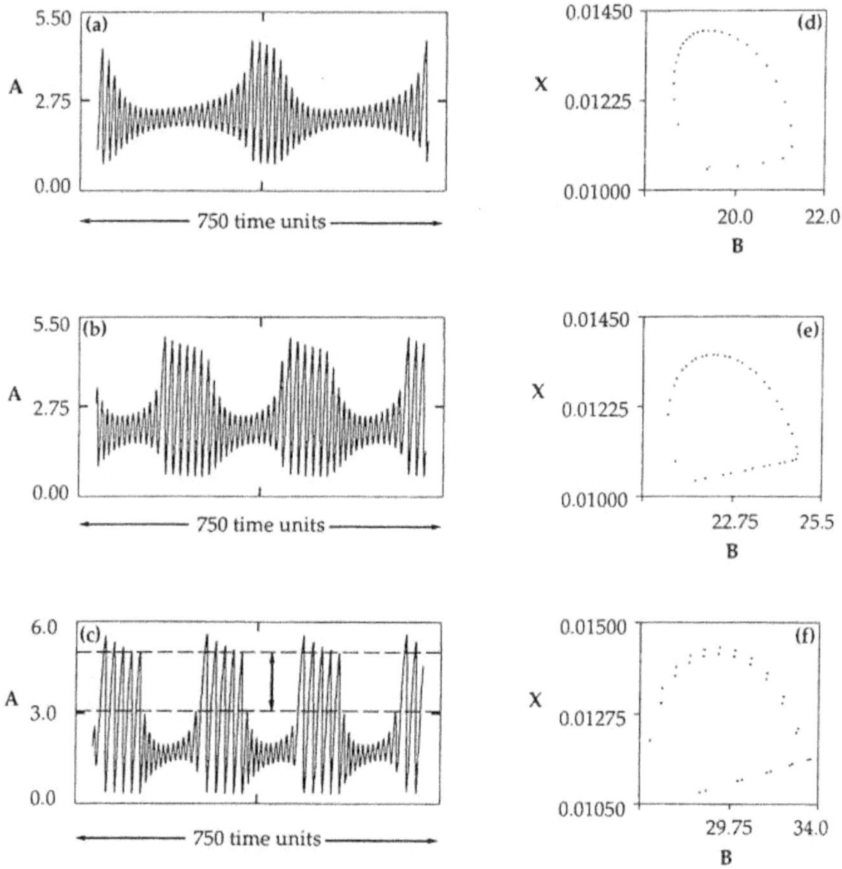

Figure 7.10. Phase-locked and mixed-mode oscillations in the DOP model. Parts a) and b) are time series of phase-locked states on a wrinkled 2-torus, while part c) is a mixed-mode state on a broken torus. Parts d) through f) are the corresponding surfaces of section. The dashed lines in part c) show the gap between the largest small oscillation and the smallest large oscillation - this gap is about as large as the small oscillations themselves. There is no such gap in parts a) and b). The k_1' values are: a) 0.012 237 795 b) 0.0439, and c) 0.1027. All other parameter values are as given in Fig. 7.5. Reproduced by permission from R. Larter and C.G. Steinmetz, *Phil. Trans. R. Soc. Lond.* **337A** (1991) 291.

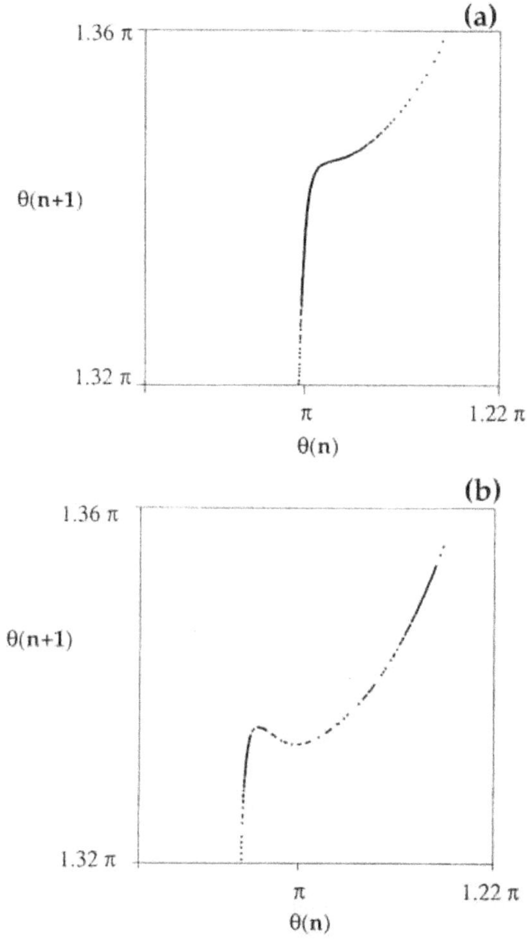

Figure 7.11. Circle maps for a highly wrinkled torus and a fractal torus. Part a) shows the map for the highly wrinkled torus, with $k_1' = 0.036$, while part b) shows the map for the fractal torus, with $k_1' = 0.0426$. All other parameter values are as in Fig. 7.5. Since the circle map in part a) is monotonic there is no chaos. Part b) clearly shows a local maximum and minimum; thus the circle map is nonmonotonic and chaos is therefore possible. Note that both parts a) and b) are close-ups of a small portion of the complete circle maps. Reproduced by permission from C.G. Steinmetz and R. Larter, *J. Chem. Phys.* **94** (1991) 1388.

corresponds to the "circle" that is the cross-section of the torus. A typical map for a quasiperiodic state on a wrinkled torus (such as that shown in Fig. 7.9b) is shown in Fig. 7.11a. It can be seen to be a simple, one-dimensional map. The bifurcation from the wrinkled torus of Fig. 7.9b to the fractal torus of Fig. 7.9c is associated with the development of an inflection point in this circle map (See Fig. 7.11b). The inflection point is significant because its existence requires that the inverse of the circle map be multiple-valued. As we will see below, this means that the rotation number which characterizes the periodicity of the state, becomes undefined. Speaking in practical terms, an undefined rotation number means that chaos becomes possible. So, the development of the inflection point shown in Fig. 7.11b, which heralds the appearance of a fractal torus attractor, means that chaos becomes a possibility, whereas it was not a possibility prior to this transition.

As alluded to above, the rotation number is critical to understanding the existence of chaos in the DOP model. A general definition of the rotation number which can be applied to quasiperiodic as well as periodic states is:

$$\rho = \lim_{n \to \infty} \frac{\theta(i + n) - \theta(i)}{2\pi n} \tag{7.6}$$

The rotation number as given by Eq. 7.6 is the long-term average of the individual ($n = 1, 2, \ldots$) rotation numbers and is independent of i if it is independent of the choice of initial conditions. If ρ is rational, then the dynamics are periodic; if it is irrational, then they are quasiperiodic. These are the only two possibilities if the rotation number is well-defined. However, if the rotation number is not well-defined, the dynamics need be neither periodic nor quasiperiodic, *i.e.*, chaos becomes a possibility. There are two ways the rotation number can be undefined: one occurs when the limit in Eq. 7.6 is dependent on initial conditions; in this case, birhythmicity is seen. On the other hand, the limit may fail to converge even for large n irrespective of the initial conditions; chaos exists in this case. So, an undefined rotation number is a necessary but not sufficient condition for chaos.

Figure 7.12 shows a plot of the rotation number over the range of values studied. The last three stages of the torus shown in Fig. 7.9 are labeled on this plot. The transition between undistorted torus (Fig. 7.9a) and wrinkled torus (Fig. 7.9b) is marked by the minimum at small k_1' values in the rotation number plot; the interval corresponding to the undistorted torus is very narrow and is not labeled in this plot. The transition between the wrinkled and fractal torus (Fig. 7.9c) is not easily seen in the rotation number plot. However, it is easily seen in Fig. 7.11, which shows the development of the inflection point in the circle map at this transition. The final transition, that between a fractal and a broken torus (Fig. 7.9d), is associated with the appearance of the six Devil's staircases that are located in this region of the rotation number plot. The Devil's staircases correspond to sequences of mixed-mode, periodic oscillations, *i.e.*, complex oscillations in which a clear distinction between large-amplitude and small-amplitude oscillations can be made (as in Fig. 10c). No such distinction can be made for the complex oscillatory states found in the wrinkled or fractal torus stages; hence, the appearance of these mixed-mode states marks the bifurcation from the fractal torus to a broken torus.

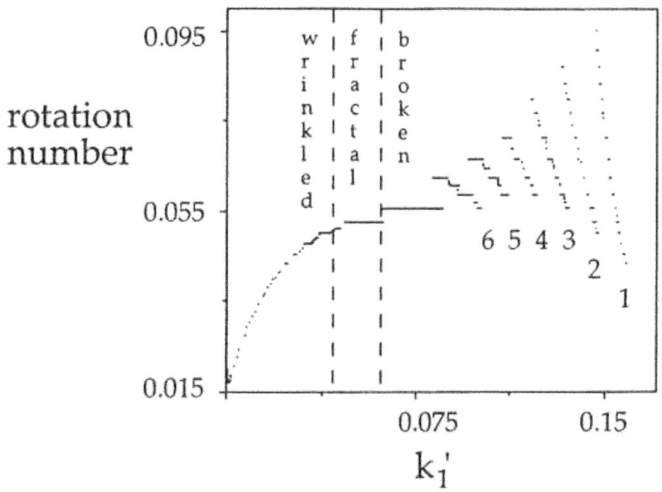

Figure 7.12. The rotation number as a function of k_1'. The number of large oscillations in each of the cascades on the broken torus is indicated. The region of the undistorted torus is too narrow to be visible. Reproduced by permission from C.G. Steinmetz and R. Larter, *J. Chem. Phys.* **94** (1991) 1388.

The staircases in Fig. 7.12 are labeled with sequence numbers (6, 5, ... , 1) which correspond to the number of large-amplitude oscillations in each repeating unit of the periodic oscillation. Each staircase is referred to as an *L*-sequence. Figure 7.13 shows an example of the detailed structure of the 5-sequence. If we use the notation L^S for the mixed-mode state with L large-amplitude oscillations and S small-amplitude oscillations, then many of the steps in the staircases are associated with secondary, tertiary, *etc.* concatenated states of the form $(L^S)^m(L^{S+1})^n$. The concatenated states are formed by combining the primary states L^S and L^{S+1} and it is found that only one of the primary states occurs more than once in any concatenated state; that is, either m or n is always found to be unity in the concatenated states in the DOP model of the PO reaction.

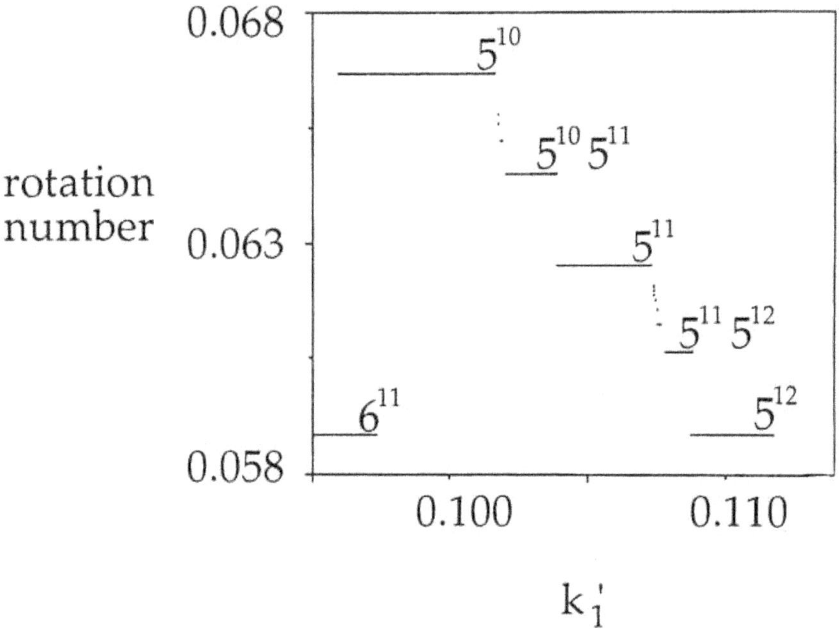

Figure 7.13. The rotation number, $\rho(k_1')$, for the 5-sequence (that is, the part of plot in Fig. 7.12 labeled "5"). The primary and secondary states are labeled, while some of the tertiary states are shown but not labeled. Notice the absence of tertiary states to the right of the secondary states ($5^{10}\,5^{11}$ and $5^{11}\,5^{12}$). Reproduced by permission from C.G. Steinmetz and R. Larter, *J. Chem. Phys.* **94** (1991) 1388.

The steps within each staircase follow a Farey sequence, which has been observed to be associated with chemical chaos in other reactions. For a mixed-mode state of the form $(L^S)^m(L^S+1)^n$ the rotation number is:
A Farey sum of two fractions p_1/q_1 and p_2/q_2 is defined as:

$$\rho = \frac{m + n}{m(L + S) + n(L + S + 1)} \tag{7.7}$$

$$\frac{p_1}{q_1} \oplus \frac{p_2}{q_2} = \frac{p_1 + p_2}{q_1 + q_2} \tag{7.8}$$

The rotation number, as given by Eq. 7.7, for the mixed mode state L^S is $(L + S)^{-1}$ and for the state L^{S+1} is $(L + S + 1)^{-1}$. The secondary concatenated states thus have a rotation number which is the Farey sum of the primary state rotation numbers of which they are composed. For example, the state $L^S L^{S+1}$ (*i.e.*, $m = n = 1$) has a rotation number according to Eq. 7.7 that is the Farey sum of the individual rotation numbers:

$$\frac{2}{L + S + L + S + 1} = \frac{1}{L + S} \oplus \frac{1}{L + S + 1} \tag{7.9}$$

This relationship holds for the higher-order concatenated states as well. However, only certain members of the Farey sequence are found to exist within the L-sequences; the states which exist are shown in boxes in the Farey tree of Fig. 7.14. The states which are missing are apparently the same as those which are found to be missing in several experimental systems totally unrelated to the PO reaction: the BZ reaction and the electrodissolution of copper.

The steps within each L-sequence occasionally overlap, resulting in small intervals of hysteresis in which two oscillatory states co-exist. For example, the 5^{11} and $5^{10}5^{11}$ states shown in Fig. 7.13 overlap; similarly, the 5^{12} and $5^{11}5^{12}$ states overlap. Alternating with these overlap regions are intervals (such as that between the 5^{10} and $5^{10}5^{11}$ states) which contain states that do not overlap. Instead, these regions are filled with small staircases of highly complex mixed-mode states; the interval between the 5^{10} and $5^{10}5^{11}$ step, for example, contains mixed mode states of the form $(5^{10})^n5^{11}$ alternating with chaotic states. The chaotic states are found to arise via a sequence of period doublings from these highly complex mixed-mode states. Figure 7.15 shows a bifurcation diagram that illustrates the period-doubling cascade from the $5^{10}5^{11}$

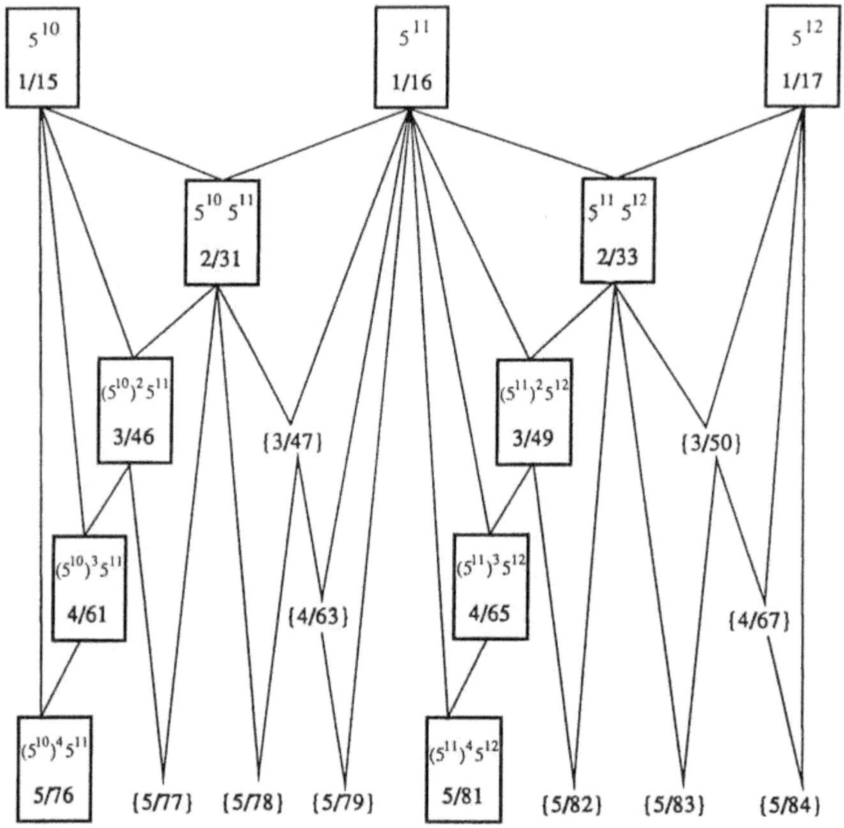

Figure 7.14. Farey tree for the 5-sequence down to the 5th level. The function $\rho(k_1')$ for the 5-sequence was shown in Fig. 7.13. All of the states shown are predicted by Farey arithmetic, but only the boxed-in states are actually observed. The states in brackets are not observed even in high-resolution calculations. The symbolic representations for the observed mixed-mode states also are given in the boxes. Reproduced by permission from C.G. Steinmetz and R. Larter, *J. Chem. Phys.* 94 (1991) 1388.

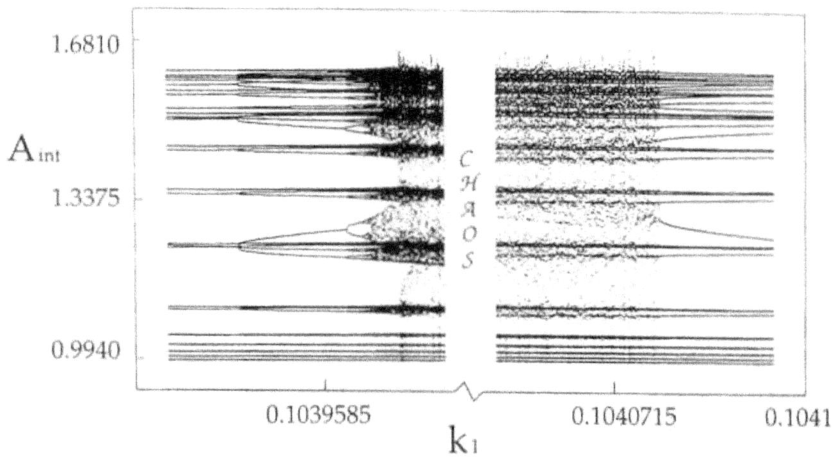

Figure 7.15. Bifurcation diagram showing the local transition to chaos. The periodic state on the left is the $5^{10}5^{11}$ mixed-mode state, while that on the right is the $5^{10}5^{10}5^{11}$ state, and in between is chaos. The ordering of states is reversed from that in Figure 7.13 because the bifurcation is shown as a function of k_1 rather than k_1'. The value plotted, A_{int}, is the value of the variable A in the Poincaré section. Reproduced by permission from R. Larter and C.G. Steinmetz, *Phil. Trans. R. Soc. Lond.* **337A** (1991) 291.

mixed-mode state that results in chaos. The chaotic region abruptly disappears and is replaced by the next mixed-mode state, $5^{10}5^{10}5^{11}$. The latter transition is a tangent bifurcation which occurs when the strange attractor collides with the basin boundary of a coexisting limit cycle attractor. A tangent bifurcation is associated with Type-I intermittency.

This section contains just a summary of the results given in references 40 and 41; the reader is referred to these papers for more details and further references to the literature.

7.3.2.3 *Other Investigations of the DOP Model*

a. *Fast/Slow-Variable Analysis*

A 1988 study of the DOP model[34] used an approach based on singular-perturbation theory in which the complete system of four independent variables was subdivided into a slowly varying subset and a more rapidly varying subset. This division allowed the origin of complex oscillatory behavior, and hence chaotic behavior, to be explained in terms of the underlying fast subsystem bifurcations.

A de-dimensionalization of the full set of governing equations reveals that the species B varies on a much slower time scale than do the other three variables. This information also can be ascertained by a visual inspection of the time series for the four species during a typical complex oscillation. Since B varies much more slowly than the other variables, it is approximately valid to assume that B is a constant and to treat it as a parameter of the equations governing the remaining three variables. The behavior of the three variable fast subsystem can then be studied separately.

Three positive and real steady states are found for the fast subsystem over certain ranges of parameter values. It is interesting to consider the behavior of the fast subsystem as a function of the special parameter B. Figure 7.16 shows the S-shaped steady state curve for the logarithm of the X variable as a function of B. The lower branch of the S-shaped curve corresponds to a stable node over most of its extent while the middle branch is always a saddle node. The upper branch is interesting because a Hopf bifurcation occurs at one point along it. A mechanism by which complex oscillations can arise exists if the Hopf bifurcation occurs at a value of B midway between the upper and lower knees of the S-shaped curve. A phase portrait corresponding to a complex oscillatory state of the full system has been superimposed on the S-shaped curve in Fig. 7.16. If the Hopf bifurcation were to fall outside of the multiple steady state region, the dynamics of the middle and upper branches would have no effect on the dynamics of the lower branch and a simple (one-frequency) oscillation would be observed. In the case shown, however, complex oscillations become possible because two oscillatory mechanisms now exist. One mechanism corresponds to relaxation type oscillations which carry the system from the upper branch to the lower branch and back again; the other mechanism involves the Hopf bifurcation point on the upper branch. The complexity of the oscillations shown is due to the interaction of these two mechanisms which carry the fast variables over the multiple steady state surface as the slow variable B swings back and forth between its limits. A plot of the bifurcation points as a function of the parameter B and a second parameter k_1 shows that these bifurcations enclose the

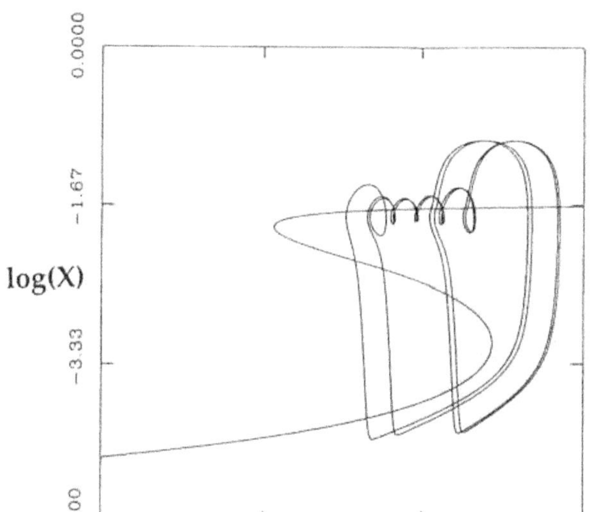

Figure 7.16. Fast/slow-variable description of the DOP model. A phase portrait computed from the full system of equations is superimposed on the steady state curve for the fast subsystem with corresponding parameter values. The parameter values are as in Fig. 7.5 except k_1 = 0.071 312 5; these parameter values correspond to a $2^4 2^5$ mixed mode state, as shown.

region in which complex oscillations are found in the full system (See the shaded region in Fig. 7.17).

b. Boundary Crises and Super-Persistent Chaotic Transients

Another investigation of the DOP model[42] found behavior consistent with a boundary crisis and associated intermittency in a region of parameter space different from that described in the preceding sections. In this range of parameter values, highly complex oscillatory states were found consisting of concatenations of many primary periodic states. For example, chaos was found very close to the $2^1 1^2 2^4 3^4 3^4$

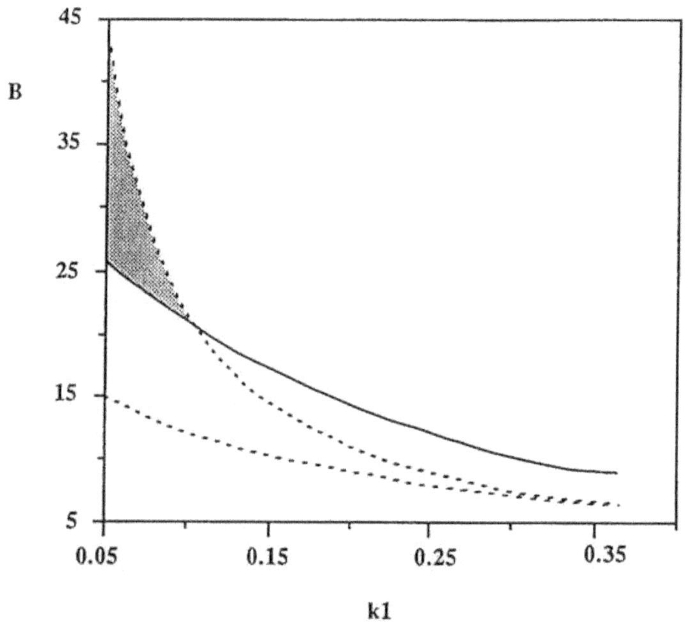

Figure 7.17 Parameter plane showing the locations of the upper and lower knees (dotted curves) and the Hopf bifurcation on the upper branch (solid curve) of the S-shaped steady state curve of Fig. 7.16. The dashed curves meet at a codimension-two bifurcation point. The shaded region corresponds to parameter values for which complex oscillations and chaos are found in the full system. Reproduced by permission from R. Larter, C.G. Steinmetz and B.D. Aguda, *J. Chem. Phys.* **89** (1988) 6506.

state in this region. This state's chaotic nature was identified by its power spectrum.

Many states identified as super-persistent chaotic transients were found close to the parameter interval in which chaos was observed. After many thousands of apparently chaotic oscillations these states would suddenly and irreversibly (but at unpredictable times) transform into a fairly simple 1^1 periodic state. The dynamical interpretation of this observation is that the strange attractor associated with chaotic states collides with the boundary separating it from the basin for the 1^1 limit cycle. The basin boundary is itself fractal, so the time required for the trajectory to penetrate the boundary is unpredictable. This type of behavior is referred to as a boundary crisis. It is interesting to note that chaotic transients have been associated with a separation of time scales, so the ability to carry out a fast/slow separation of this model may, in fact, be related to the existence of super-persistent chaotic transients and intermittency.

c. Oscillation-Steady State Bistability

The experimental observation of bistability between an oscillatory state and a co-existing steady state was described in a previous section. This bistability (variously known as hard excitation and O-S bistability) had been predicted from simulations with an eight-variable detailed model. Recent calculations at Stanford[20] have shown that the DOP model might also be consistent with this observation. The method used to induce a transition from the steady state to the oscillatory state was different from that used in the study of the eight-variable model and grew out of studies of the effect of sinusoidal and non-sinusoidal perturbations of the oxygen flow. When a sinusoidal perturbation was imposed on a steady state solution of the DOP model, an oscillatory state persisted even after the perturbation was turned off. The O-S bistability observed for the DOP model agreed with experimental observations in one way that the simulations with the more detailed model did not: the steady-state oxygen concentration was found mid-way between the minimum and maximum of the oscillatory state, as was observed experimentally. The steady state associated with the O-S bistability was found in the DOP calculations to be a locally stable focus.

7.3.2.4 *The Olsen 83 Model*

The slightly modified model proposed by Olsen in 1983, Eqs. 7.5, also reproduces the periodic and chaotic behavior observed experimentally, but there are some differences between the results found for the DOP model, Eqs. 7.4, and the modified 83 model. Many of these differences can be characterized as merely quantitative, rather than important qualitative differences. However, a few of the differences are, in fact, qualitative in nature; these differences will be highlighted in the following.

The parameter intervals over which chaotic behavior was first observed in the 83 model are somewhat broader than those described above for the DOP model. In addition, the nature of the transition from periodic to chaotic behavior for the parameter values initially studied is quite different. The transition to chaos was found to follow the period-doubling route; no quasiperiodicity, *i.e.*, no evidence of a torus attractor, is found near the region in which chaotic behavior was first observed. However, recent results[43] indicate that quasiperiodicity also can be found in the 83 model in other regions of parameter space and that the period-doubling route changes smoothly into the quasiperiodic route in this model.

The next-amplitude map, which can be formed from the chaotic solutions originally found by Olsen in 1983, bears a remarkably close quantitative resemblance to the next-amplitude maps formed from experimental data (see Figure 7.6). The computed solutions form a map which is multi-branched and which, in fact, has a fractal character.[44] The multiple branching is probably indicative of the highly developed nature of this chaotic state. It is likely that maps corresponding to chaotic states found just past the transition to chaos would have a less elaborate branching structure.

Olsen (1985) also studied the effects of imposing periodic and stochastic perturbations on this model.[45] The purpose of this investigation was to determine whether the nonlinear PO system would amplify the small but inherent fluctuations in the experimental constraints, such as the pumping rate, stirring rate, *etc.* To study the effect of a small periodic fluctuation in the NADH pumping rate, the 83 model was modified by changing the k_8 term to $k_8 + \mathrm{C}\sin(2\pi t)$. Perturbations in the inlet concentration of oxygen also were studied. The bifurcation diagrams in the presence and absence of the periodic perturbation were largely the same, regardless of the value of C (the amplitude of the perturbation); hence, the periodic perturbations were judged to be relatively unimportant in their effect on the dynamics of the 83 model. However, the imposition of random (as opposed to periodic) perturbations did affect the structure of the bifurcation diagram. Specifically, the sharp bifurcation points became more and more blurred as the random perturbation increased in amplitude. It was still possible to distinguish between the truly chaotic behavior and the irregular behavior due to the imposed noise, though, by inspection of the next-return maps. These show a distinct structure (as shown in Fig. 7.6) for the unperturbed chaos, whereas a blurred, nondistinct "map" is seen when large-amplitude random perturbations are imposed on either the periodic or chaotic solutions of this model. One conclusion of this study that has experimental significance is that periodic motion can be totally obscured by the imposition of large-amplitude random fluctuations. The identification of chaotic motion in the presence

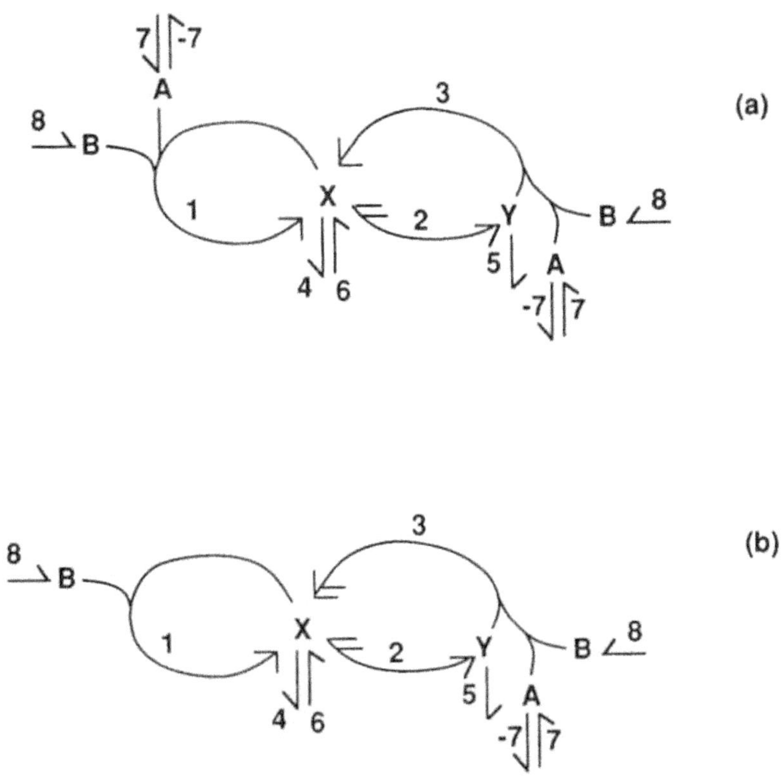

Figure 7.18. Network diagrams showing the reactions of a) the DOP model, Eqs. 7.4, and b) the 83 Model, Eqs. 7.5. In both models, $A = O_2$, B = NADH and X and Y are free-radical intermediates. The notation uses arrow heads and tail feathers to indicate stoichiometry: a single-headed arrow indicates a stoichiometry of 1, a tail with 2 feathers indicates a stoichiometric coefficient of 2, *etc.*; a tail with no feathers implies a stoichiometric coefficient of 1. Reproduced by permission from B.D. Aguda and R. Larter, *J. Am. Chem. Soc.* **112** (1990) 2167.

of these random fluctuations is possible, however, if the amplitude of the fluctuation is not too large.

Aguda, Larter and Clarke[46] studied the dynamical behavior of the 83 model by inspecting the flow through the reaction network (see Fig. 7.18). The arrows in this diagram correspond to the steps in the mechanism (see Eqs. 7.2 and 7.3). The first, fourth, sixth and eighth steps form a subnetwork that, on its own, will produce damped oscillations. This subnetwork was dubbed a dynamic element and given the designation DE-1. The damped oscillations of the dynamic element DE-1 tend to pull the trajectories of the full network toward the single steady state. The existence of other dynamic elements in the full reaction network allow these damped oscillations to become sustained oscillations and also permit chaotic behavior in this model. A second dynamic element, DE-2, corresponds to simply the third step of the reaction; DE-2 provides for the existence of a second feedback loop and serves as a switch between the large-amplitude and small-amplitude oscillations. Only damped oscillations would be observed without this second dynamic element. In addition, chaotic oscillations also would be impossible without DE-2. A third dynamic element, DE-3, was identified as the reversible feed of A, the seventh step. This element is important because it is essentially the only reaction occurring for a significant length of time during each cycle of the oscillation. DE-2 is responsible for "turning on" DE-1 after DE-3 has operated for a sufficient length of time.

7.3.2.5 The Dunford Model

Alexandre and Dunford recently proposed a new model[39] for the PO reaction which emphasizes the mechanistic role of the enzyme. The native form of the enzyme, Per^{3+}, and the crucial enzyme species coIII are included specifically in the mechanism:

$$\begin{aligned} P + A + X &\xrightarrow{k_1} Y + 2X \\ Y + X &\xrightarrow{k_2} P + 2X \\ B &\xrightarrow{k_3} X \\ X &\xrightarrow{k_4} \text{Product} \\ A_0 &\overset{k_1}{\rightleftarrows} A \end{aligned} \tag{7.10}$$

Here, A is O_2 as in the other small models, but the NADH concentration is assumed to be constant. P is taken to be the native form of the enzyme (Per^{3+}) while Y is coIII. The third step models the spontaneous generation of *X*, the NAD• free radical, and the fourth step its auto-decomposition. The first and second steps are both autocatalytic in X and involve changing the enzyme from the Per^{3+} form (P) to the coIII form (Y) and back again. The existence of two coupled autocatalytic cycles makes this mechanism similar to the DOP and 83 mechanisms, and oscillatory behavior is expected. The advantage of mechanism 7.10, according to Alexandre and Dunford, is that the role of the enzyme is stated explicitly. However, the role which the two forms P and Y fulfill in mechanism 7.10 does not agree entirely with the role that the analogous forms Per^{3+} and coIII play in the detailed mechanisms. We will return to this point later.

Alexandre and Dunford assumed while studying the dynamics of this model that the total enzyme concentration was fixed, *i.e.*,

$$E_{tot} = P + Y \tag{7.11}$$

Using Eq. 7.11 to eliminate the enzyme variable *P* and substituting the result into the differential equations for model 7.10 yields a system of differential equations for the three remaining variables *X*, *Y* and *A*. This system is found to possess a single steady state which undergoes a Hopf bifurcation at a certain point in parameter space. Alexandre and Dunford found that k_4, corresponding to the termination reaction for NAD•, was the most critical variable affecting the oscillatory behavior. Simple oscillations were found in numerical simulations, but no complex oscillations or chaos were observed.

7.3.3 *Detailed Models*

A number of more detailed PO models have been proposed and investigated over the years. Recently, some variations of these models have been shown to be capable of explaining the existence of oscillatory behavior, both regular and chaotic. All of these models have a common "skeleton" of the steps shown in Fig 7.19 and listed in the top half of Table 7.2. Note that only four of the five known enzyme-substrate complex intermediates appear in this skeleton mechanism. The fifth, Per^{2+}, is present in only one of the models to be described below; it is interesting that this latter model is the only detailed model known to exhibit chaotic oscillations, as of this writing. The intermediate NAD• is critical in these schemes,

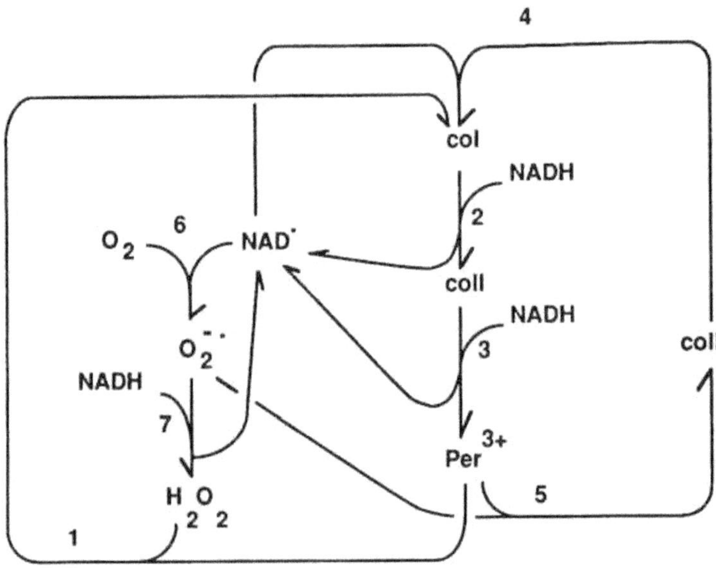

Figure 7.19. Network diagram showing the steps which are common to the detailed reaction mechanisms discussed in the text. These steps are also listed in conventional chemical equation form in the top portion of Table 7.2.

Table 7.2 Reactions Included in Detailed Models.

Skeleton

R_1			Per^{3+}	+	H_2O_2	→	coI				
R_2			coI	+	NADH	→	coII	+	NAD•		
R_3			coII	+	NADH	→	Per^{3+}	+	NAD•		
R_4			coIII	+	NAD•	→	coI	+	NAD^+		
R_5			Per^{3+}	+	O_2^{-}•	→	coIII				
R_6			NAD•	+	O_2	→	NAD^+	+	O_2^{-}•		
R_7	O_2^{-}•	+	NADH	+	H^+	→	H_2O_2	+	NAD•		

Model A

R_8					2NAD•	→	$(NAD)_2$				
R_9						→	O_2				
R_{10}					O_2	→					

Model C

R_{11}						→	NADH				
R_{12}			Per^{3+}	+	NAD•	→	Per^{2+}	+	NAD^+		
R_{13}			Per^{2+}	+	O_2	→	coIII				

YY Model

R_{14}			$2O_2^{-}$•	+	$2H^+$	→	H_2O_2	+	O_2		
R_{15}	H^+	+	H_2O_2	+	NAD•	→	NAD^+	+	H_2O	+	HO•
R_{16}			HO•	+	NADH	→	NAD•	+	H_2O		
R_{17}	H^+	+	O_2	+	NADH	→	H_2O_2	+	NAD^+		
R_{18}					coIII	→	coII	+	H_2O_2		

primarily because it is involved in at least one autocatalytic cycle, *i.e.*, a cycle in which the consumption of one NAD• radical results in the production of three NAD• radicals. This autocatalytic cycle corresponds to reactions 1, 2, 3, 6 and 7. The differences among the mechanisms to be described below involve the nature of the remaining reactions.

7.3.3.1 The YY Model

Yokota and Yamazaki proposed in 1977 the detailed model for the PO reaction shown in Fig. 7.20 and henceforth called the YY model.[36] This mechanism includes the steps in the top and bottom portions of Table 7.2. The bimolecular reaction between coIII and NAD• is considered to be the dominant decomposition pathway for coIII in the YY mechanism, even though a second pathway involving the production of coII and H_2O_2 from coIII is included. Simulation studies with the YY mechanism by Yokota and Yamazaki reproduced the four phases in the coIII kinetics they had observed experimentally in a closed system: initial burst, induction, steady state and abrupt decomposition. These studies did not reveal any oscillatory behavior in this model, however. Subsequent investigations by Aguda and Larter[25] showed several mathematical inconsistencies in the rate equations used by Yokota and Yamazaki in their simulations; this may be a partial explanation for the absence of oscillatory behavior in the YY model. However, the YY mechanism models a closed system since it does not allow for input of reactants or a termination pathway for the autocatalytic species; as such, it would not be expected to yield sustained oscillations.

7.3.3.2 The FAB Model

A model proposed by Fed'kina, Ataullakhanov and Bronnikova[37] in 1984 is simpler than the YY model but adds a unimolecular termination reaction for the NAD• radicals; this addition corrects a problem with the YY model which caused the concentration of NAD• to increase monotonically throughout the YY simulation.

Simulations with the FAB mechanism were carried out by applying certain simplifying assumptions; these assumptions were, in fact, so broad in scope that the eight-variable FAB mechanism was reduced to a two-variable approximate mechanism. The assumptions involved taking the other 6 species to be at steady or quasi-steady state. The two-variable approximation does show oscillatory behavior and has other features which are partially consistent with experimental observations.

Quantitative comparison with experiment is not possible, however, because of the sweeping assumptions made in the reduction of this mechanism.

7.3.3.3 *Model A*

Aguda[47] showed that the molecularity of the termination reaction in the FAB model (assumed to be unimolecular) results in only a single steady state for this mechanism. A bimolecular termination reaction was shown to be necessary for the existence of multiple steady states in the FAB mechanism and, hence, bistability; the mechanism by which oscillations arise may also be affected by this if the oscillations are of the relaxation type, *i.e.*, related to the existence of multiple steady states. A model identical to the FAB model in all respects except the molecularity of the NAD• termination reaction was thus proposed. This model, known as Model A, is shown in Fig. 7.21 and consists of the steps in the top two portions of Table 7.2; it can be seen that a bimolecular termination reaction is included as step 8. Aguda and Clarke[38] showed that Model A could explain the existence of bistability in the PO reaction via the multiple steady states referred to above. They also found damped oscillatory behavior for certain parameter values used in their study. Later, Aguda and Larter[25] found sustained oscillations in simulations with this model for somewhat different parameter values. This oscillatory behavior was of the type characterized as "simple," *i.e.*, the oscillations could be described by a single frequency. No complex oscillatory behavior or chaos was found in the simulations with Model A.

7.3.3.4 *Model C*

By comparing the dynamical behavior of the smaller models (the DOP model and the 83 model), Aguda and Larter[25] predicted that chaos would be found if Model A was modified to include a reaction in which oxygen could serve to "force" the non-autocatalytic feedback loop, reactions 6, 5 and 4 in Figs. 7.21 and 7.22. This prediction was based on the fact that both the DOP model and the 83 model include oxygen as a reactant in the autocatalytic loop involving species Y. Because of the similarities in behavior of the smaller four-variable models and Model A, a suggestion was made that species Y in the four-variable models played the same role as coIII in the detailed model, whereas species X was NAD•. Using these analogies, and the fact that chaos is observed in both the DOP model and the 83 model, led to the addition of a pair of reactions which allowed a constant feed of oxygen to force the

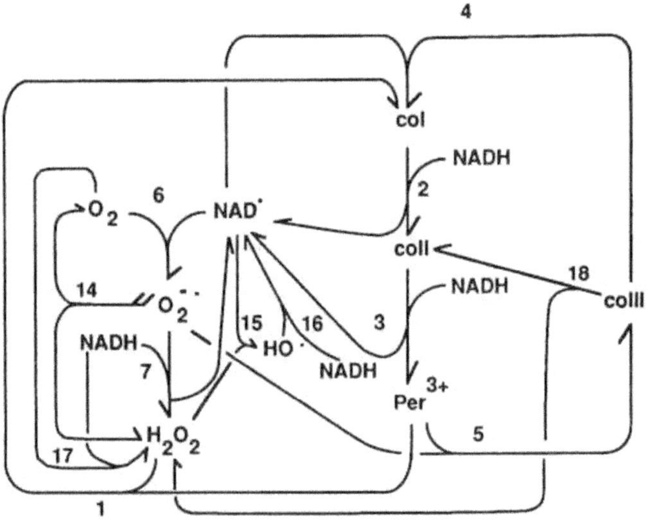

Figure 7.20. The Yokota-Yamazaki mechanism for the PO reaction. This mechanism includes all steps in the top and bottom portions of Table 7.2. Reproduced by permission from B.D. Aguda and R. Larter, *J. Am. Chem. Soc.* **112** (1990) 2167.

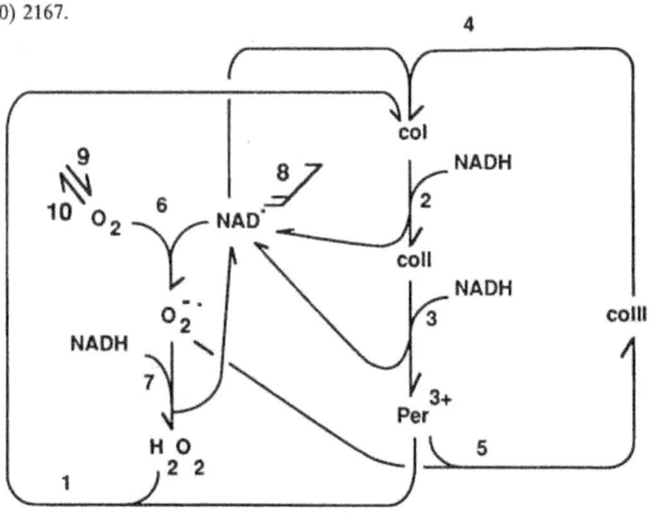

Figure 7.21 Model A for the PO reaction. This mechanism includes the steps in the top two portions of Table 2. Reproduced by permission from B.D. Aguda and R. Larter, *J. Am. Chem. Soc.* **112** (1990) 2167.

Table 7.3. Rate Constant Values.

R_i	Rate Expression	k_i, exp'l[(a)]	ref	k_i, ref 18	k_i, ref 48
R_1	$k_1[Per^{3+}][H_2O_2]$	1.2×10^7	12,36,53-56	1.0×10^6	1.2×10^6
R_2	k_2[coI][NADH]	5.3×10^3	36, 53-56	1.0×10^6	5.4×10^3
R_3	k_3[coII][NADH]	8.0×10^2	53-55	1.0×10^6	8.0×10^2
R_4	$k_4[NAD^{\cdot}]$[coIII]	1.3×10^8	36,61	1.0×10^6	1.3×10^6
R_5	$k_5[Per^{3+}][O_2^{\cdot -}]$	1.9×10^6	36,59,62	6.0×10^7	1.9×10^7
		1.5×10^5	50		
R_6	$k_6[NAD^{\cdot}][O_2]$	2.0×10^9	36,52,59	1.0×10^6	2.0×10^6
R_7	$k_7[O_2^{\cdot -}]$[NADH]	5.9×10^{3}[(b)]	36,59	1.0×10^6	5.9×10^3
		$<3.5 \times 10^{4}$[(c)]	50		
R_8	$k_8[NAD^{\cdot}]^2$	5.6×10^7	52	1.0×10^6	3.0×10^6
		2.7×10^6	51		
R_9	k_9[(d)]	1.0×10^{-7} Ms^{-1}	1,12,17[(h)]	1.0×10^{-7}	1.0×10^{-7}
R_{10}	$k_{10}[O_2]$[(d)]	0.1 s^{-1}	1,12,17[(h)]	0.1 s^{-1}	0.1 s^{-1}
R_{11}	k_{11}[(e)]	1.0×10^{-7} Ms^{-1}	1,12,17[(h)]	1.0×10^{-7}	1.0×10^{-7}
R_{12}	$k_{12}[Per^{3+}][NAD^{\cdot}]$	very slow	36,59	1.0×10^6	1.0×10^6
R_{13}	$k_{13}[Per^{2+}][O_2]$	5.8×10^4	59,60	1.0×10^6	5.8×10^6
R_{14}	$k_{14}[O_2^{\cdot -}]^2$	1.1×10^7	36,64		
		2.0×10^{7}[(f)]	50,63		
R_{15}	$k_{15}[NAD^{\cdot}][H_2O_2]$	8.6×10^{8}[(g)]	36		
R_{16}	$k_{16}[NADH][HO^{\cdot}]$	unknown			
R_{17}	k_{17}[NADH]	3.0×10^{-6} s^{-1}[(g)]	36,61		
R_{18}	k_{18}[coIII)]	2.0×10^{-3} s^{-1}	58		
		3.0×10^{-4} s^{-1}	36		

a) Units are $M^{-1}s^{-1}$ unless otherwise indicated; b) Calculated by fitting simulation to data at pH = 5.6; c) pH dependent, measured at pH = 5.1; d) $k_9 = k_{10}[O_2]_{eq}$ where $[O_2]_{eq}$ is the concentration of O_2 in the gas phase; e) $k_{11} = k_{11}'[NADH]_0$; f) pH dependent, calculated for pH = 5.1 from data given in reference 63; g) calculated by fitting simulation to data at pH = 5.6; h) calculated from typical experimental conditions used in referenced studies.

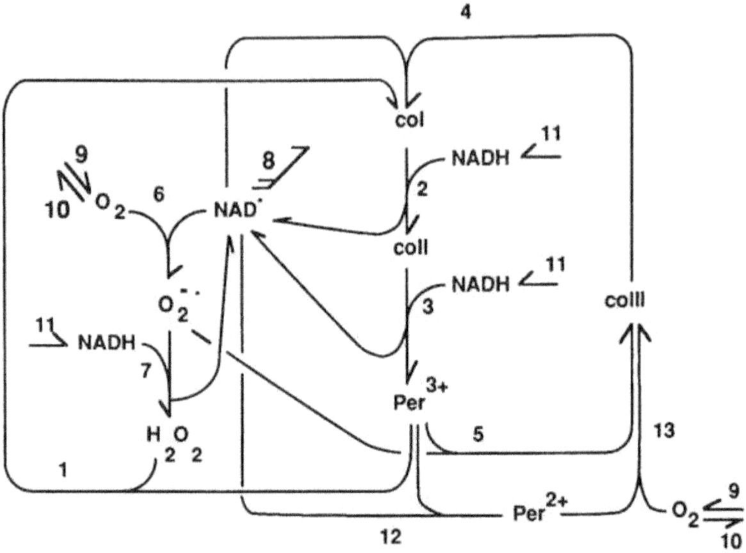

Figure 7.22. Model C for the PO reaction. The steps in the top three portions of Table 7.2 are included. Reproduced by permission from B.D. Aguda and R. Larter, *J. Am. Chem. Soc.* **113** (1991) 7913.

nonautocatalytic loop in model A involving coIII. The resulting model consists of the steps in the top three portions of Table 7.2 and is shown in Fig. 7.22. This model is referred to as Model C (for "chaos") and is seen to include the fifth, and last, enzyme species, Per^{2+}. An additional difference between Models A and C is the inclusion of reaction 11, which allows NADH to be fed into the system and to be, therefore, a time-dependent variable. Simulations with Model C revealed[18] the existence of simple oscillations followed by a period-doubling sequence of bifurcations that occurred when the total enzyme concentration was varied. This sequence of period doublings resulted in chaos (see Fig. 7.23) with a next-amplitude map that was quite similar to that seen experimentally.

One area of concern about the complex oscillations and chaos found in these simulations with Model C involves the values of the rate constants used to generate Fig. 7.23. Table 7.3 shows a comparison between the experimentally determined rate constants (third column) and the rate constants used in the simulations of reference 18 (fifth column). In some cases, no experimental values for the rate constants are known. In addition to the absence of information about some rate constants, it is also known that the experimental conditions (such as pH, temperature, presence of

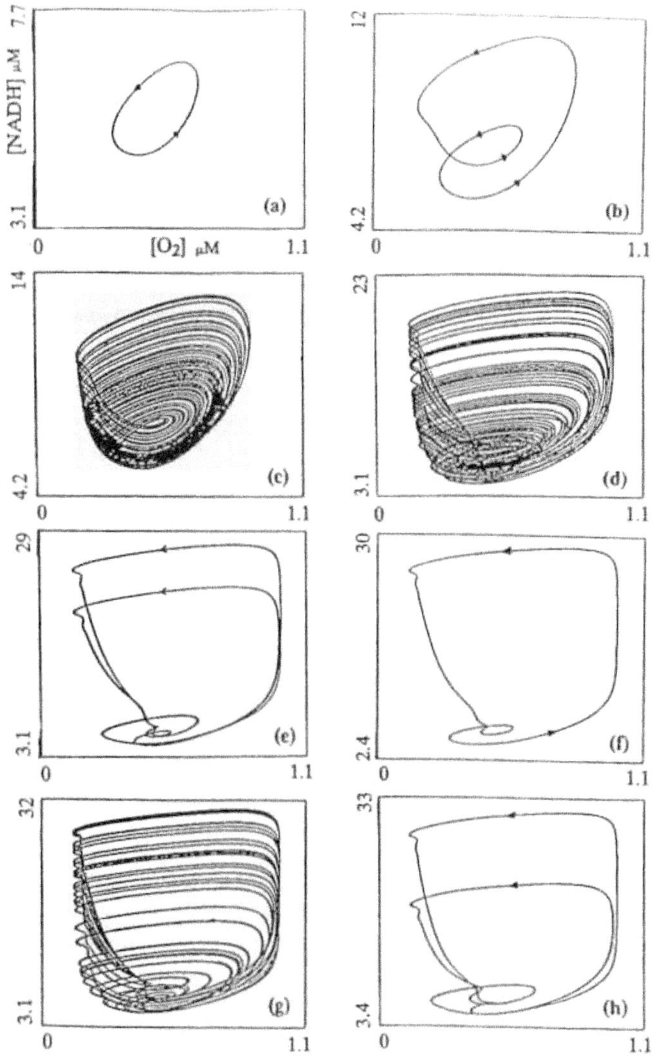

Figure 7.23. Alternating periodic-chaotic sequence seen in simulations with Model C using the parameter values in the next-to-last column of Table 7.3. The total enzyme concentration, E_{tot} = [Per^{3+}] + [Per^{2+}] + [coI] + [coII] + [coIII], is fixed for each run at the following values (all in μM): a) 3.0; b) 4.0; c) 4.2; d) 4.3; e) 4.8; f) 4.9; g) 5.1; and h) 5.2. The chaotic state in (c) yields a nearly one-dimensional next-amplitude map that is quite similar to that seen experimentally. Reproduced by permission from B.D. Aguda and R. Larter, *J. Am. Chem. Soc.* **113** (1991) 7913.

additives, etc.) were not the same for all rate constant measurements. Nevertheless, the values of the rate constants used in the 1991 simulation are not completely reasonable, so an attempt was made to look at the dynamical behavior of Model C for rate constants that were more experimentally reasonable. Values closer to those determined experimentally were chosen[48] and are given in column six of Table 7.3. Simulations using these values revealed another period-doubling sequence into chaos as the enzyme concentration was varied, verifying the robustness of the result of reference 18. The next-amplitude map associated with the chaotic state found for the values in column six is shown in Figure 7.24; although the map is quite beautiful and elaborate, its form does not agree as well with experiment as the next-amplitude

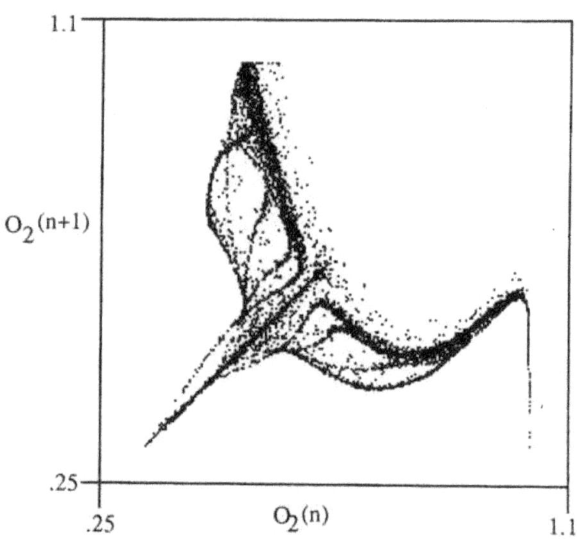

Figure 7.24. Next-amplitude map for a chaotic state calculated with Model C using the experimentally more reasonable parameter values given in the last column of Table 7.3. Here, the total enzyme concentration is 5.1 μM.

map found for the rate constants in the fifth column, or, indeed, with the map found in simulations with the smaller 83 model. This result does, however, show that chaos can be achieved with Model C when experimentally reasonable rate constants are used.

7.3.4 *Comparison of Detailed Models with Four-Variable Models*

The extensive theoretical and computational studies of the detailed models described above have made it possible to begin assessing the identity of the unknown species X and Y in the DOP model and Olsen's 1983 model. These small models have been remarkably accurate in reproducing the shape of the experimentally observed oscillations and bifurcation sequences. The first of these models was constructed by forming a branched quadratic structure with a mechanism similar to the Lotka-Volterra model.[49,50] It now appears that the mechanistic structure determined in this rather *ad hoc* way is remarkably similar to the overall structure of the detailed models which are based more firmly in experimental observation.

The species X is most likely NAD•, which is produced autocatalytically through a cycle involving superoxide radical, hydrogen peroxide and the native form of the enzyme. The autocatalysis comes about when coI and coII react with NADH to produce additional NAD• free radicals. These pathways are fairly well-established.

The identity of the species Y is still somewhat uncertain, but it is likely to be related to one of the other enzyme species. Alexandré and Dunford's recently proposed[39] simple four-variable model for the PO reaction was the first of the small models to explicitly assign the mechanistic roles of any of the enzyme species; this model was discussed in Section 7.3.2.5. Two critical enzyme species, Per^{3+}, the native form of the enzyme, and coIII were specifically included, and a variable similar to the species *Y* in the DOP model was assigned to coIII. In the Alexandré and Dunford mechanism, Per^{3+} is converted to coIII via a reaction which simultaneously produces NAD• autocatalytically. CoIII is converted back to Per^{3+} by a second autocatalytic cycle. However, only one autocatalytic cycle is present in all of the detailed models described above, and it does not involve either of these enzyme conversion reactions. This discrepancy may not be important at the level of simplification taken by Alexandré and Dunford, however.

It is still possible that further study will reveal the existence of two autocatalytic cycles and clarify the roles of the key enzyme species in the mechanism and their relation to the simple four-variable mechanisms originally proposed.

Recent experimental interest in this system also promises to reveal more details about the mechanistic origins of oscillatory behavior, both regular and chaotic, in the PO reaction, and to help clarify the relationships between the small models and the more detailed mechanisms proposed above.

7.4 Conclusions

Study of the PO reaction has been undertaken using both detailed theoretical and computational investigations as well some experimental approaches. However, the experimental study lags, by far, the theoretical and computational studies, and the time is now ripe to carry out a new round of detailed experimental investigations. Achieving a higher degree of understanding of this reaction is important since the PO reaction is only the second known example of a homogeneous chemical oscillator that undergoes a well-characterized transition to chaos. The ability of the nonlinear dynamics community to make general statements about chemical chaos hinges on having more than one well-characterized example of this phenomenon.

Some of the remaining questions involving this reaction focus on the details of the mechanism involved with the generation of regular and chaotic oscillations. The question of the roles played by the various enzyme species, especially the native form, Per^{3+} and the crucial intermediate form coIII, remains unconfirmed. Recent theoretical studies of possible detailed mechanisms bring us closer to answering these questions. However, these answers can be only conjectures without further experimental work. Other unresolved questions involve the role of the critical additives methylene blue and dichlorophenol. At a dynamical level, the origin of the oscillations and the route to chaos have only begun to be studied. Further extensive investigation of this system is necessary to complete our understanding of the nonlinear dynamics of the PO reaction.

Since it is an enzyme-catalyzed reaction, the PO reaction is especially intriguing and brings us back to the question of whether chaos is an important phenomenon *in vivo*. Further experimental study is necessary to determine whether this phenomenon is found in other autonomous enzyme reactions. The fact that enzyme reactions are usually found *in vivo* within complex networks of nonlinear processes suggests that chaos is an inescapable feature of living systems. Whether or not this is true awaits further investigation.

Literature Cited

1. L.F. Olsen and H. Degn, *Nature* **267** (1977) 177.
2. R.A. Schmitz, K.R. Graziani and J.L. Hudson, *J. Chem. Phys.* **67** (1977) 3040.
3. M. Mäder and V. Amberg-Fisher, *Plant Physiol.* **70** (1982) 1128.
4. M. Mäder and R. Füssl, *Plant Physiol.* **70** (1982) 1132.
5. O. Pantoja and C.M. Willmer, *Planta* **174** (1988) 44.
6. H.B. Dunford, in *Peroxidases in Chemistry and Biology, Vol. II*, J. Everse, K.E. Everse and M.B. Grisham, Eds. (CRC Press, Boca Raton, 1991).
7. P. Gray and S.K. Scott, *Chemical Oscillations and Instabilities; Nonlinear Chemical Kinetics* (Clarendon Press, Oxford, 1990). p. 15.
8. I. Yamazaki, K. Yokota and R. Nakajima, *Biochem. Biophys. Res. Commun.* **21** (1965) 582.
9. H. Degn, *Nature* **217** (1968) 1047.
10. H. Degn, *Biochim. Biophys. Acta* **180** (1969) 271.
11. S. Nakamura, K. Yokota and I. Yamazaki, *Nature* **222** (1969) 794.
12. L.F. Olsen and H. Degn, *Biochim. Biophys. Acta* **523** (1978) 321.
13. T.-Y. Li and J. Yorke, *Am. Math. Mon.* **82** (1975) 985.
14. O.E. Rössler, *Z. Naturforsch.* **31A** (1976) 259.
15. M. Markus, D. Kuschmitz and B. Hess, *FEBS Lett.* **172** (1984) 235.
16. O. Decroly and A. Goldbeter, *Proc. Natl. Acad. Sci. USA* **79** (1982) 6917.
17. T. Geest, C.G. Steinmetz, R. Larter and L.F. Olsen, *J. Phys. Chem.* **96** (1992) 5678.
18. B.D. Aguda and R. Larter, *J. Am. Chem. Soc.* **113** (1991) 7913.
19. T. Geest, L.F. Olsen and W.M. Schaffer, "*Lyapunov exponents and the fractal dimension of chaos in the peroxidase-oxidase reaction*," preprint (1992).
20. M. Samples, Ph.D. thesis, Stanford University (1992).
21. G. Baier, P. Urban and K. Wegmann, *Z. Naturforsch.* **43A** (1988) 987.
22. G. Baier, P. Urban and K. Wegmann, *Z. Naturforsch.* **43A** (1988) 995.
23. G. Baier, P. Urban and K. Wegmann, *Z. Naturforsch.* **44A** (1989) 1107.
24. G. Ertl, *Science* **254** (1991) 1750.
25. B.D. Aguda and R. Larter, *J. Am. Chem. Soc.* **112** (1990) 2167.
26. B.D. Aguda, L.L. Hofmann Frisch and L.F. Olsen, *J. Am. Chem. Soc.* **112** (1990) 6652.
27. J.G. Lazar and J. Ross, *Science* **247** (1990) 189.
28. J.G. Lazar and J. Ross, *J. Chem. Phys.* **92** (1990) 3579.
29. B. Halliwell, *Planta* **140** (1978) 81.
30. P. Sevcîk and H.B. Dunford, *J. Phys. Chem.* **95** (1991) 2411.
31. M. Burger and R.J. Field, *Nature* **307**, 720 (1984).
32. H. Degn, L.F. Olsen and J.W. Perram, *Ann. N.Y. Acad. Sci.* **316** (1979) 623.
33. R. Larter, C.L. Bush, T.R. Lonis and B.D. Aguda, *J. Chem. Phys.* **87** (1987) 5765.
34. R. Larter, C.G. Steinmetz and B.D. Aguda, *J. Chem. Phys.* **89** (1988) 6506.
35. L.F. Olsen, *Phys. Lett.* **94A** (1983) 454.
36. K. Yokota and I. Yamazaki, *Biochem.* **16** (1977) 1913.
37. V. Fed'kina, F. Ataullakhanov and T. Bronnikova, *Biophys. Chem.* **19** (1984) 259.
38. B.D. Aguda and B.L. Clarke, *J. Chem. Phys.* **87** (1987) 3461.

39. S. Alexandre and H. B Dunford, *Biophys. Chem.* **40** (1991) 189.
40. C.G. Steinmetz and R. Larter, *J. Chem. Phys.* **94** (1991) 1388.
41. R. Larter and C.G. Steinmetz, *Phil. Trans. R. Soc. Lond.* **337A** (1991) 291.
42. C.G. Steinmetz, Ph.D. thesis, Indiana University-Purdue University at Indianapolis (1991).
43. C.G. Steinmetz, "*Period-doubling and quasiperiodicity in a model for chemical chaos*," manuscript in preparation (1992).
44. L.F. Olsen, in *Stochastic Phenomena and Chaotic Behavior in Complex Systems*, P. Schuster, Ed. (Springer-Verlag, Berlin, 1984). p. 116.
45. L.F. Olsen, *Z. Naturforsch.* **40A** (1985) 1283.
46. B.D. Aguda, R. Larter and B.L. Clarke, *J. Chem. Phys.* **90** (1989) 4168.
47. B.D. Aguda, Ph.D. thesis, University of Alberta (1986).
48. M. Klein and R. Larter, "*Chaotic solutions for an experimentally reasonable model of the peroxidase-oxidase reaction*," manuscript in preparation (1992).
49. H. Degn and D. Mayer, *Biochim. Biophys. Acta* **180** (1969) 291.
50. L.F. Olsen, *Biochim. Biophys. Acta* **527** (1978) 212.
51. Z. Samec, W.T. Bresnahan and P.J. Elving, *J. Electroanal. Chem.* **133** (1982) 1.
52. E.J. Land and A.J. Swallow, *Biochem. Biophys. Acta* **234** (1971) 34.
53. B. Chance, *Arch. Biochem.* **22** (1949) 224.
54. B. Chance, *Arch. Biochem. Biophys.* **41** (1952) 416.
55. P. George, *Biochem. J.* **54** (1953) 267.
56. D. Dolman, G.A. Newell, M.D. Thurlow and H.B. Dunford, *Can. J. Biochem.* **53** (1975) 495.
57. P. George, *J. Biol. Chem.* **201** (1953) 427.
58. M. Tamura and I. Yamazaki, *J. Biochem. (Tokyo)* **71** (1972) 311.
59. K. Yokota and I. Yamazaki, *Biochim. Biophys. Acta* **105** (1965) 301.
60. J.B. Wittenberg, R.W. Noble, B.A. Wittenberg, E. Antonini, M. Brunori and J. Wyman, *J. Biol. Chem.* **242** (1967) 626.
61. I. Yamazaki and K. Yokota, *Mol. Cell. Biochem.* **2** (1973) 39.
62. T. Odajima and I. Yamazaki, *Biochim. Biophys. Acta* **284** (1972) 355.
63. D. Behar, G. Czapskim, J. Rabani, L.M. Dorfman and H.A. Schwarz, *J. Phys. Chem.* **74** (1970) 3209.
64. J. Rabani and S.O. Nielsen, *J. Phys. Chem.* **73** (1969) 3736.

CHAPTER 8: A NEW THEORETICAL APPROACH TO THE DESCRIPTION OF PHYSICO-CHEMICAL REACTION DYNAMICS WITH CHAOTIC BEHAVIOR

V. Gontar

The International Group for Scientific and
Technological Chaos Studies
Ben Gurion University of the Negev
Israel

ABSTRACT

A major feature of systems exhibiting deterministic chaos is their extreme sensitivity to initial conditions and parameter values. There is thus a practical and conceptual problem with simulating chaotic systems by numerical integration of their governing differential equations, which necessarily introduces very small perturbations into the behavior of the system. A new algebraic, iterative approach to the theoretical description of chaotic systems is developed here which does not involve the use of differential equations.

8.1 Introduction

Strange attractor, deterministic chaos, self-organization -- interpretation of these concepts, begun relatively recently, is still underway.[1-7] Lorenz's discovery of the extraordinary behavior of the numerical solution of a system of ordinary differential equations[8] marks the dawn of the era of chaotic dynamics, a field which is the product of classical physics and computer-aided mathematics. Chaotic dynamics has very substantially broadened our conceptions of space and time, determinism and randomness.

By chaotic behavior is meant a substantial, arbitrarily large change in the character of the behavior of a system as a result of an arbitrarily small change in the external parameters or initial conditions. Such a definition makes it problematic to use the apparatus of differential calculus to describe chaotic motion mathematically.

This is because the requirement that the solutions of differential equations must be continuous and robust leads to the condition that only infinitesimal dispersal of trajectories can result from an infinitesimal change in the initial conditions of any parameter. The appearance of computerized calculations on the scene, while significantly extending the possible range of numerical integration, cast doubt on the robustness of iterative methods and raised the question of the finite accuracy of computer calculations. Whether Lorenz's attractor is a product of the finite accuracy of calculations on a computer, or whether it represents a new class of solutions of differential equations, is a question which, in our view, calls for analysis. We are faced here with a special situation -- the traditional chain linking a system of differential equations with an analytic solution cannot be automatically applied to cases where there is no analytic solution. The finite accuracy of calculations, which forms the definition of chaotic behavior and cannot fail to influence the solution, has severed this link and made of the computer an essential element of the chain: mathematical model -- computer (numerical method, accuracy) -- solution and analysis of solution. This problem leads us to the paradoxes of the ancient Greeks: is it mathematically and logically possible to formulate a contradiction-free description of the process of approach to an object if the distance to this object contains an infinite number of segments? Newton's answer, the calculus of infinitesimals forming the basis of the theory of differential equations has been to date the only mathematical apparatus for describing dynamics. It seemingly will have to be modified before it can be used to describe the nonlinear dynamics of systems with chaotic behavior. The problem of the existence of a lim $\Delta x/\Delta t$ for $\Delta t \rightarrow 0$ is particularly crucial in the study of chaos. Is the language of differential calculus which appears in the mathematical formulation of all dynamical laws truly so universal? Or is it possible to construct a mathematical tool or even a new calculus for describing dynamics without using derivatives and free of the contradiction between the continuous time and space of differential equations and the discrete process of calculation? Equally urgently, how is one to define the role and place of chaotic behavior in the deterministic world of Newton and the probabilistic-statistical world of Boltzmann?

Below we present a new theory and concordant mathematical model for describing the dynamics of complex chemical transformations. As it is algebraic and iterative, it is free of the contradictions mentioned above. It describes all known types of dynamic behavior in chemical reactions, including chaotic behavior.

8.2 The π Theorem of the Theory of Similarity and Dimensionality and the Stoichiometry of Chemical Reactions

From the point of view of the theory of dimensionality, all physical quantities fall into two categories, primary and secondary. Primary quantities are those that are determined by direct measurement, yielding a numerical value: $u = V/c_o$, where V is the primary quantity and c_o is the standard unit of measure. Take the average height of a man, $V = 175$ cm; the numerical value is obviously $u = 175$ ($c_o = 1$ cm) and depends on the choice of unit. On the other hand, the ratio of similar quantities, *i.e.*, those having the same dimensionality, is independent of the unit. Thus the ratio of the numerical value of a man's height to the length of his arm does not depend on the choice of unit of measurement. This concept of direct measurement is called the "absoluteness of relation."

Secondary physical quantities are expressed in terms of primary quantities. To properly construct a secondary quantity, one has to determine what primary quantities it depends on and what the form of the dependence is. The equations with whose help secondary quantities are expressed in terms of primary ones are called determination equations.

It is natural to demand that the concept of absolute "absoluteness of relation" ratio also hold for secondary quantities. Hence the determination equations should have the wholly concrete form:

$$X = f(u_1, u_2, \ldots u_M) \qquad (8.1)$$

where u_1, $u_2, \ldots, u_M$ are primary quantities, X is a secondary quantity and f is a homogeneous function. The property of homogeneity is defined by the condition that any transformation of the variables will entrain a similar transformation of the function. In other words, if we change the system of units (multiply u_j by some arbitrary coefficients $1/t_i$), then X will also change, as a result of the homogeneity of f:

$$f(u_1/t_1, u_2/t_2, \ldots, u_M/t_M) = f(t_1, t_2, \ldots, t_M)\, f(u_1, u_2, \ldots, u_M) \qquad (8.2)$$

The condition of homogeneity of function f leads to the function $f(u)$:

$$f(u) = Cu_1^{a_1},\ u_2^{a_2}, \ldots, u_M^{a_M} \qquad (8.3)$$

where C is a constant. Thus to obtain a new physical characteristic (secondary quantity), one need only multiply the primary quantities by each other and raise them to some power.

Having defined the concepts of primary and secondary quantities characterizing the process under investigation, we shall now look for relations between these quantities which would ensure invariance under transformations of primary quantities of the type:

$$u_j' = \frac{u_j}{t_j} \qquad t_j > 0$$

where t_j are arbitrary coefficients.

The solution of this problem is given by the π theorem of the theory of similarity and dimensionality, and in Brandt's formulation[9] it is as follows. Consider the system of secondary quantities:

$$X_i = u_1^{a_{i1}} \ u_2^{a_{i2}} \dots u_M^{a_{iM}} \qquad\qquad i = 1, 2,\dots,N,$$

where $\| a_{ij} \|$ is a deterministic matrix of dimension N x M and of rank r. Let the function $F(X)$ be homogeneous with respect to the quantities u_j and X_i. Then the equation

$$F(X_1, X_2,\dots, X_N) = 0 \tag{8.4}$$

is equivalent to the following

$$F(1, 1,\dots,1, \pi_1, \pi_2,\dots, \pi_L) = 0, \tag{8.5}$$

where the first r arguments are units and $\pi_1, \pi_2,\dots, \pi_L$ (L = N − M) are L independent products formed by $X_1, X_2,\dots, X_N$. The theorem has the following consequence: let the quantities $X_1, X_2,\dots, X_N$ have a determinant matrix $\| a_{ij} \|$ of rank M:

$$\| a_{ij} \| = \| \frac{R}{Q} \| \tag{8.6}$$

where R is a nondegenerate matrix of dimension M × M, and Q is a matrix of dimension (N − M) × M. If F is a homogeneous function, then Eq. 8.4 is equivalent to Eq. 8.5, and

$$\pi_l = X_1^{\nu_{l1}} X_2^{\nu_{l2}} \ldots X_N^{\nu_{lN}} \quad l = 1, 2, \ldots, N\text{-}M \tag{8.7}$$

N − M are independent quantities with the deterministic matrix $\| \nu_{li} \|$:

$$\nu_{li} = \| -R^{-1} \times Q \vdots I \|, \tag{8.8}$$

where I is the unit matrix (N-M) × (N-M). From Eq. 8.6 and Eq. 8.8,

$$\| \nu_{li} \| \times \| a_{ij} \| = 0 \tag{8.9}$$

which expresses the condition of independence of π_l from u_j.

Thus the π theorem enables us to describe systems using the quantities π_l, which are invariant under transformations of the primary quantities u_j, while Eq. 8.8 makes it possible to construct invariant complexes of π_l from the quantities X_i on the given deterministic matrix $\| a_{ij} \|$.

The invariance of the complexes of π_l reflects the following fact about modeling the behavior of systems: since the choice of primary quantities is arbitrary, for any given set of the quantities, u_j, the system of secondary quantities, X_i, will depend on the values of the quantities u_j; this is not a reflection of the essence of the phenomena taking place in the system but rather a consequence of the definite dependence of the primary and secondary quantities expressed by Eq. 8.3. To construct quantities that do not depend on the choice of the quantities u_j, one must construct quantities characterizing the internal properties of the system under investigation. The quantities π_l, being criteria of similarity, will be the same for similar systems.

Let us illustrate this with a simple example. In describing the kinematics of motion, let us specify a set of four quantities: the distance S, time T, velocity V and acceleration W. The determinant matrix $\| a_{ij} \|$ of the dimensionalities of these quantities is:

	s[m]	t[c]	
S	1	0	N = 4
T	0	1	M = 2
V	1	−1	L = 2
W	1	−2	

From Eq. 8.8

$$R = \begin{Vmatrix} 1 & 0 \\ 0 & 1 \end{Vmatrix}, \quad Q = \begin{Vmatrix} 1 & -1 \\ 1 & -2 \end{Vmatrix}, \tag{8.10}$$

we obtain the determinant matrix of the invariants π_l:

$$\|v_{li}\| = \begin{Vmatrix} -1 & 1 & 1 & 0 \\ -1 & 2 & 0 & 1 \end{Vmatrix}$$

and

$$\pi_1 = S^{-1} T^1 V^1 W^0 = \frac{V T}{S}$$

$$\pi_2 = S^{-1} T^2 V^0 W^1 = \frac{W T^2}{S} \tag{8.11}$$

The meaning of these expressions is obvious and is a consequence of the very general postulates of the theory.

Now that we have formulated the principal results of the theory of similarity and dimensionality, let us attempt to extend them to analysis of chemical transformations. Take M substances in the known initial amounts $u_1^0, u_2^0, \ldots, u_M^0$. In the terminology of the theory of dimensionality, the concentrations corresponding to these chemical substances or elements will be "primary" quantities. Having mixed these substances and assuming their chemical transformation -- accompanied by the formation of a series of new substances -- we must now introduce the concept of "secondary" quantities formed as a result of the reaction with the concentrations

X_i. The determination matrix for the substances formed as a result of the reaction will look like this:

$$\begin{array}{ccccc} & u_1 & u_2 & \cdots & u_M \\ X_1 & a_{11} & a_{12} & \cdots & a_{1M} \\ X_2 & a_{21} & a_{22} & \cdots & a_{2M} \\ \vdots & \vdots & \vdots & & \vdots \\ X_N & a_{N1} & a_{N2} & \cdots & a_{NM} \end{array} \tag{8.12}$$

a_{ij} being a real number indicating the number of elements of type j in a constituent of type i. After specifying the set of u_j and correspondingly the set of X_i with the help of the matrix $\| a_{ij} \|$, we specify an infinite multitude of similar chemical systems differing only in the set of initial concentrations u_j^0 of the original substances. The values of X_i of these systems will naturally be different as well. When applied to the problem of description of chemical systems, the π theorem makes it possible to create similar criteria of similarity, π_l, for all chemical systems differing in initial concentrations, u_j^0. By analogy with the π theorem, the number of independent quantities, π_l, needed to describe the properties of such a chemical system is given by L = N − M. To illustrate this, let us look at the chemical system resulting from the interaction of the primary elements iodine and hydrogen. Assuming that I, H, I_2, H_2, HI are present in the system, let us construct the determinant matrix $\| a_{ij} \|$ (henceforth referred to as the molecular matrix)

$$\begin{array}{cccc} & & H & I \\ X_1 & H & 1 & 0 \\ X_2 & I & 0 & 1 \\ X_3 & HI & 1 & 1 \\ X_4 & H_2 & 2 & 0 \\ X_5 & I_2 & 0 & 2 \end{array} \qquad \begin{array}{l} N = 5 \\ M = 2 \\ L = 3 \end{array} \tag{8.13}$$

From Eq. 8.8, the determinant matrix for the quantities π_1, π_2, π_3 is

$$\|\nu_{li}\| = \left\| \begin{array}{ccccc} -1 & -1 & 1 & 0 & 0 \\ -2 & 0 & 0 & 1 & 0 \\ 0 & -2 & 0 & 0 & 1 \end{array} \right\| \tag{8.14}$$

and

$$\pi_1 = X_1^{-1} \cdot X_2^{-1} \cdot X_3 = \frac{X_3}{X_1 X_2}$$

$$\pi_2 = X_1^{-2} \cdot X_4 = \frac{X_4}{X_1^2}$$

$$\pi_3 = X_2^{-2} \cdot X_5 = \frac{X_5}{X_2^2} \qquad (8.15)$$

The requirement that the quantities π_l be invariant (be independent of the initial concentrations of the quantities u_j) is fulfilled, since the matrix product Eq. 8.9 is equal to zero.

Now it is an easy matter to recognize in the quantities π_l the equilibrium constants of the corresponding chemical reactions:

$$H + I \rightleftarrows HI$$

$$2I \rightleftarrows I_2$$

$$2H \rightleftarrows H_2$$

The matrix $\| v_{li} \|$ coincides with the stoichiometric matrix, while Eq. 8.9 defines the mass balance under chemical transformations. Indeed, any system of physico-chemical reactions can be described by means of the matrix $\| v_{li} \|$ in the form

$$\sum_{i=1}^{N} \nu_{li} A_i = 0 \qquad (8.16)$$

where A_i is a symbol denoting the i-th constituent of the physico-chemical system: this could be atoms, molecules, atomic groupings, radicals, elementary particles, quasiparticles or any objects whose interactions could be expressed, by analogy with ordinary chemical reactions, by some scheme such as $A_1 + A_2 \rightleftarrows A_3$; $A_3 \rightleftarrows A_4$, and so forth. As is apparent from this definition, the form of expression of the transformations of A_i (Eq. 8.16) is universal. Another important element of the formal description of chemical reactions is the laws of conservation. If X_i is a certain quantitative measure of A_i (number of particles, concentration, charge, *etc.*), and b_j is the total number of particles of the j-th constituent of the system, the laws of conservation may then be written with the help of the molecular matrix $\| a_{ij} \|$ as:

$$\sum_{i=1}^{N} a_{ij} X_i = b_j , \tag{8.17}$$

The constituents of the system are defined as the M components if the laws of conservation can be written down.

Equations 8.7 and 8.15 are identical with the mass-action thermodynamic law for the system of chemical reactions. This law, as is well known, was obtained for equilibrium chemical reactions in ideal gas mixtures within the bounds of the laws of thermodynamics and only holds for ideal gases.

In deriving the relations within the bounds of the theory of similitude and dimensionality, we used notions such as "absoluteness of relation ratio" of quantities and the π theorem of the theory of similarity and dimensionality. The fact that Eq. 8.7 is identical in form with the mass-action law, which is obtained in chemical thermodynamics from the minimum free energy condition, supports the analogy of this law with the relations obtained from the theory of similarity and dimensionality. Unlike dimensionless invariants, Eq. 8.12, chemical invariants possess dimensionality, hence in this case we are speaking of a purely formal analogy. However, if we assume that invariance -- in the sense of independence from the concentrations of the components in the chemical systems -- is just as fundamental a property of physico-chemical systems as is nondimensionality, we arrive at the following ideas:

1. The form of Eq. 8.7 must be conserved whether one is describing reactions in the ideal gas phase or in any condensed phase (liquid, solid, plasma). It is necessary and sufficient to write down the entire set of physicochemical reactions (constituents) in order to adequately describe equilibrium states in these phases. These descriptions must include defects, clusters, structural characteristics, *etc.* in the form of a system of reactions (Eq. 8.16). Instances of the use of this approach are known and have been successful.[11,12]

2. The form of Eq. 8.7 must be conserved not merely in the equilibrium state but also at any arbitrary instant. It would in fact be strange if the requirement we used regarding independence from the initial concentrations had to be obeyed only in the equilibrium state at some definite instant $t \to \infty$. This condition

must be met independently of time; that is, according to the logic of the theory of similarity and dimensionality, it is necessary that our "invariants" be independent of time, just as they are independent of the initial concentrations of the components. For this condition to be realized, it is necessary to introduce time -- or a function of time -- among the "elements" and obtain a system of equations of the type of Eq. 8.7 in which time appears -- that is, use Eq. 8.7 to describe the kinetics of chemical transformations.

8.3 A New Model for the Kinetics of Chemical Transformations

Equations 8.7 and 8.17 form a system of N nonlinear algebraic equations having a unique solution for $X_i > 0$ and defining a complex chemical equilibrium. This system has been thoroughly studied and admits of reduction of dimensionality. Equations 8.7 and 8.17 reduce to the system of M equations:[13]

$$\sum_{i=1}^{N} a_{ij} \exp [\sum_{k=1}^{M} a_{ik} Z_k + f_i] = b_j$$

$$X_i = \exp [\sum_{k=1}^{M} a_{ik} Z_k + f_i]$$

$$\ln \pi_l = - \sum_{i=1}^{N} v_{li} f_i \qquad (8.18)$$

Since usually M << N, the computational advantages of 8.18 are obvious.

As concluded in the preceding section, when formulating π_l when X_i depends on time, one must include an additional element -- the time function $\theta(t)$ among the components I and H of the chemical system. Then the matrix $\| a_{ij} \|$ becomes

		H	I	$\theta(t)$
X_1	H	1	0	0
X_2	I	0	1	0
X_3	$\theta(t)$	0	0	1
X_4	HI	1	1	γ_1
X_5	H_2	2	0	γ_2
X_6	I_2	0	2	γ_3

(8.19)

From Eq. 8 we obtain the matrix $\| v_{li} \|$, and therefore

$$\| v_{li} \| = \begin{Vmatrix} -1 & -1 & -\gamma_1 & 1 & 0 & 0 \\ -2 & 0 & -\gamma_2 & 0 & 1 & 0 \\ 0 & -2 & -\gamma_3 & 0 & 0 & 1 \end{Vmatrix}$$

$$\pi_1 = \frac{X_4}{X_1 X_2 \theta^{\gamma_1}}$$

$$\pi_2 = \frac{X_5}{X_1^2 \theta^{\gamma_2}}$$

$$\pi_3 = \frac{X_6}{X_2^2 \theta^{\gamma_3}} \tag{8.20}$$

Like $\theta(t)$, the concentrations X_i are functions of time t. Several properties of the function $\theta(t)$ for closed chemical systems become apparent when Eq. 8.20 is rewritten in the form:

$$\frac{X_4(t)}{X_1(t)\, X_2(t)} = \pi_1 \, \theta^{\gamma_1}(t)$$

$$\frac{X_5(t)}{X_1^2(t)\,X_2(t)} = \pi_2\theta^{\gamma_2}(t)$$

$$\frac{X_6(t)}{X_2^2(t)} = \pi_3\theta^{\gamma_3}(t) \tag{8.21}$$

Assuming that the initial values of the concentrations are $X_1(0) = b_1$, $X_2(0) = b_2$, $X_3(0) = O$, $X_4(0) = 0$ and $X_5(0) = O$, the function $\theta(t)$ should satisfy the conditions $\theta^{\gamma}(0) = 0$ and $\theta^{\gamma}(\infty) = 1$. Based on these restrictions, the following function was proposed for $\theta(t)$:[14]

$$\theta^{\gamma}(t) = \exp\left[-\frac{w_l}{\gamma^l t}\right] = \exp\left[-\frac{\alpha_l}{t}\right], \tag{8.22}$$

the

$$\frac{w_l}{\gamma_l} = \alpha_l$$

being empirical parameters characterizing the rate of the l-th reaction.

Thus, assuming that the parameters π_l, α_l and b_j are given, to obtain the kinetic curves, $x_i(t)$, of the system of a complex reaction specified by the stoichiometric matrix $\| v_{l\,i} \|$, it is necessary to solve the following system of equations (or the equivalent system, Eqs. 8.18):

$$\prod_{i=1}^{N} X_i^{v_{il}}(t_s) = \pi_l \exp\left(-\frac{\alpha_l}{t_s}\right)$$

$$\sum_{i=1}^{N} a_{ij}\,X_i(t_s) = b_j \tag{8.23}$$

for all values of t_s, s = 0, 1, 2... for the specified initial conditions

$$X_i(0) = \left[\begin{array}{ll} b_j & i = 1, 2, ..., M \\ 0 & i = M+1, ..., N \end{array} \right] \tag{8.24}$$

The procedure for determining the parameters π_l and α_l from a statistical analysis of kinetic experiments is known.[15]

It is interesting to compare the solutions obtained for diverse chemical reaction mechanisms (υ_{li}) with the corresponding solutions obtained by integrating the system of differential equations of the mass-action kinetic law. These calculations have been carried out and indicate good qualitative correspondence between the solutions for all types of reaction in closed systems.[16]

Additional independent support for this theory is provided by the simple time dependence proposed by M. Garfinkle in 1982 for the chemical affinity, A. After analyzing experimental data on the kinetics of over 100 reactions in closed systems, he demonstrated that all can be described satisfactorily by introducing the form below for the time dependence of A:[17,18]

$$A(t) = A_r \ln \left[\frac{t}{t_k} \exp\left(1 - \frac{t}{t_k}\right)\right], \tag{8.25}$$

A_r and t_k being empirical parameters. It is easy to show that the function $\theta(t)$ is related to the chemical affinity, A, by the simple expression, $ln\ \theta(t) = A(t)$. The results of analysis of real experimental kinetic data, as well as the ideas we have formulated regarding "chemical" invariance, support the validity of the ideas put forward here for closed chemical systems.

8.4 The Dynamics of Chemical Reactions

We have constructed in the preceding sections a new mathematical model for describing the kinetics of chemical reactions and used it successfully to describe kinetics in closed systems.

However, problems arise when we seek to use the new kinetic model (NKM) to describe oscillating chemical reactions. As is well known, the mass-action kinetic law (KMAL) equations leading to oscillations are obtained when the reaction mechanism is made to include reciprocal relations (auto-catalysis, or mutual influence of the concentrations of the component parts of the system on the acceleration of deceleration of a given reaction). Take for example:

$$A + B \rightarrow C \tag{8.26}$$

This means that [C] accelerates the process of its own formation.

How should the equations of the new kinetic model be changed to allow description of processes of the type illustrated by Eq. 8.26 ?

The answer lies in the method of construction of chemical invariants π_l. Every time we turn our attention to a new factor (new chemical element, time, *etc.*) we must change the deterministic matrix $\| a_{ij} \|$ so as to assure that the invariants, π, subsequently constructed in accordance with Eq. 8.8, will not depend on this factor, which appears here as a new variable. As we did for t, for which we introduced the function $\theta(t_s)$ in $\| a_{ij} \|$, we should introduce a new function of the concentrations, $\Psi(X_i)$, which exerts a definite influence on the course of the given reaction ([C] in Eq. 8.26), as functions of the times t_{s-p} (henceforth written as (s − p)):

		A	B	$\theta(t_s)$	$\Psi(C(t_{s-p}))$
X_1	A	1	0	0	0
X_2	B	0	1	0	0
	θ	0	0	1	0
	Ψ	0	0	0	1
X_3	C	1	1	γ	β

(8.27)

Taking into account Eqs. 8.8 and 8.22, the determinant equation for the invariant π of Eq. 8.26 is:

$$\frac{X_3(t_s, (s-p))}{X_1(t_s, (s-p))\, X_2(t_s, (s-p))\psi^{\beta}(X_1(s-p))} = \pi \exp\left[-\frac{\alpha}{t_s}\right]$$

$$X_1(t_s, (s-p)) + X_3(t_s, (s-p)) = b_1 \qquad p = s-1,\ s-2$$

$$X_2(t_s, (s-p)) + X_3(t_s, (s-p)) = b_2 \tag{8.28}$$

Determining the explicit form of the Ψ function is not the present task facing the theory. Here we will limit ourselves to the simple empirical dependence which, for any arbitrary mechanism of reaction, can be written in the form:

$$\psi = \exp\left[\sum_{i=1}^{N} \frac{\delta_{il} X_i(s-p)}{t_s}\right] \quad \begin{matrix} s = 1, 2, \ldots, \\ p = s-1, s-2 \end{matrix} \tag{8.29}$$

δ_{il} being empirical parameters characterizing the degree of influence of X_i on the l-th reaction.

Generalizing Eq. 8.28 to the case of an arbitrary stoichiometry, and taking into account the equations of conservation of mass, we obtain:

$$\prod_{i=1}^{N} X_i^{\nu_{il}}(t_s, (s-p)) = \pi_l \exp\left(-\frac{\alpha_l}{t_s}\right)\psi$$

$$\sum_{i=1}^{N} a_{ij} X_i(t_s, (s-p)) = b_j \tag{8.30}$$

Let us confine ourselves to the special case of the function $X_i(t_{s-p})$, p = s − 1, *i.e.*, to preceding instants of time. Clearly, for $t_s \rightarrow \infty$, Eqs. 8.30 also define the equilibrium condition of a closed chemical system. An open system is characterized by a time of establishment of the stationary state (T). Processes in open systems can be modeled using Eqs. 8.30 by modifying $0 < t_s < T$ = constant. Our introduction of an explicit function of time, θ, and discrete time, $\Psi(s-p)$, has opened up the possibility of using a system of algebraic equations to describe the kinetics of chemical reactions. It is possible to employ Eqs. 8.30 to simulate the complex (and also chaotic) dynamics of chemical reactions by using the special delayed function, Ψ. Since Eqs. 8.30 represent an inexplicit system of multidimensional iterations for a given chemical reaction mechanism (matrix υ_{li}), and by taking into account the reciprocal influence of the component parts of the system (Ψ function), Eqs. 8.30 fully determine $X_i(t_s)$. A particular case of Eqs. 8.30 is the following representation of the form of a system of multi-dimensional interations:

$$X_i^{n-1} = F(X^n) \tag{8.31}$$

The next step in the development of the new dynamic model is to apply it to describe the spatial behavior of chemical reactions.[25] In order to be able to apply Eqs. 8.30 to the construction of the spatial distribution, $X_i(r_g)$, one must add to the right-hand side of Eqs. 8.30 the function $\phi(X_i(r_{g-f}))$ of all the concentrations $X_i(r_{g-f})$, calculated at points in space r_f in the vicinity of r_g. By analogy with the Ψ function, function ϕ must be introduced by means of an additional column in matrix $\| a_{ij} \|$. The entire spatial region in which the chemical reaction takes place is divided into cells with linear dimensions Δr_g ($g = 1,2...$) in order to do this. Assuming that the chemical reaction in a cell with coordinates r is influenced by the concentrations $X_i(r_{g-1})$ from the cells nearest to r, we write by analogy with Eqs. 8.29

$$\phi = \exp\left(\frac{\sum_{i=1}^{N} e_{i1} X_i(r_{g-1})}{t_s} \right) \tag{8.32}$$

where the ϵ_{i1} are empirical parameters characterizing the influence of the concentrations in the neighborhood of $x_i(r_g)$.

8.5 Modeling of a Complex Reaction with Chaotic Behavior within the New Dynamics Model

To illustrate the use of the formalism set forth above, let us consider the chemical reaction dynamics represented by the following mechanism:

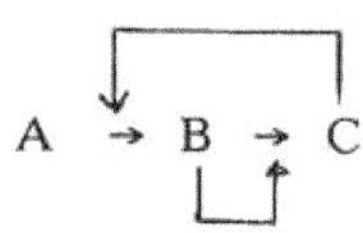

(8.33)

Element A transforms into B, and B transforms into C, while C accelerates the rate of reaction A → B, and B slows its own transformation into C. In accordance with the formalism outlined above (Eqs. 8.29 and 8.30), the corresponding equations describing the dynamics of this complex reaction become

$$\frac{X_B(t_s, s)}{X_C(t_s, s)} = \pi_1 \psi_1 (X_C(s-1))$$

$$\frac{X_C(t_s, s)}{X_B(t_s, s)} = \pi_2 \psi_2 (X_B(s-1))$$

$$X_A(t_s, s) + X_B(t_s, s) + X_C(t_s, s) = b \tag{8.34}$$

$X_{A,B,C}$ being a collective measure of the constituents A, B, C in the system (particle number, concentration). Let us confine ourselves to the influence of X_B and X_C on the process at the instants (s − 1) preceding the time t_s. Figures. 8.1-8.4 show

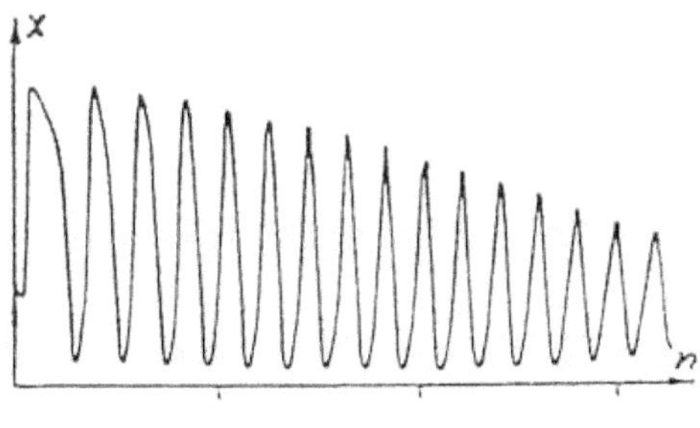

Figure 8.1. Damped oscillations in a closed system.

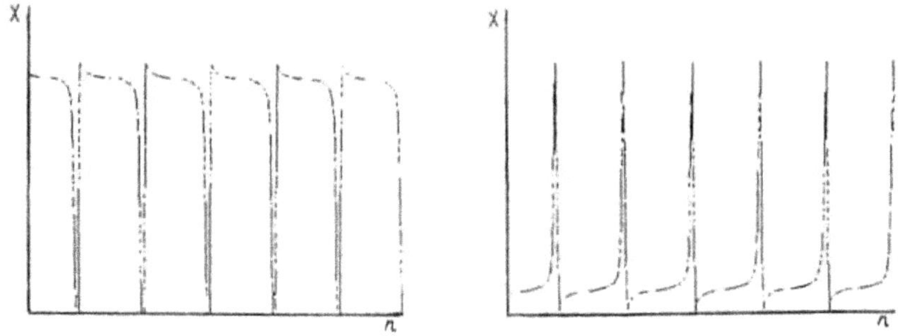

Figure 8.2. Quasiperiodic oscillations in an open system.

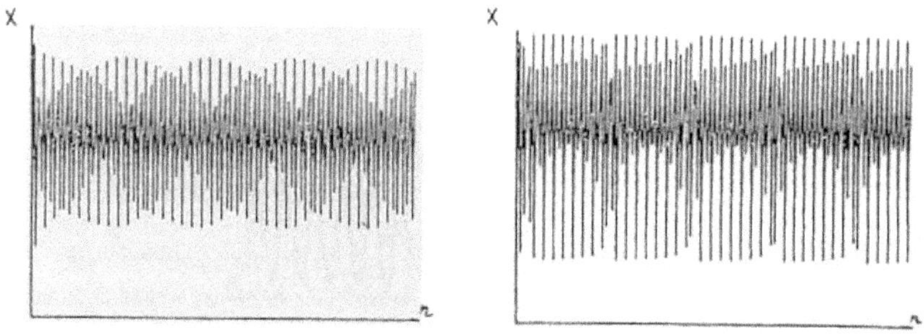

Figure 8.3. Various forms of quasiperiodic oscillation.

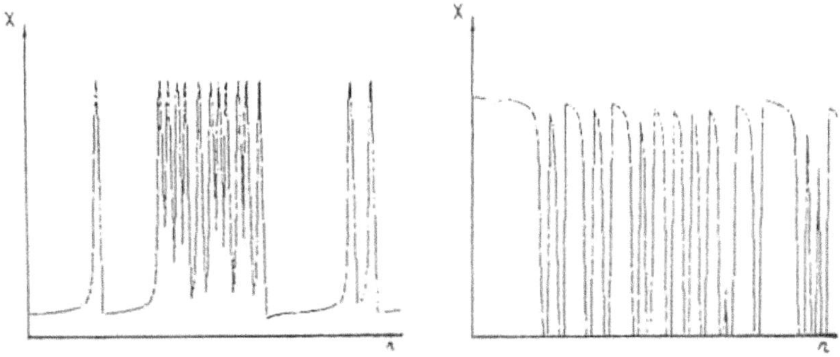

Figure 8.4. Time Series - Belousov-Zhabotinsky (BZ) reaction modeling.

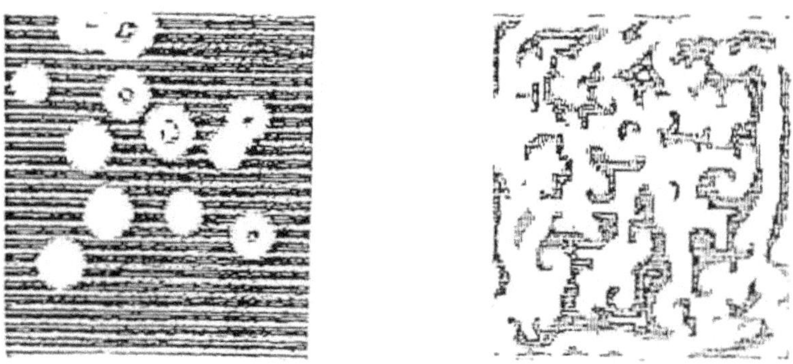

Figure 8.5. Trigger waves (left) and spiral waves in the BZ reaction.

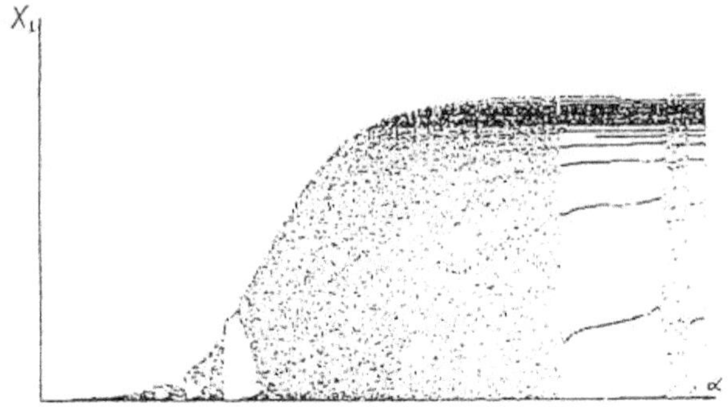

Figure 8.6. Calculated bifurcation diagram for the BZ reaction.

different kinds of oscillatory solutions of Eqs. 8.34. Of particular interest to us is the quasi-continuous solution (Fig. 8.2) which we discovered for $t_s = T$ (open system). As it is discrete (does not have a derivative at any point), this solution imitates oscillatory processes. As is well known, obtaining oscillatory regimes of this kind has hitherto been the prerogative of differential equations and special functions associated with them. In our view, the construction of an oscillatory process using a system of iteration that lacks a continuous variable (time, coordinate) points to the possibility of creating a discrete wave mechanics with applications in various fields of science. The model given by Eqs. 8.34 should make it possible to obtain diverse types of chaotic behavior, including different regimes which could be compared as required with the experimental data.[20-23] We will not address this task here and instead will confine ourselves to a single example: the experimental bifurcation diagram obtained in reference 19 and that calculated by us using Eq. 8.34 (See Fig. 8.6). While not precisely identical, they are similar enough to encourage us in our hopes of applying our approach to the quantitative description of the Belousov-Zhabotinsky reaction.[25]

To obtain the spatial behavior of a chemical reaction according to Eq. 8.32 we must treat the concentrations of the component parts of the system as functions of the spatial coordinates r. The corresponding equations and solutions, in the form of ring and spiral waves characteristic of the BZ reaction, are described in reference 24 and shown here in Fig. 8.5.

8.6 Conclusions

We have formulated a new mathematical model of the dynamics of physico-chemical transformations. However, it is natural to question the necessity of such a model: why is it that the classsical equations of the mass-action kinetic law become problematic when it comes to describing complex chaotic systems? There are several reasons:

1. The fact that computers are actively used to generate chaotic regimes characterized by high sensitivity to numerical procedure and finite computational accuracy makes it impossible to carry out constructive analysis and interpretation of the solutions obtained. The contradiction between the continuous space-time variables of differential equations and the discrete method of calculation with computers makes it necessary to perform special analysis, which does not overcome this contradiction. The new dynamic model may be regarded as a contradiction-free consequence of such analysis.

2. From their inception, modern theoretical physics and chemistry have naturally maintained that systems have small responses to small perturbations. Perturbation theory, the distinction between rapid and slow variables, the various simplification procedures and similar problems -- all traditionally use the concept of a "small" parameter with negligible effect on the system. These conditions cannot be fulfilled when one turns to the description of systems with chaotic behavior. There is no such thing as a "smaller" parameter in such systems. An infinitesimal change in the initial conditions or parameters can lead to the exponential dispersal of trajectories. What earlier appeared to be an instability or peculiarity of solution from the view point of differential calculation, assumes an entirely natural, regular character in the study of chaotic systems. This leads to the necessity of constructing a new calculus free of the above contradictions for systems with chaotic behavior.

Without claiming universality, we see the foregoing presentation as an attempt to construct an alternative language for describing the dynamics of systems of chemical reactions with complex (including chaotic) behavior. To be accepted, our theory will have to change our way of conceiving time. Our model has a minimum of two types of time: t_s and $s-p$ ($p = s-1$, $s-2$). t_s could compare with the astronomic time traditionally used in differential equations. In principle, this time is not very different from continuous time. But we obtain new time scales in our model when we introduce "internal" clocks by using the s−p iteration. The oscillations we obtained in Fig. 8.2 actually have a period which depends only on the mechanism of the interactions and on some parameters, unrelated to traditional time, frequency, *etc*. The system can only be characterized by its mechanism -- its temporal behavior must be calculated. We are led to the same conclusion when we look at the spatial behavior: the geometry of the waves depends only on the mechanism involved. We obtained *ring* waves even when we started from a square. This means that the spatio-temporal behavior is a result of the time of non-linear interactions present, and can be expressed by an algebraic interational model.

Of course we are only beginning to understand this new conception of space and time, but it looks very attractive, especially when we compare our results with the experimental data.

To conclude this article, it is necessary to remark that the results reported here should lead to fundamental changes in our conception of randomness in physics. In reality the chaotic behavior we obtained using iterations is nothing more than the "random numbers generator" well known to all programmers engaged in simulations of stochastic processes. This means that if we succeed in constructing theories like the above (namely, a dynamics that leads to difference rather than to differential equations), it will no longer be necessary to take into consideration probabilistic concepts. Recent successes in the application of modern chaos theory give us grounds to hope that our dream will become a reality in the near future.

Acknowledgement. I would like to express my gratitude to the Hebrew University of Jerusalem, the Lady Davis Trust, Prof. R. Levine and Ms N. Blouinfield, as well as to Ben-Gurion University of the Negev, its President, Dr. A. Braverman, and the Doron Foundation for Education and Welfare, for their support of my activities.

Literature Cited

1. I. Prigogine and R. Lefever, *J. Chem. Phys.* (1968) 1695.
2. O.E. Rössler, *AMS Lectures in Applied Maths*. **17** (1979) 141.
3. R.M May, *Nature*, **261** (1976)459.
4. H.L Swinney, *Physica* **7D** (1983) 3.
5. D. Avnir, *The Fractal Approach to Heterogeneous Chemistry* (Wiley, New York, 1989).
6. G. Nicolis, J.S. Rao and G.S. Rao, *Chaos: A Possible Role in Prebiotic Evolution*, Preprint, Department of Chemical Physics, Free University of Brussels (1984).
7. N. Samardzija, L.D. Greller and E. Wasserman, *J. Chem. Phys.* **90** (1989) 4.
8. E.N. Lorenz, *J. Atmos. Sci.* **20** (1963) 130-41.

9. L. Brandt, *Arch. Rat. Mech. Anal.* **1** (1957) 35.
10. V. Gontar, *Russian J. Phys. Chem.* **56** (1982) 861.
11. A. Men, V. Fetisov, and G. Chufrov, *Dokl, Acad. Nauk SSSR.* **247** (1979)327.
12. L. Nicolaeva, Prikhodko, *Russian J. Phys. Chem.* **58** (1984) 349.
13. V. Gontar and A. Evseev, *Russian J. Phys. Chem.* **50** (1976) 744.
14. V. Gontar, *Vest. Mosk. Universiteta, Chem.* **16**(5) (1975) 520.
15. V. Gontar, *Russian J. Phys. Chem.* **50** (1976) 2154.
16. V. Gontar, *Russian J. Phys. Chem.* **55** (1981) 2301.
17. M. Garfinkle, *Materials Chemistry* **7** (1982) 359-594.
18. M. Garfinkle, *J. Chem. Phys.* **93** (1989) 2158.
19. F.W Schneider and A.F. Münster, *J. Chem. Phys.* **95** (1991) 2130.
20. V. Gontar, A. Ilin and N. Prichodko, *Russian J. Phys. Chem.* **63** (1989) 1799.
21. V. Gontar, A. Ilin and N. Timashava, *Russian J. Phys. Chem.* **63** (1989) 3221.
22. V. Gontar, A. Ilin, V. Klimova and M. Kolpakova, *Russian J. Phys. Chem.* **64** (1990) 1547.
23. V. Gontar in *International Conference on Unsteady State Processes in Catalysis* (North Holland Press, Amsterdam, 1990).
24. V. Gontar and A. Ilin, *Physica*, **52D** (1991) 528.
25. V. Gontar and A. Ilin, *Contrib. Plasma Phys.* **31** (1991) 6.

CHAPTER 9: FROM PERIODIC BEHAVIOR TO CHAOS IN BIOCHEMICAL SYSTEMS

ALBERT GOLDBETER

Faculté des Sciences, Université Libre de Bruxelles,
Campus Plaine, C.P. 231, B-1050 Brussels, Belgium

ABSTRACT

This chapter begins with an overview of some of the most important periodic phenomena in biochemical systems. Addressed in turn are glycolytic oscillations in yeast and muscle, intracellular Ca^{2+} oscillations triggered in a variety of cells by hormones or neurotransmitters, and the mitotic oscillator which controls the onset of cell division in eukaryotic cells. How simple periodic behavior gives way to more complex oscillatory phenomena is then examined in a series of regulated enzymatic models. Particular attention is devoted to the conditions in which two or three stable limit cycles coexist (bi- and tri-rhythmicity). The analysis of a multiply regulated biochemical model shows how complex periodic oscillations (bursting) and aperiodic oscillations (chaos) occur as a result of the interplay between two instability-generating mechanisms. The model shows that despite the complexity of its regulatory structure, simple periodic oscillations remain the most common type of dynamic behavior in this biochemical system. The periodic properties of the cyclic-AMP (c-AMP) signaling system which controls aggregation of *Dictyostelium* amoebae after starvation are considered next. The analysis of a model for c-AMP signaling based on receptor desensitization predicts the possible occurrence of autonomous chaos in *Dictyostelium*. The relevance of these results to the behavior of some "aperiodic" mutants is examined. The failure to detect chaos in suspensions of these mutants could result from the suppression of chaos by the strong coupling with periodic oscillations in cellular suspensions.

9.1 Introduction

Although it is also encountered in nonbiological, chemical systems, particularly those which contain an autocatalytic reaction, rhythmic behavior can be viewed as a basic property of living organisms. Oscillations indeed occur at all levels of biological organization,[1] with periods ranging from milliseconds (neurons) to seconds (cardiac cells), minutes (oscillatory enzymes), hours (pulsatile hormone secretion), 24 h (circadian rhythms), weeks (ovarian cycle) and years (circannual rhythms, epidemiological processes and predator-prey interactions in ecology). When the neurophysiologist Alfred Fessard published in 1936 a book entitled *Propriétés rythmiques de la matière vivante*,[2] he had in mind the dynamics of neurons; it is clear, however, that rhythmic properties extend to all eukaryotic cells and cover a huge range of frequencies as shown by the short, non-exhaustive list of biological rhythms drawn above.

The reason why oscillatory phenomena are so common in biological systems stems from their regulatory properties, which were developed and selected in the course of evolution. Positive as well as negative regulatory feedback provide a source of nonlinearity which, in conjunction with cooperative processes, gives rise to instabilities associated with oscillatory behavior. Such a source of nonlinearity due to regulatory feedback is lacking in chemical systems, which are not subjected to evolutionary pressure. However, this does not rule out the existence of nonlinear, activating or inhibitory processes in chemical kinetics, but these processes are less common and remain "gratuitous" as they do not have any physiological role.

Besides periodic behavior, other, more complex oscillatory phenomena have been identified and increasingly studied in recent years. Among these phenomena are complex periodic oscillations (bursting) and aperiodic oscillations (chaos). From a mechanistic point of view, two major routes leading to bursting and chaos have been identified. These two kinds of oscillations arise either from the periodic forcing of an oscillatory system, or from the interaction of at least two instability-generating mechanisms within the same system. In contrast to the former scenario, which has been followed in many experimental and theoretical studies devoted to chaos in biology, particularly in biochemical and neurobiological systems,[3-5] chaos obtained following the second route is autonomous as it occurs in the absence of any periodic forcing.

This chapter will be devoted to an overview of simple and complex oscillatory phenomena in biochemical reactions. After considering in Section 9.2 some of the best-known examples of periodic behavior in biochemical systems, the emphasis will shift in Section 9.3 to the passage from simple to complex oscillations, including chaos. The peroxidase reaction, which was among the first biochemical oscillators to be demonstrated *in vitro*[6] and which later provided the first example of autonomous chaos in an enzyme reaction,[7] will not be considered here, as it is discussed in Chapter 7. Section 9.4 will be devoted to the periodic properties of the cyclic-AMP signaling system that controls intercellular communication in *Dictyostelium* amoebae. The dynamics of c-AMP signaling in these cells also will be discussed in terms of the possible occurrence of autonomous chaos.

9.2 Periodic Behavior in Biochemical Systems

The number of experimentally observed oscillators in biochemical systems did not significantly change during the decade following 1975, when cyclic-AMP (c-AMP) oscillations were observed in the slime mold *Dictyostelium discoideum*.[8] Some ten years before, around 1965, oscillations were demonstrated in glycolysis, first in intact yeast cells and then in yeast (and later muscle) extracts (See, *e.g.*, ref. 9 and refs. 10-13 for reviews). Around the same time oscillations were also found in the peroxidase reaction (See ref. 6 and Chapter 7) and in mitochondria.[11,14] Studies of the latter oscillations were, however, not pursued much beyond their initial characterization. Lately, the investigation of glycolysis, the peroxidase reaction, or c-AMP oscillations in *Dictyostelium* has turned from periodic behavior to chaos.

In 1985,[15] oscillations in intracellular Ca^{2+} were added to the list of periodic phenomena observed at the cellular level (See refs. 16-18 as well as a special issue of *Cell Calcium*[19] for reviews). These widespread oscillations in cytosolic Ca^{2+} differ from those involving voltage-dependent membrane conductances in electrically excitable cells.

The three dates — 1965, 1975, 1985 — seem to point themselves to some inherent periodicity in the experimental progress on biochemical oscillations! However, less than a decade elapsed before the next important example of biochemical oscillatory behavior was characterized. Recent experimental advances have indeed thrown light on the oscillator which controls the onset of mitosis in eukaryotic cells. Evidence which has accumulated in the last four years points to the existence of a continuous biochemical oscillator underlying the cell-division cycle in embryonic cells.[20]

Before addressing the transition of simple to complex oscillatory phenomena in Section 9.3, experimental and theoretical studies of the most important examples of periodic behavior in biochemical systems, besides peroxidase, will be discussed. Considered in turn will be glycolytic oscillations in yeast, cytosolic Ca^{2+} oscillations, and the mitotic oscillator. The case of c-AMP oscillations in *Dictyostelium* cells will be dealt with separately in Section 9.4; besides representing a prototype for pulsatile signaling in intercellular communication, this system indeed provides an example for the transition from periodic behavior to autonomous chaos.

9.2.1 Glycolytic Oscillations

To this date, glycolytic oscillations remain the prototype of periodic behavior originating from the regulation of an enzyme reaction. In yeast, glycolytic oscillations occur with a period of 5 to 10 minutes when a substrate such as glucose or fructose is injected at an appropriate rate.[21] Oscillations also occur in muscle extracts with a somewhat longer period of the order of 20 minutes.[22,23]

It was demonstrated soon after this observation that glycolytic oscillations originate from the reaction catalyzed by phosphofructokinase (PFK), a key regulatory enzyme controlling the glycolytic flux.[9-11] The PFK oscillations stem from the peculiar regulation of this allosteric enzyme: PFK is indeed activated by ADP, one of the reaction products. In muscle, autocatalysis is primarily exerted by the other product of the reaction, fructose-1,6-P_2.[22,23]

The most conspicuous property of glycolytic oscillations is their control by the substrate injection rate. The range of constant input rates producing oscillations is bounded by two critical values.[21] The period of the oscillations diminishes while the amplitude passes through a maximum as the input rate increases from the first up to the second critical value.

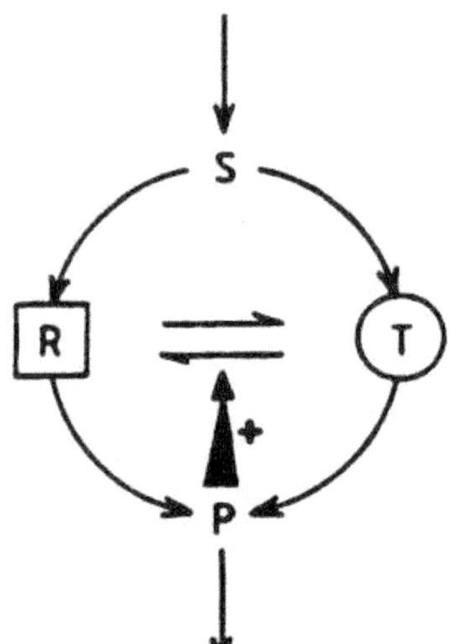

Figure 9.1. Schematic representation of a product-activated, allosteric enzyme reaction. The substrate S, injected at a constant rate, binds to the states R and T of the enzyme and is subsequently transformed into product P. The product is removed at a rate proportional to its concentration and also promotes the transition from the less reactive (T) to the more reactive (R) enzyme state. The enzyme contains multiple subunits which undergo the conformational transition between the R and T states in a concerted manner.

Following the proposal of phenomenological, enzymatic models for glycolytic oscillations based on the positive feedback exerted by the reaction product,[24,25] a model taking explicitly into account the allosteric nature of the product-activated enzyme was analyzed for the PFK reaction.[26-28] The model (See Fig. 9.1) is governed by two differential equations which describe the time evolution of the substrate, α, and product γ normalized concentrations in continuously stirred yeast extracts:

$$\frac{d\alpha}{dt} = \nu - \sigma\phi \tag{9.1a}$$

$$\frac{d\gamma}{dt} = q\sigma\phi - k\gamma \tag{9.1b}$$

with

$$\phi = \frac{\alpha(1+\alpha)(1+\gamma)^2}{L + (1+\alpha)^2(1+\gamma)^2} \tag{9.1c}$$

In these equations, ν and σ are the normalized substrate input and maximum enzyme reaction rate, respectively; q is a dimensionless parameter, and L is the allosteric constant closely related to the degree of cooperativity of allosteric interactions between the enzyme subunits. Equation 9.1c gives the simplest form of rate function for the product-activated allosteric enzyme, assuming that the enzyme consists of two identical subunits obeying the concerted transition model of Monod, Wyman and Changeux.[29]

Linear stability analysis of Eqs. 9.1 indicates that for appropriate values of the other parameters, the unique steady state becomes unstable when the substrate input rate is between two critical values:

$$\nu_{1c} \leq \nu \leq \nu_{2c} \tag{9.2}$$

Such a result can also be reached by means of phase-plane analysis (Fig. 9.2). For oscillations to occur, the nullcline $d\gamma/dt = 0$ must possess a sigmoidal, N shape. The steady state, located at the intersection of the two nullclines, is unstable as soon as it lies on a region of sufficiently negative slope on the γ nullcline, such that

$$(d\alpha/d\gamma) < -1/q \tag{9.3}$$

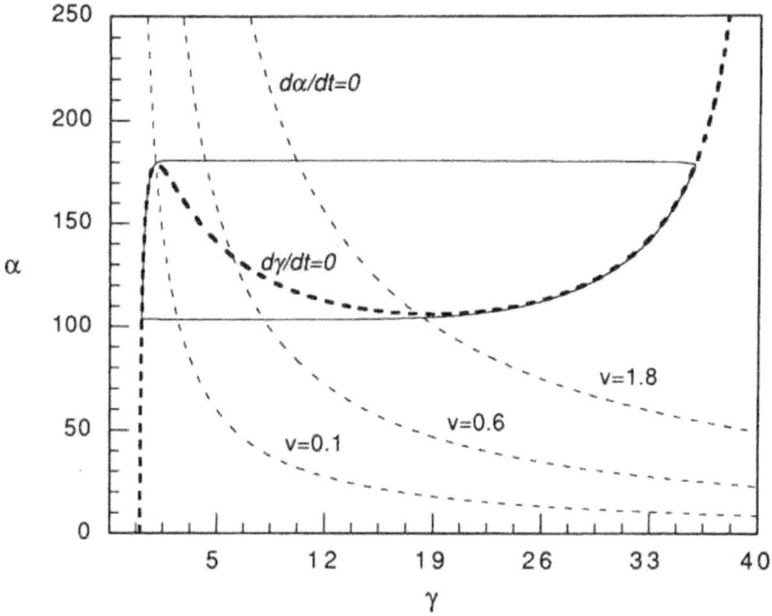

Figure 9.2. Typical phase-plane dynamics observed for the product-activated, allosteric enzyme reaction proposd for glycolytic oscillations. Shown are the sigmoidal product nullcline (heavy dashed line) and three substrate nullclines obtained at three different values of the substrate input ν (in m^{-1}). The steady state is unstable when the two nullclines intersect in the region of negative slope on the γ nullcline (*e.g.*, for $\nu = 0.6\ m^{-1}$); then the system evolves toward a stable limit cycle (solid line). Typical parameter values yielding oscillations with Eqs. 9.1 are $\sigma = 4\ min^{-1}$, $k_s = 0.1\ m^{-1}$, $L = 5 \times 10^6$, $q = 1$; relaxation oscillations similar to those shown here are obtained at larger values of q and k_s.

Increasing the substrate injection rate, ν, from a low initial value, one sees from Fig. 9.2 how the steady state is at first stable, being in a region of positive slope ($d\alpha/d\gamma$) on the product nullcline, and then becomes unstable as soon as condition 9.3 is satisfied. Further increase in ν brings the steady state again into a region of positive slope and the system leaves the oscillatory domain. The latter is thus delimited by two critical values of the substrate input, as expressed by the instability condition 9.2.

How more complex modes of oscillatory behavior may arise in the model for glycolytic oscillations will be considered in Section 9.3 below.

9.2.2. *Intracellular Ca^{2+} Oscillations*

In a large variety of cells, stimulation by an external signal such as a hormone or a neurotransmitter triggers a train of cytosolic Ca^{2+} spikes.[16-19] The period of these oscillations generally ranges from seconds (cardiac cells) to minutes (endothelial cells, fibroblasts, or hepatocytes to cite but a few examples). The frequency of Ca^{2+} oscillations rises with the degree of stimulation. Below a critical magnitude of stimulation, cytosolic Ca^{2+} settles at a low steady-state level; above a second, higher critical value of the external stimulus, a high steady-state level of cytosolic Ca^{2+} is established. In some cells such as cardiac myocytes or pituitary cells, Ca^{2+} oscillations can occur spontaneously in the absence of stimulation.

While the number of experimental studies of Ca^{2+} oscillations has rapidly increased in the last few years, interest also has been extended to the spatial aspects of Ca^{2+} signaling.[30] The propagation of Ca^{2+} waves had long been observed in amphibian eggs after fertilization.[31] In hepatocytes[32] and endothelial cells,[33] a link between Ca^{2+} oscillations and the wave propagation of Ca^{2+} signals has since been established. The same link has been demonstrated in cardiac myocytes where Ca^{2+} waves propagate as a sharp band along the cell, at a rate[34] of the order of 100 ms^{-1}. In contrast, Ca^{2+} waves in hepatocytes and endothelial cells propagate as "tides,"[17] with a progressive increase in Ca^{2+} all over the cell, at a rate[32,33] close to 20 ms^{-1}.

The mechanism of Ca^{2+} oscillations and Ca^{2+} wave propagation involves the synthesis of inositol 1,4,5-trisphosphate (IP_3).[16,30] The level of this intracellular messenger increases after stimulation, owing to the activation of phosphoinositidase C. The role of IP_3 is to mobilize Ca^{2+} from an intracellular store. Models for Ca^{2+} oscillations primarily differ according to whether or not they rely on concomitant oscillations in IP_3.[35]

In the model proposed by Meyer and Stryer,[36] oscillations originate from the elevation of the cytosolic Ca^{2+} level by IP_3 and from the activation of IP_3 synthesis by Ca^{2+}. Such a cross-feedback loop results in a global process of self-amplification; Ca^{2+} oscillations are necessarily accompanied by a periodic variation in IP_3. Extensions of the original version of this model have been proposed.[18]

A second class of models relies on the process of Ca^{2+}-induced Ca^{2+} release (CICR) to account for Ca^{2+} oscillations.[35,37,38] Here (See Fig. 9.3), the rise in IP_3 triggers a constant release of Ca^{2+} from an IP_3-sensitive store into the cytosol.

Cytosolic Ca^{2+} is transported into a second store, insensitive to IP_3, from which it is eventually discharged in a process activated by cytosolic Ca^{2+}; the latter CICR process has been demonstrated only in cardiac and muscle cells but some evidence for its occurrence in other cell types has been obtained.

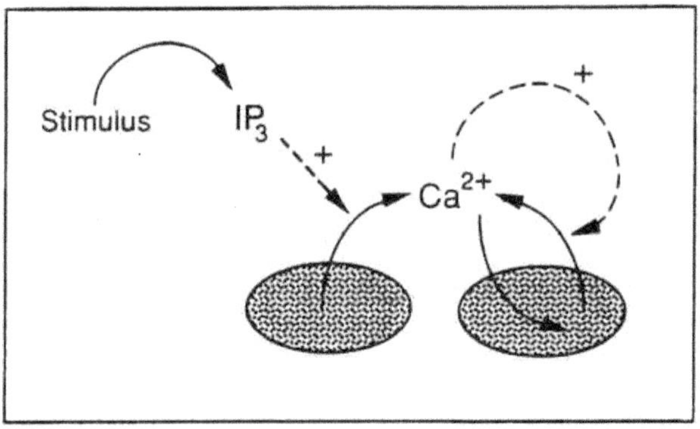

Figure 9.3. Schematic representation of the model for Ca^{2+} oscillations based on Ca^{2+}-induced Ca^{2+} release. An extracellular stimulus elicits the synthesis of inositol 1,4,5-trisphosphate (IP_3) which mobilizes Ca^{2+} from an intracellular store. Cytosolic Ca^{2+} is pumped into an IP_3-insensitive Ca^{2+} store from which it is released in a process activated by cytosolic Ca^{2+}. The latter regulation is capable of producing sustained Ca^{2+} oscillations (Redrawn from ref. 35).

The model based on CICR, in its simplest form, is governed by two differential equations which describe the time evolution of cytosolic Ca^{2+} (Z), and

$$\frac{dZ}{dt} = v_0 + v_1\beta - v_2 + v_3 - kZ$$

$$\frac{dY}{dt} = v_2 - v_3 \tag{9.4}$$

with

$$v_2 = V_2 \frac{Z^n}{(K_2^n + Z^n)}, \quad v_3 = V_3 \frac{Y^m}{(K_R^m + Y^m)} \cdot \frac{Z^p}{(K_A^p + Z^p)}. \tag{9.5}$$

In the above equations, V_2 and V_3 denote the maximum rates of Ca^{2+} pumping into the IP_3 insensitive intracellular store and of Ca^{2+} release into the cytosol. The rates v_2 and v_3 have been written so as to allow for positive cooperativity in pumping and release, as well as in the activation of the latter process by cytosolic calcium; K_2, K_R and K_A denote the threshold constants for these processes, while n, m and p represent the Hill coefficients characterizing their degree of cooperativity ($n, m, p \geq 1$). The threshold constant K_R and concentration Y are both defined with respect to the total intracellular volume.

One of the most salient results of the model based on the CICR mechanism is represented in Fig. 9.4. There, the temporal evolution of cytosolic Ca^{2+} is shown at different values of the parameter β associated with increasing levels of stimulation. In the absence of stimulation, a stable steady state is established, corresponding to a low level of cytosolic Ca^{2+} (panel A). Upon increasing the value of β, the steady state becomes unstable and oscillations appear, with a frequency that rises with the degree of stimulation (panels B and C). Finally, above a critical degree of stimulation, oscillations disappear and the system evolves towards a stable steady state corresponding to a high level of cytosolic Ca^{2+} (panel D). Similar results are obtained upon raising the level of extracellular Ca^{2+}.[37,38] The sequence of dynamic behavior predicted by the model in response to increasing levels of stimulus is observed in many cell types.[16-19]

In some cells it appears that the distinction between the two pools of Ca^{2+}, one sensitive to IP_3 and the other to Ca^{2+}, is not so clear-cut; in these cells, indeed, Ca^{2+} and IP_3 behave as co-agonists for the induction of Ca^{2+} release.[39] The analysis of such a dual regulation of the Ca^{2+} channel indicates that sustained oscillations of cytosolic Ca^{2+} may still occur in a one-pool model in these conditions.[40]

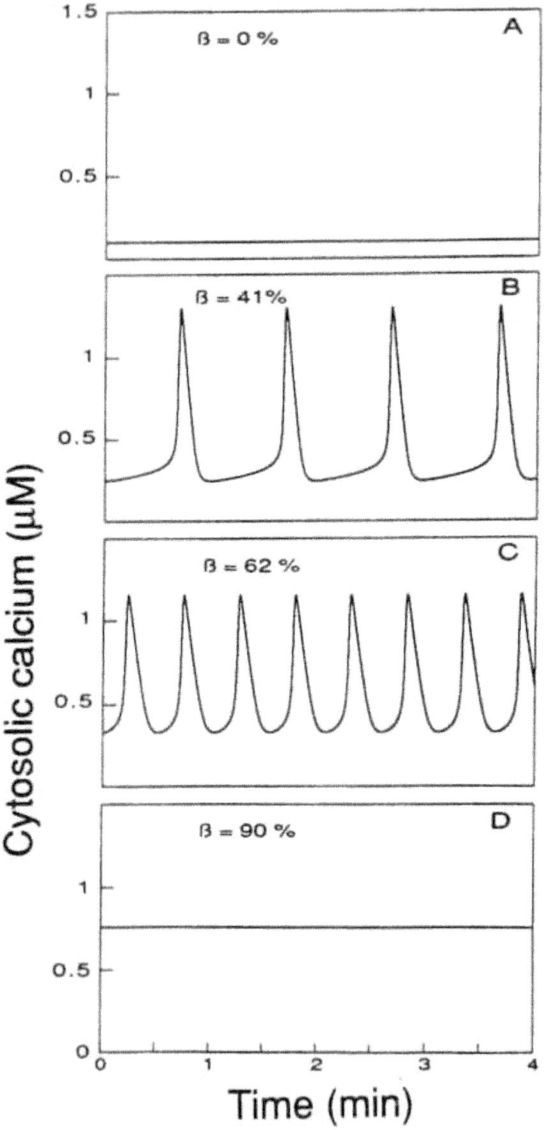

Figure 9.4. Sustained oscillations in cytosolic Ca^{2+} generated by means of Eqs. 9.4-9.5 for increasing values of the parameter β measuring the stimulation level in the model based on Ca^{2+}-induced Ca^{2+} release (redrawn from ref. 35).

The model based on CICR can be applied to Ca^{2+} wave propagation once the diffusion of cytosolic Ca^{2+} is taken into account. The analysis shows[35] that waves similar to those observed in cardiac cells or in hepatocytes and endothelial cells can occur, with the observed rates, depending on whether the period of Ca^{2+} oscillations is of the order of seconds, as in cardiac cells, or minutes, as in the latter types of cells. The model also has been used to generate spatial patterns similar to those observed in oocytes.[41]

Most Ca^{2+} oscillations observed in the experiments are of the simple rather than complex periodic (or chaotic) type. In some cells, however, stimulation by certain agonists triggers a train of Ca^{2+} spikes riding on a slower oscillation (See, *e.g.*, ref. 30); such behavior, reminiscent of bursting, suggests the interplay between at least two regulatory mechanisms. It is likely that the slower oscillation involves the variation of IP_3 while the faster spikes originate from CICR.

9.2.3 *The Mitotic Oscillator*

Besides Ca^{2+} oscillations, the most important oscillatory process discovered in cell biology during the last decade is the biochemical oscillator controlling the onset of mitosis in eukaryotic cells. Advances in the molecular identification of the mitotic control system originate, primarily, from biochemical studies performed in frog, sea urchin and starfish embryos on the one hand,[20] and in yeast on the other hand.[42] While the experimental evidence points to the existence of a continuous biochemical oscillator driving the onset of mitosis in rapidly dividing embryonic cells with a periodicity of the order of 30 min,[20] the situation is more complex in yeast and somatic cells.[42] There, additional feedback processes exist; the mechanism regulating the onset of mitosis is blocked until a sufficient cell size is reached or processes such as DNA replication or mitotic spindle formation are completed. Evidence for a continuous mitotic oscillator has also been obtained in *Drosophila* where the 13 first nuclear divisions occur with a period close to 8 min.

In embryonic cells, the cell-division cycle is driven by a protein named cyclin.[20,43] The progressive accumulation of cyclin leads, through a cascade of biochemical reactions, to the activation of an enzyme called cdc2 kinase (the name originates from genetic studies in yeast where a number of cell division cycle — or cdc — mutations have been characterized). The periodic activation of cdc2 kinase initiates in turn the various events associated with cell division, such as breakdown of the nuclear envelope, chromosome condensation or spindle assembly.

The regulation of cdc2 kinase is achieved by a cascade of phosphorylation-dephosphorylation reactions.[44-46] Several models have recently been proposed for the mitotic oscillator. Most of them rely[47-49] on the nonlinear, autocatalytic activation of cdc2 kinase for which some experimental evidence exists. However, the analysis of a minimal cascade model[50] based on experimental observations[51] shows that autocatalysis is not required for producing oscillations in cdc2 kinase activity. In this model (Fig. 9.5) cyclin is synthesized at a constant rate and activates enzyme E_1 (a protein phosphatase, product of the gene cdc25) which dephosphorylates the inactive form of cdc2 kinase (M^+) and thereby brings it into the active form (M); a protein kinase (E_2) reverts this activation step. The active form of cdc2 kinase in turn

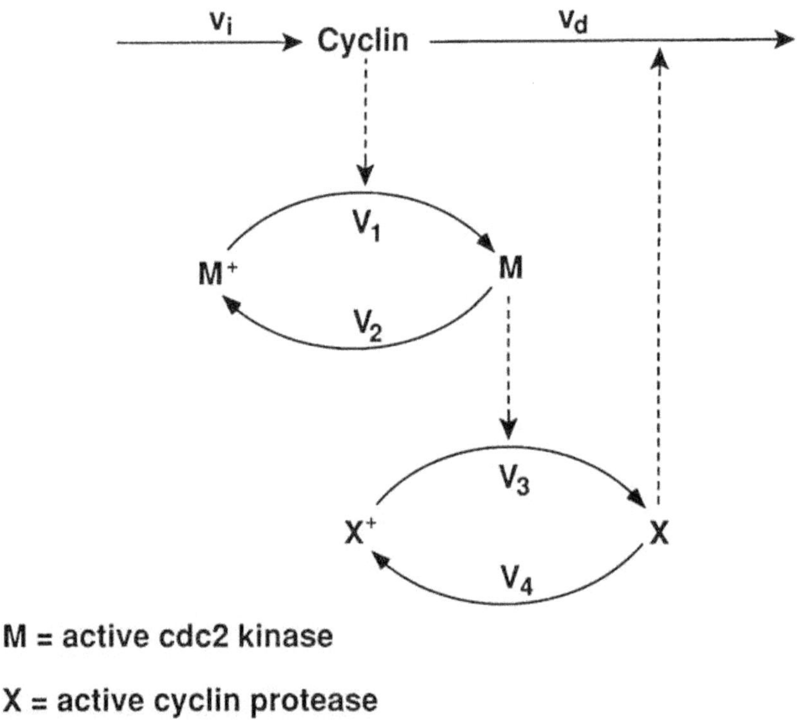

Figure 9.5. Minimal cascade model for the mitotic oscillator. The model incorporates cyclin synthesis, activation of cdc2 kinase by cyclin through a dephosphorylation-phosphorylation cycle, activation of cyclin protease through a second phosphorylation-dephosphorylation cycle involving cdc2 kinase, and destruction of cyclin by the active form of protease X (redrawn from ref. 50).

phosphorylates the inactive form of a cyclin protease (X^+) and thereby brings it into its active state (X); this activation step is reversed by a phosphatase (E_4). The protease X degrades cyclin, and this step triggers the inactivation of cdc2 kinase. It is possible, and even likely, that the activation of cyclin protease by cdc2 kinase involves additional phosphorylation-dephosphorylation cycles;[51] such enlargement of the cascade would, however, not significantly alter its dynamic behavior.

It has been suggested[20,51] that the activation of cdc2 kinase by cyclin and the activation of cyclin degradation by cdc2 kinase constitute a negative feedback loop which may result in sustained oscillations. The analysis of the model of Fig. 9.5 shows that this conjecture holds, provided that thresholds exist in the dependence of cdc2 kinase activation on cyclin, and in the activation of cyclin protease by cdc2 kinase.[50]

The model of Fig. 9.5 is governed by a set of three differential equations describing the time evolution of cyclin concentration (C), the fraction of active cdc2 kinase (M), and the fraction of active cyclin protease (X):

$$\frac{dC}{dt} = v_i - v_d X \frac{C}{K_d + C} - k_d C$$

$$\frac{dM}{dt} = V_1 \frac{(1 - M)}{K_1 + (1 - M)} - V_2 \frac{M}{K_2 + M}$$

$$\frac{dX}{dt} = V_3 \frac{(1 - X)}{K_3 + (1 - X)} - V_4 \frac{X}{K_4 + X} \tag{9.6}$$

with,

$$V_1 = \frac{C}{K_c + C} V_{M1}\,, \quad V_3 = M V_{M3} \tag{9.7}$$

Parameters V_i and K_i denote the maximum rate and normalized Michaelis constants of enzyme E_i (i = 1,..., 4); v_i and v_d denote the input rate and maximum rate of cyclin degradation by protease X, while constant k_d relates to a relatively negligible, nonspecific cyclin degradation. It is noteworthy that the only type of nonlinearity in Eqs. 9.5-9.6 is of the Michaelian type. These equations nevertheless admit periodic solutions over a wide range of parameter values, once thresholds occur in the activation of M by C and of X by M (See refs. 50 and 52 for further details). A typical example of oscillatory behavior is shown in Fig. 9.6.

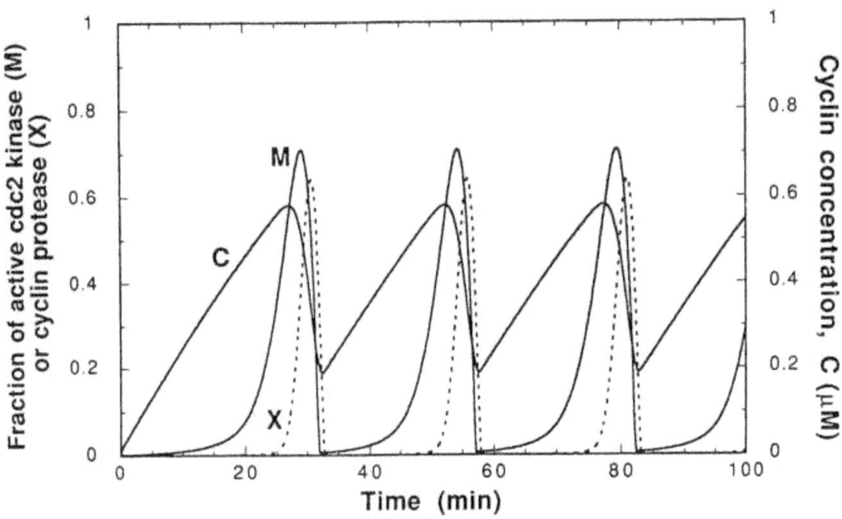

Figure 9.6. Sustained oscillations generated by means of Eqs. 9.6-9.7 in the minimal cascade model for the mitotic oscillator, schematized in Fig. 9.5 (from ref. 50).

It can be seen that the progressive rise in cyclin (C) leads to the abrupt activation of cdc2 kinase (M), the latter brings about a sharp rise in cyclin protease (X) after a short delay, and the latter event triggers the rapid decline in cyclin. As a result M drops precipitously, followed by a rapid decrease in X. When the rate of cyclin synthesis exceeds its rate of degradation, the level of cyclin may rise again and a new cycle of oscillations begins.

Although based on a number of simplifying assumptions, such as disregarding the formation of complexes between cyclin and cdc2 kinase and the effect of a second phosphorylation-dephosphorylation reaction involved in cdc2 kinase regulation, the model described by Eqs. 9.5 nevertheless shows that the negative feedback loop built in the cascade suffices, in principle, to produce sustained oscillations. The model indicates the importance of time delays in generating limit cycle behavior; these delays are not introduced in an *ad hoc* manner into Eqs. 9.5 but result from the existence of thresholds in the activation curves of cdc2 kinase and cyclin protease.[50]

Autocatalysis can be incorporated into the cascade model of Fig. 9.5 by assuming that the phosphatase E_1 is itself activated through phosphorylation by cdc2 kinase. The analysis indicates[52] that such an autocatalytic regulation, although not essential for oscillations, renders them more abrupt and enlarges the domain of instability in parameter space.

As the experimental study of the cell cycle is undergoing rapid developments, further insights into the molecular mechanism of the mitotic oscillator can be expected in the near future. With regard to the main theme of this volume, it should be noted that the current evidence in this biochemical system points to the occurrence of simple periodic behavior rather than more complex oscillatory phenomena.

9.3 Complex Oscillatory Phenomena in a Multiply Regulated Biochemical System

To analyze the transition from simple to complex oscillatory phenomena, two complementary ways exist *a priori*. The first is to try to account for complex oscillations observed in the experiments. The second is to rely on the analysis of more or less abstract models whose primary goal is not so much to account for experimental results but rather to explore the realm of possible modes of complex behavior. To this approach belong a series of simple biochemical models which were developed to investigate the occurrence of birhythmicity (coexistence of two stable limit cycles), bursting and chaos in biochemical systems.[28] The interest of these studies is of course not limited to the biochemical field because the phenomena predicted by these models can occur in other, nonbiological, chemical systems. Thus birhythmicity was demonstrated in a chemical system[53] following its theoretical prediction in one of these biochemical models.[54]

9.3.1 Birhythmicity

The vast majority of oscillation in chemical and biological systems corresponds to the evolution toward a stable limit cycle. Such a limit cycle is generally unique for a given set of parameter values or for given experimental conditions. To investigate the conditions in which multiple, stable limit cycles may coexist, the model for the product-activated allosteric enzyme analyzed for glycolytic oscillations (Fig. 9.7a) has been extended (Fig. 9.7b). The extension considered[55] was dictated by a conjecture based on phase-plane analysis: so as to transform the sigmoidal product nullcline which admits a single region of negative slope ($d\alpha/d\gamma$) into one admitting two regions

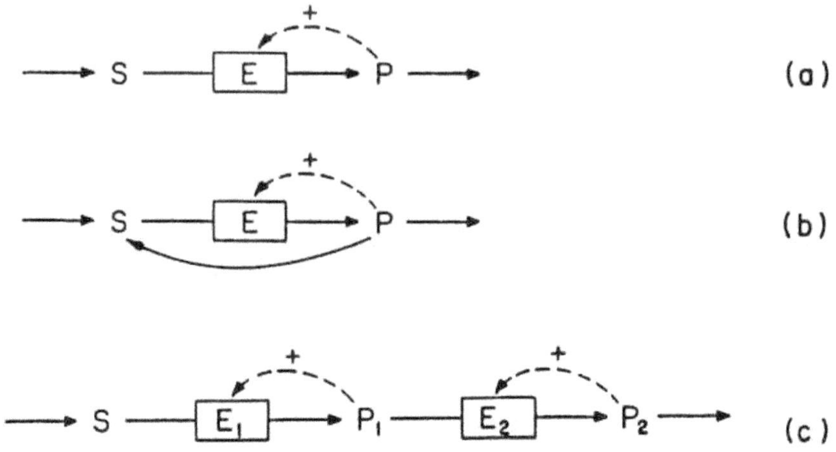

Figure 9.7. Regulated enzymatic models showing increasingly complex modes of oscillatory behavior. (a) product-activated, allosteric enzyme model for glycolytic oscillations, admitting simple periodic behavior; (b) same model with an added reaction of product recycling into substrate, admitting simple periodic behavior and also birhythmicity; (c) coupling in series of two product-activated, allosteric enzyme reactions. In addition to simple periodic behavior and birhythmicity, the latter model admits, a.o., complex periodic oscillations (bursting) and chaos.

of negative slope, a step of nonlinear recycling of product into substrate was added.[28,55] The resulting set of equations, modified from Eqs. 9.1, takes the form:

$$\frac{d\alpha}{dt} = \nu + \frac{\sigma_i \gamma^n}{K^n + \gamma^n} - \sigma \phi$$

$$\frac{d\gamma}{dt} = q \sigma \phi - k_s \gamma - \frac{q \sigma_i \gamma^n}{K^n + \gamma^n} \tag{9.8}$$

The new term with respect to Eqs. 9.1 relates to the recycling reaction whose maximum rate is given by the parameter σ_i, while K denotes a threshold constant and n gives the degree of cooperativity of the enzyme.

In addition to simple periodic behavior corresponding to the evolution toward a unique limit cycle, the system governed by Eqs. 9.8 can admit the coexistence between two stable limit cycles; these two cycles are separated by an unstable cycle (see Fig. 9.8).

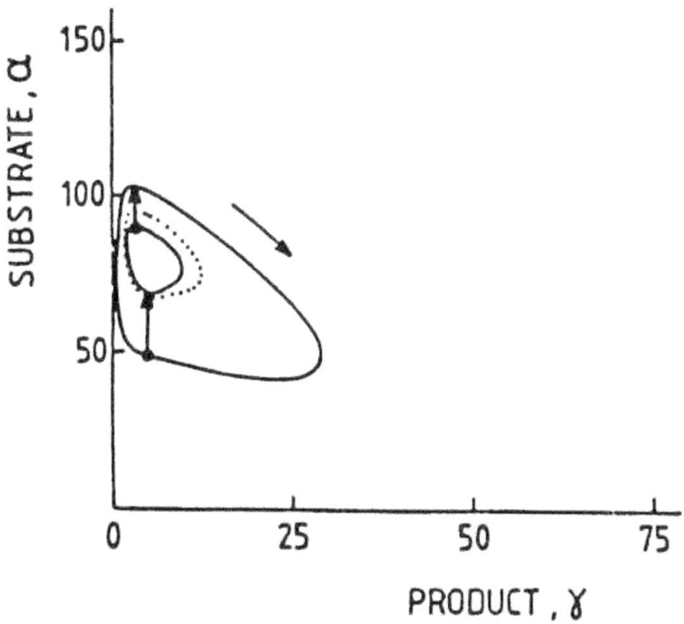

Figure 9.8. Birhythmicity. For certain parameter values, the model of Fig. 9.7b governed by Eqs. 9.8 admits a coexistence between two stable limit cycles (solid lines) separated by an unstable cycle (dotted line). The arrows indicate how the addition of an appropriate amount of substrate at the right phase allows one to switch reversibly between the two stable cycles (from ref. 55).

This phenomenon, referred to as birhythmicity,[54] opens up the possibility of reversibly switching between the two stable cycles upon appropriate perturbations. Such perturbations, in the form of instantaneous increases in substrate, are indicated by the vertical arrows in Fig. 9.8.

9.3.2 Bursting and Chaos

Because the system governed by Eqs. 9.8 contains only two variables, it cannot admit complex periodic oscillations (bursting) or aperiodic oscillations (chaos). To investigate the conditions for the occurrence of the latter types of oscillatory behavior, a three-variable model has been developed[28,54,56] which is again based on the allosteric model proposed for glycolytic oscillations. Here, two allosteric enzymes are coupled in series; each of these enzymes is activated by its reaction product (Fig. 9.7c).

The first substrate (S) is injected at a constant rate (ν) and transformed into product P_1 by enzyme E_1; product P_1 is transformed by enzyme E_2 into product P_2 which leaves the system at a rate proportional to its concentration, with an apparent first-order rate constant k_s. This three-variable system is governed by the following set of kinetic equations:

$$\frac{d\alpha}{dt} = \nu - \sigma_1 \phi(\alpha, \beta)$$

$$\frac{d\beta}{dt} = q_1 \sigma_1 \phi(\alpha, \beta) - \sigma_2 \eta(\beta, \gamma)$$

$$\frac{d\gamma}{dt} = q_2 \sigma_2 \eta(\beta, \gamma) - k_s \gamma \tag{9.9}$$

where the rate functions ϕ and η of the allosteric enzymes E_1 and E_2 are given, respectively, by expressions 9.10a and 9.10b.

$$\phi(\alpha, \beta) = \frac{\alpha(1 + \alpha)(1 + \beta)^2}{L_1 + (1 + \alpha)^2(1 + \beta)^2} \tag{9.10a}$$

$$\eta(\beta, \gamma) = \frac{\beta(1 + \gamma)^2}{L_2 + (1 + \gamma)^2} \tag{9.10b}$$

In these equations, α, β and γ denote the normalized concentrations of S, P_1 and P_2, respectively (See refs. 28 and 54 for further details).

The dynamic behavior of System 9.9 has been studied in detail as a function

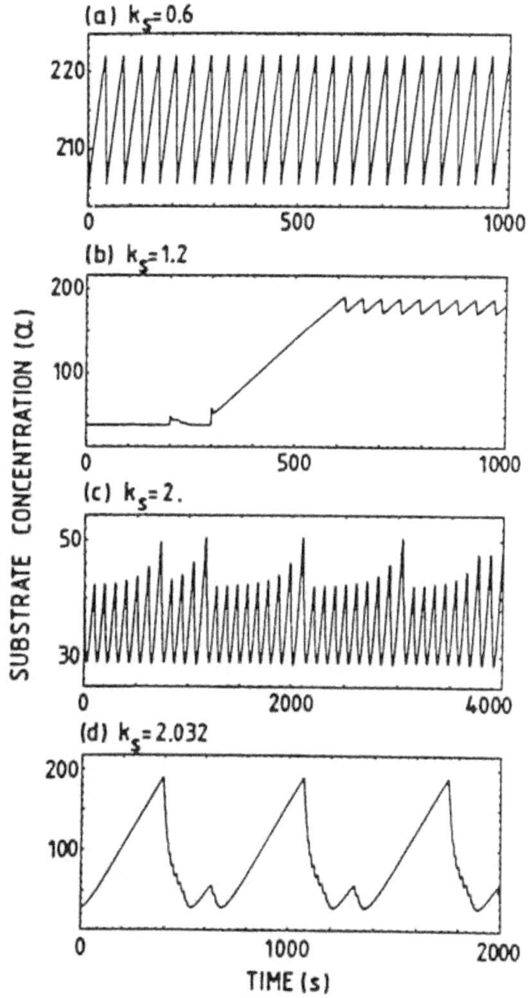

Figure 9.9. From simple to complex oscillatory phenomena in the multiply regulated biochemical model of Fig. 9.7c governed by Eqs. 9.9-9.10. For increasing values of parameter k_s, the following modes of oscillatory behavior are shown: (a) Simple periodic oscillations; (b) hard excitation; (c) aperiodic oscillations, *i.e.*, chaos; (d) complex periodic oscillations (other phenomena observed in this model, not shown in this figure, include birhythmicity or trirhythmicity). Only the variable α is represented as a function of time (Redrawn from ref. 54).

of two control parameters, v and k_s. The variety of dynamic behavior is much larger than in the system comprising a single positive feedback loop. Here, as the parameter k_s is increased, one can observe the following modes of oscillatory behavior (see Fig. 9.9): (i) simple periodic oscillations of the limit cycle type; (ii) hard excitation, *i.e.* coexistence of a stable limit cycle with a stable steady state; (iii) chaos, which is reached after a sequence of period-doubling bifurcations; (iv) complex periodic oscillations in the form of bursting. Multiple periodic regimes also can be obtained in this model: up to three coexisting, stable limit cycles (trirhythmicity) have thus been found, but such multiplicity is only seen in a small domain in parameter space. In contrast to bursting, which corresponds to the evolution toward a closed curve, aperiodic oscillations are associated with a strange attractor in the (α, β, γ) space (Fig. 9.10).

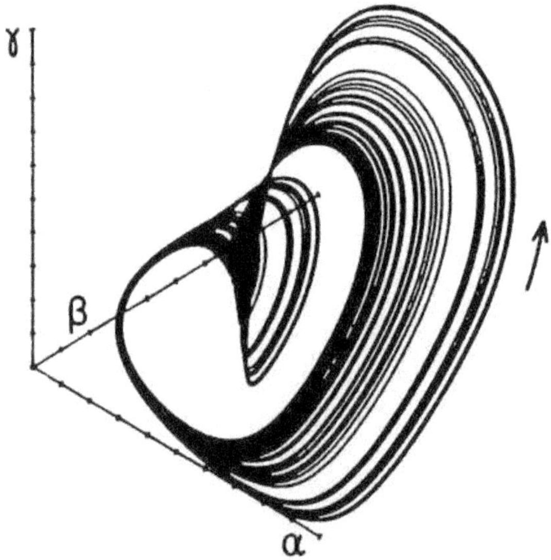

Figure 9.10. Strange attractor associated with chaos in the multiply regulated biochemical system of Fig. 9.7c (From ref. 54).

Bursting and chaos can be conveniently studied in this model by means of a one-dimensional map.[56] Similar modes of dynamic behavior have been observed in another biochemical model containing an autocatalytic step.[57]

Bursting and chaos clearly originate in this multiply regulated biochemical system from the interplay between two instability-generating mechanisms. Each of the two positive feedback loops can indeed produce on its own sustained oscillations for appropriate parameter values. Complex oscillatory phenomena result from the interaction of the two instability mechanisms when these become active at the same time. An empirical method for finding domains of complex oscillations and chaos in parameter space has been developed based on this conjecture:[26,58] It consists of identifying two distinct domains of instability in a given parameter space, each of which is associated with one of the two destabilizing feedback mechanisms. Then, by changing some other parameter, the two domains of instability are brought closer to each other until they overlap. It is often in the region of overlap that complex oscillatory phenomena occur in the form of birhythmicity, bursting or chaos.

A detailed mapping of the different modes of dynamic behavior in the two-parameter space v-k_s shows the relative sizes of the associated domains, and hence the relative likelihood of occurrence of these various oscillatory phenomena.[26,54] This study indicates that simple periodic oscillations are the most common type of dynamic behavior encountered in the model; then come complex periodic oscillations (bursting) and hard excitation. Much less common are birhythmicity and chaos. While the former phenomenon occurs in two regions of the v-k_s parameter space, chaos occurs in at least three distinct regions. Trirhythmicity occurs in the tiny region where the two domains of birhythmicity overlap. Despite the complex regulatory structure of the model, simple periodic oscillations thus remain by far the most common type of oscillatory behavior.

9.4 From Periodic Behavior to chaos in *Dictyostelium* cells

The cellular slime mold *Dictyostelium discoideum* represents a prototype of spatio-temporal organization at the cellular level. These cells indeed aggregate after starvation, by a chemotactic response to pulses of cyclic-AMP (c-AMP) emitted by centers with a periodicity of 5-10 min; as a result of such a periodicity in c-AMP secretion, cells collect around the aggregation centers in a wavelike manner.[59-61] From the point of view of temporal organization, the periodic synthesis of c-AMP in *Dictyostelium* represents a remarkable example of biological rhythm involved in intercellular communication. With regard to the main theme of this volume, of particular interest is the possibility that in some cases, the signaling system in *Dictyostelium* may function in an aperiodic, chaotic manner. The periodic and aperiodic oscillatory properties of the c-AMP signaling system will be examined in turn below.

9.4.1. *Model for c-AMP Signaling Based on Receptor Desensitization*

The mechanism of c-AMP oscillations in *Dictyostelium* rests on the positive feedback exerted by extracellular c-AMP on its intracellular production. Upon

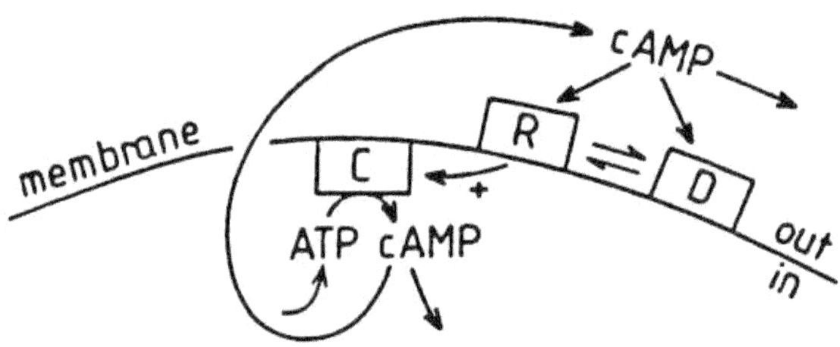

Figure 9.11. Model for c-AMP signaling in *Dictyostelium* based on receptor desensitization. Extracellular c-AMP binds to the active (R) state of the receptor and thereby elicits the activation of adenylate cyclase (C), which synthesizes c-AMP from ATP. Binding of c-AMP to the receptor in the R state induces its transition to the desensitized (D) state which cannot activate the cyclase. Other arrows refer to ATP synthesis, intracellular c-AMP transport into the extracellular medium, and c-AMP hydrolysis by phosphodiesterase (from ref. 64).

binding to a cell surface receptor, c-AMP triggers the activation of adenylate cyclase which produces c-AMP from ATP. Intracellular c-AMP thus synthesized is transported into the extracellular medium where it binds to the receptor and further enhances its own production, while part of it is hydrolyzed by the enzyme phosphodiesterase present on the cell membrane and in the extracellular medium,[60,61]

As important for oscillations as is the positive feedback loop described above, the factor that limits autocatalysis is equally important. A major role here is played by desensitization of the c-AMP receptor through reversible phosphorylation.[62,63] Holding with these observations, a model for c-AMP signaling based on receptor desensitization (See Fig. 9.11) has been proposed by Martiel and Goldbeter.[28,64] In its simplest version,it describes the time evolution of intracellular and extracellular c-AMP and of the fraction of active c-AMP receptor; ATP is treated as a parameter in that version of the model since the amplitude of ATP variations remains reduced in the course of c-AMP oscillations. The resulting three-variable system is governed by the following kinetic equations:

$$\frac{d\varrho_T}{dt} = -f_1(\gamma)\,\varrho_T + f_2(\gamma)(1 - \varrho_T)$$

$$\frac{d\beta}{dt} = q\,\sigma\,\phi(\varrho_T, \gamma, \alpha) - (k_i + k_t)\,\beta$$

$$\frac{d\gamma}{dt} = \left(\frac{k_t\,\beta}{h}\right) - k_e\,\gamma \qquad (9.11)$$

where,

$$f_1(\gamma) = \frac{k_1 + k_2\gamma}{1 + \gamma}, \quad f_2(\gamma) = \frac{k_1 L_1 + k_2 L_2 c\,\gamma}{1 + c\,\gamma}$$

$$\phi(\varrho_T, \gamma, \alpha) = \frac{\alpha\,(\lambda\,\theta + \epsilon\,Y^2)}{1 + \alpha\,\theta + \epsilon\,Y^2(1 + \alpha)}, \quad Y = \frac{\varrho_T\,\gamma}{1 + \gamma} \qquad (9.12)$$

In these equations, β and γ denote the normalized concentrations of intracellular and extracellular c-AMP while ρ_T denotes the fraction of active receptor; α is the normalized level of ATP and other parameters relate to the kinetic characteristics of the receptor, adenylate cyclase, c-AMP transport, and c-AMP hydrolysis by phosphodiesterase (See ref. 64 for further details).

A recent extension of the model incorporates G proteins which act as intermediates between c-AMP binding to the receptor and activation (or inhibition) of adenylate cyclase (J. Halloy and A. Goldbeter, manuscript in preparation). The results obtained in that extended model are qualitatively similar to those obtained in the simpler model governed by Eqs. 9.11-9.12.

9.4.2 *Oscillations and Relay of c-AMP Signals*

The model decribed by Eqs. 9.11-9.12 generally admits a unique steady state solution which can become unstable for appropriate values of the parameters. In these conditions, the system evolves toward a stable limit cycle corresponding to sustained oscillations in c-AMP. These oscillations (Fig. 9.12), whose period is of the order of 5-10 min for the parameter values collected from the experiments, are accompanied by a periodic alternation of the receptor between its active and desensitized states.[64] A similar alternation is observed in the experiments, since the receptor oscillates between its phosphorylated and dephosphorylated forms in the course of c-AMP oscillations.[63]

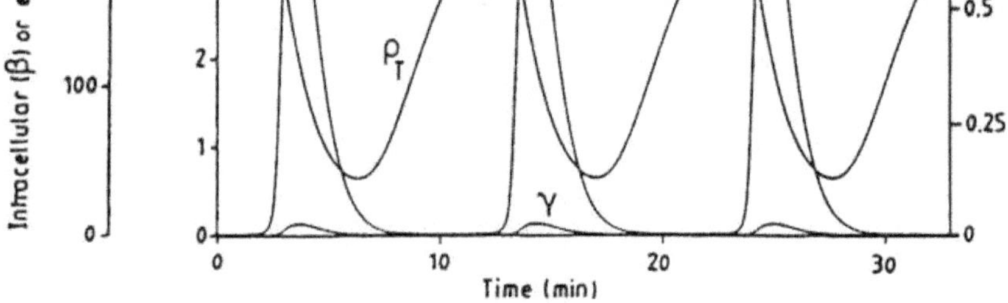

Figure 9.12. Sustained oscillations of c-AMP in the model based on receptor desensitization. Shown are the normalized concentrations of intracellular (β) and extracellular c-AMP (γ), ATP (α), and the fraction of active c-AMP receptor (ρ_T) (From ref. 64).

Relay of c-AMP signals in the course of aggregation consists in the pulsatory amplification of suprathreshold c-AMP pulses. The analysis of the model in the phase plane indicates that a close link exists between relay and oscillatory behavior.[64] The former type of phenomenon reflects the excitability of the c-AMP signaling system: for parameter values close to those producing oscillations, the steady state is indeed stable but excitable; a suprathreshold increase in extracellular c-AMP then leads to the synthesis of a large pulse of c-AMP before the system returns to the stable steady state.

9.4.3 *Birhythmicity, Bursting and Chaos*

To observe more complex modes of oscillatory behavior, it is necessary to take into account the time variation of the substrate ATP which, in the above equations, was treated as a constant parameter. The reason why this variation is required for the occurrence of complex oscillatory phenomena can be related to the results obtained in the multiply regulated biochemical system discussed in the preceding section. The consumption of ATP acts as a second factor limiting autocatalysis, besides receptor desensitization. Thus, two oscillatory mechanisms can operate in parallel within the c-AMP signaling system; both mechanisms share the same positive feedback loop exerted by c-AMP on its intracellular production, but differ by the factor limiting this autocatalysis.

Taking into account the variation of ATP, the signaling system is governed by a set of four kinetic equations. A quasi-steady state hypothesis for intracellular c-AMP (β) permits one, however, to reduce the number of variables so that the system is governed by the following set of three kinetic equations:[65]

$$\frac{d\varrho_T}{dt} = -f_1(\gamma)\varrho_T + f_2(\gamma)(1 - \varrho_T)$$

$$\frac{d\alpha}{dt} = v - \sigma\,\phi(\varrho_T, \gamma, \alpha)$$

$$\frac{d\gamma}{dt} = \frac{q k_t \sigma\,\phi(\varrho_T, \gamma, \alpha)}{h(k_i + k_t)} - k_e\gamma \tag{9.13}$$

where,

$$f_1(\gamma) = \frac{k_1 + k_2\gamma^2}{1 + \gamma^2}, \quad f_2(\gamma) = \frac{k_1 L_1 + k_2 L_2 c^2 \gamma^2}{1 + c^2\gamma^2}$$

$$\Phi(\varrho_T, \gamma, \alpha) = \frac{\alpha(\lambda\theta + \epsilon Y)}{1 + \alpha\theta + \epsilon Y(1 + \alpha)}, \quad Y = \frac{\varrho_T \gamma^2}{1 + \gamma^2} \tag{9.14}$$

Besides the inclusion of α as a variable and the quasi-steady state assumption for β, what distinguishes Eqs. 9.14 from Eqs. 9.11 is the hypothesis that the nonlinearity required for oscillations occurs in the activation step between the receptor and the cyclase rather than at the level of c-AMP binding to the receptor. Both kinds of nonlinearity yield similar types of dynamic behavior.

The system governed by Eqs. 9.13-9.14 presents new modes of oscillatory dynamics, in addition to simple periodic behavior and relay of suprathreshold c-AMP signals. Thus, bursting and birhythmicity can occur in this system,[65] as well as aperiodic oscillations;[66,67] the latter again appear through a cascade of period-doubling bifurcations, both with respect to parameter v and to parameter k_e which relate, respectively, to the input of ATP and to the hydrolysis of extracellular c-AMP by phosphodiesterase. The domains of birhythmicity and chaos in parameter space are again much more reduced than those where simple periodic oscillations or bursting are found.[67]

The prediction of chaotic behavior in the model for c-AMP signaling raises the possibility that autonomous chaos may occur in the course of *D. discoideum* aggregation.[66] Of particular interest is the observation by Durston[68] that in contrast to wild type amoebae which aggregate in a periodic manner on agar after starvation, aggregation is "aperiodic" in the mutant *Fr17*: whereas the interval between successive waves remains close to 5 min in the wild type, it varies from 4 to 24 min in *Fr17*. Preliminary evidence from experiments in cell suspensions of the mutant *HH201* derived from *Fr17* indicates that the synthesis of c-AMP in that mutant is "erratic".[69] To test the occurrence of chaos in *HH201*, cells placed in stirred suspensions were studied by light scattering so as to monitor the dynamic behavior in a continuous manner.[70] Instead of aperiodic behavior, rather regular oscillations were recorded, with a progressive drift in the period from 8 to 6 min, as observed in suspensions of wild type cells.

9.4.4 *Suppression of Chaos by Periodic Oscillations*

There are several ways to explain the occurrence of regular rather than aperiodic oscillations in the putatively chaotic mutant *HH201*. First, it is possible that cells were grown in the experiments so as to reach after starvation a domain of periodic oscillations in parameter space; such an explanation is plausible since in the model for c-AMP signaling the domain of periodic behavior is much larger than the domain of chaos. Alternatively, the fact that chaos did not occur in the experiments might be due to the interaction of cells endowed with different dynamic properties within the stirred suspension. In particular, the question arises as to what is the dynamic behavior resulting from the mixing of cells behaving chaotically with cells behaving in a periodic manner, given that the cells are strongly coupled through extracellular c-AMP. Could it be that chaos is suppressed by the presence of periodic cells within a mixed suspension?

To test this possibility, the dynamic behavior of a system comprising variable proportions of cells from two populations, one chaotic and the other periodic, was determined.[67,71,72] For each population cells synthesize and secrete c-AMP according to Eqs. 9.13-9.14, with one equation for the fraction of active c-AMP receptor (ρ_T) and one for the substrate ATP (α), but the kinetic equation for extracellular c-AMP (γ) takes into account the production of c-AMP by the two populations present within the mixed suspension, as well as the degradation of c-AMP by phosphodiesterase produced by the two types of cells.

Starting with a homogeneous population of periodic cells ($F_1=1$) and increasing progressively the proportion F_2 of (initially) chaotic cells, a sequence of period-doubling bifurcations leading to chaos is found[71] (Fig. 9.13) as a function of F_2. The physiologically most significant result, however, is that the presence of a tiny proportion of periodic cells, *e.g.* of the order of a few %, suffices to suppress the chaotic behavior of the large majority of cells present in the suspension.[71,72] Thus, Fig. 9.13 indicates that the presence of 5% periodic cells suffices to transform the (initially) chaotic behavior of 95% of the cells into complex periodic oscillations. A detailed bifurcation analysis of the mixed system shows[67] that such a sensitivity of chaos can be related to the relative smallness of the chaotic domain as compared to the domain of periodic behavior in parameter space. Mixing the chaotic and periodic cell populations amounts to shifting the effective value of a control parameter in a homogeneous population out of the range of chaos into the range of values corresponding to simple or complex periodic oscillations.[67]

To further check the effect of mixing a small proportion of periodic cells with chaotic ones, a population of chaotic amoebae was subjected to a small-amplitude

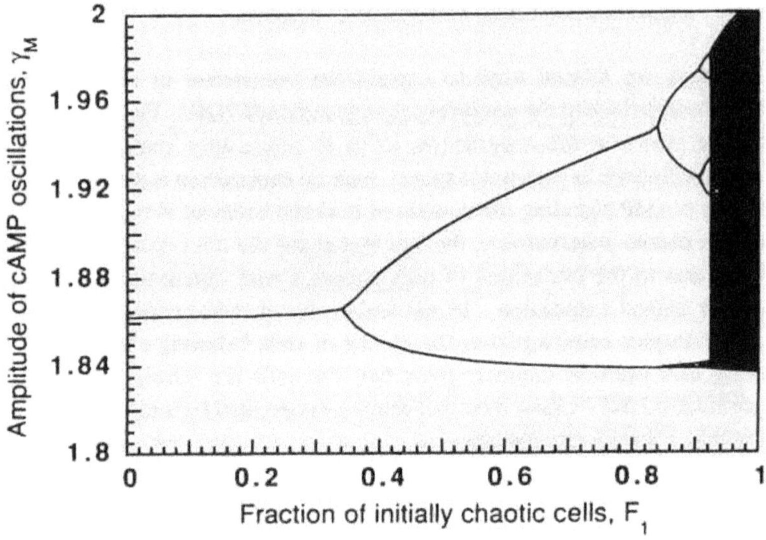

Figure 9.13. Period-doubling route to chaos observed upon mixing periodic with chaotic cells in the model for c-AMP signaling in *Dictyostelium* based on receptor desensitization. The black domain refers to chaotic oscillations (See text for details; figure redrawn from ref. 71).

periodic forcing.[67,72] This was done in the model by adding a sinusoidal input term to the kinetic equation for γ in Eqs. 9.13. Confirming the results of the mixing of periodic and chaotic cells, simulations indicate (Fig. 9.14) that a tiny periodic perturbation of the strange attractor is sufficient to transform chaotic into periodic behavior.

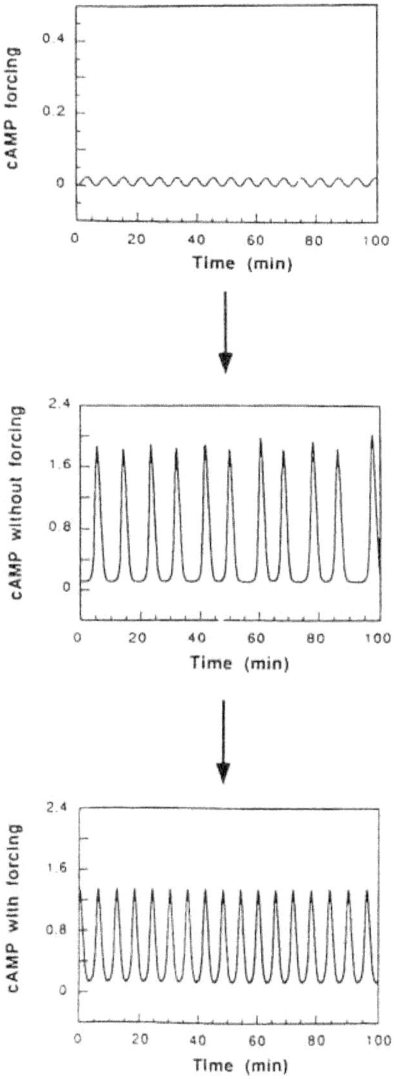

Figure 9.14. Suppression of chaos by a small-amplitude periodic forcing in the model for c-AMP signaling in *Dictyostelium* based on receptor desensitization (From ref. 72).

9.5 Concluding Remarks

Lately, the study of biochemical oscillations has undergone important advances owing to the discovery of intracellular Ca^{2+} oscillations in a large variety of cells and to the molecular characterization of the mitotic oscillator that underlies the cell division cycle in eukaryotes. These periodic phenomena have extended the list of biochemical oscillations which were, until then, restricted primarily to glycolytic oscillations in yeast and muscle, cyclic-AMP oscillations in the slime mold *Dictyostelium discoideum*, and the oscillatory peroxidase reaction. These rhythms were briefly presented in this chapter, together with models proposed for experimental observations (the peroxidase reaction was not discussed, as it is considered in detail in Chapter 7).

The physiological function of biochemical oscillations has been clarified to various degrees. The clearest and most important case is that of the cell division cycle which plays a major role in development. Moreover, perturbations of the continuous biochemical oscillator controlling mitosis may lead to cell cycle arrest or may conversely be involved in abnormal cell proliferation. As to Ca^{2+} oscillations, their effects only begin to be investigated; most of the work so far has indeed been devoted to the characterization of the oscillatory phenomenon. It appears that Ca^{2+} oscillations could be encoded in terms of their frequency:[15-19,30,37] the latter indeed rises with the level of stimulation by hormones or neurotransmitters. Oscillations in cytosolic Ca^{2+} are often accompanied by Ca^{2+} waves propagating within the cell from the site of stimulation;[31-34] these waves could serve in intracellular signalling as well as in propagating signals from cell to cell.

The function of glycolytic oscillations remains unclear, although it has been suggested that they may enhance the thermodynamic efficiency of glycolysis,[73] the adenylate energy charge,[74] or the ATP/ADP ratio[23] in this ATP-producing pathway. These oscillations could well be merely an accidental consequence of the peculiar regulation of phosphofructokinase; a similar view may also hold for the oscillatory peroxidase reaction. Even so, the virtue of these systems would be to provide highly useful models for biochemical oscillations that can be studied *in vitro*.

As to oscillations of c-AMP in *Dictyostelium*, which have not yet been characterized *in vitro*, they represent a prototype of pulsatile signaling in intercellular communication. A most significant observation is that there exists an optimal frequency of c-AMP pulses that is required for inducing the physiological response in target cells.[75,76] The analysis of the model for c-AMP signaling discussed above accounts for this result and shows that the optimal frequency closely depends on the kinetic characteristics of desensitization and resensitization of the c-AMP receptor.[77]

The existence of an optimal pattern of pulsatile stimulation by c-AMP signals in *Dictyostelium* can be related[78,79] to the optimal fequency observed in the action of a number of hormones[80] whose secretion also follows a pulsatile pattern. It is noteworthy that in the model based on receptor desensitization which applies both to c-AMP and hormonal signaling, the optimal periodic signal proves more effective than random or chaotic pulsatile stimuli.[81]

Oscillatory phenomena observed so far in biochemical systems appear to be periodic rather than irregular, although one can occasionally observe a drift of the period in intact cell preparations, *e.g.* in *Dictyostelium*. Such a variability, resulting from the interrelation with other time-dependent processes in these cells, represents a marked difference with respect to isolated chemical reactions that can be studied in the test tube.

Examples of bursting or chaos occurring spontaneously —*i.e.*, in the absence of periodic forcing— in biochemical oscillatory systems are relatively rare and reduce to the peroxidase reaction[7] and to some occasional observations of highly irregular oscillations in glycolyzing yeast extracts (See, *e.g.*, ref. 10). On the other hand, chaotic behavior has been demonstrated both theoretically[82,83] and experimentally[84] in such extracts subjected to a periodic input of substrate.

At the cellular level, the "aperiodic signaling" properties observed[68] during aggregation on agar in the *Dictyostelium* mutant *Fr17* could provide an example of autonomous, biochemical chaos. Attempts at characterizing in terms of chaos the behavior of this mutant in cell suspensions have failed, however, since rather regular oscillations instead of aperiodic behavior were observed.[70] Because this result could well be due to the suppression of chaos by the strong coupling with some periodic cells in the mixed suspension[67,71,72] further studies of the putatively chaotic mutant of *Dictyostelium* should best be carried out during aggregation on agar, as originally considered by Durston.[68] Any center behaving in a chaotic manner would then be capable of expressing its aperiodic nature in the absence of the strong coupling that takes place in cell suspensions.

The analysis of models for biochemical oscillations throws light on the molecular mechanisms of periodic behavior and clarifies the conditions in which simple periodic oscillations give way to more complex oscillatory phenomena, including chaos. With regard to periodic behavior, although endproduct inhibition can in principle give rise to limit cycle oscillations,[85,86] the most prevalent type of regulation involved in generating the phenomenon is positive feedback.[12] Such is the case for the phosphofructokinase reaction and for c-AMP signaling in *Dictyostelium*. The regulatory mechanisms which underlie Ca^{2+} oscillations also appear to include positive feedback in the form of Ca^{2+}-induced Ca^{2+} release. As to the mitotic oscillator, some models are again based on positive feedback, via the (not yet fully characterized) autocatalytic regulation of cdc2 kinase,[47-49] but other models show[50,52] that oscillations may result solely from the delayed negative feedback present in the interactions between cyclin and cdc2 kinase (See Figs. 9.5 and 9.6 above). Required in all these oscillatory mechanisms is a minimum degree of nonlinearity. Sources for nonlinearity abound in biochemical systems, the most prevalent being that arising from the cooperative properties of allosteric enzymes, transport proteins or receptors, or from the kinetics of phosphorylation-dephosphorylation cycles.

What are the biochemical ingredients needed to allow for complex oscillatory phenomena? The basic result of the study of models for regulated enzymatic reactions is that autonomous bursting and chaos can occur in systems containing (at least) two interacting oscillatory mechanisms. Thus, in the model of two product-activated enzymes coupled in series, based on an extension of the model proposed for glycolytic oscillations (See Fig. 9.7), each positive feedback loop can produce oscillations on its own; complex oscillations occur once the two instability-generating mechanisms are active for the same set of parameter values. A similar result is obtained in the model for c-AMP signaling: there, two oscillatory mechanisms are coupled in parallel; both share the same positive feedback loop but differ by the mechanism limiting the autocatalysis. These results can be related to the well-known route to bursting and chaos, which consists in subjecting an oscillatory system to some exogenous, periodic perturbation. Here, in contrast, the two sources of oscillations are endogenous, so that chaos is truly autonomous.

A recurrent finding in all models investigated for complex oscillations is that however complex the regulatory structure of the biochemical system may be, simple periodic oscillations by far remain the most common mode of dynamic behavior, followed by complex periodic oscillations in the form of bursting. The occurrence of multiple stable limit cycles and chaos is restricted to much smaller domains in parameter space. This may explain the fact that for most oscillations observed in biological systems, at least in autonomous conditions periodic behavior prevails over chaos.

Acknowledgements. I wish to thank O. Decroly, G. Dupont, J. Halloy, Y.X. Li, J.L. Martiel, F. Moran and B. Wurster with whom the various results synthesized in this chapter were obtained. This work was supported by the Belgian national incentive program for fundamental research in the life sciences (Convention BIO/08) launched by the Science Programming Policy Unit of the Prime Minister's Office (SPPS).

Literature Cited

1. A.T. Winfree, *The Geometry of Biological Time* (Springer, New York, 1980).
2. A. Fessard, *Propriétés rythmiques de la matière vivante* (Hermann, Paris, 1936).
3. L.F. Olsen and H. Degn, *Quart. Rev. Biophys.* **18** (1985) 165.
4. A. Holden, *Chaos* (Manchester Univ. Press., Manchester, UK, 1986).
5. L. Glass and M.C. Mackey, *From Clocks to Chaos: The Rhythms of Life* (Princeton Univ. Press, Princeton, NJ, 1988).
6. I. Yamazaki, K. Yokota and R. Nakajima, *Biochem. Biophys. Res. Commun.* **21** (1965) 582.
7. L.F. Olsen and H. Degn, *Nature* **267** (1977) 177.
8. G. Gerisch and U. Wick, *Biochem. Biophys. Res. Commun.* **65** (1975) 364.
9. B. Chance, B. Hess and A. Betz, *Biochem. Biophys. Res. Commun.* **16** (1964) 182.
10. E.K. Pye, *Can. J. Bot.* **47** (1969) 271.
11. B. Hess and A. Boiteux, *Annu. Rev. Biochem.* **40** (1971) 237.
12. A. Goldbeter and S.R. Caplan, *Annu. Rev. Biophys. Bioeng.* **5** (1976) 449.
13. M.J. Berridge and P.E. Rapp, *J. Exp. Biol.* **81** (1979) 217.
14. B. Chance and T. Yoshioka, *Arch. Biochem. Biophys.* **117** (1966) 451.
15. K.S.R. Cuthbertson and P.H. Cobbold, *Nature* **316** (1985) 541.
16. M.J. Berridge and A. Galione, *FASEB J.* **2** (1988) 3074.
17. R.W. Tsien and R.Y. Tsien, *Annu. Rev. Cell Biol.* **6** (1990) 715.
18. T. Meyer and L. Stryer, *Annu. Rev. Biophysics Biophys. Chem.* **20** (1991) 153.
19. K.S.R. Cuthbertson and P.H. Cobbold, Eds., *Cell Calcium* **12** (1991) 2.
20. A.W. Murray and M.W. Kirschner, *Science* **246** (1989) 614.
21. B. Hess, A. Boiteux and J. Krüger, *Adv. Enzyme Regul.* **7** (1969) 149.
22. R. Frenkel, *Arch. Biochem. Biophys.* **125** (1968) 151.
23. K. Tornheim, V. Andrés and V. Schultz, *J. Biol. Chem.* **266** (1991) 15675.
24. J. Higgins, *Proc. Natl. Acad. Sci. USA* **51** (1964) 989.
25. E.E. Sel'kov, *Eur. J. Biochem.* **4** (1968) 79.
26. A. Goldbeter and R. Lefever, *Biophys. J.* **12** (1972) 1302.
27. A. Goldbeter and G. Nicolis, in *Progress in Theoretical Biology* **4**, eds. F. Snell and R. Rosen (Academic Press, New York, 1976) p. 65.
28. A. Goldbeter, *Rythmes et chaos dans les systèmes biochimiques et cellulaires* (Masson, Paris, 1990). (English translation to be published by Cambridge Univ. Press under the title *Rhythms and Chaos in Biochemical and Cellular Systems*).
29. J. Monod, J. Wyman and J.P. Changeux, *J. Mol. Biol.* **12** (1965) 88.
30. M.J. Berridge, P.H. Cobbold and K.S.R. Cuthbertson, *Phil. Trans. R. Soc. Lond. B* **320** (1988) 325.
31. J.C. Gilkey, L.F. Jaffe, E.B. Ridgway and G.T. Reynolds, *J. Cell Biol.* **76** (1978) 448.
32. T.A. Rooney, E.J. Sass and A.P. Thomas, *J. Biol. Chem.* **265** (1990) 10792.
33. R. Jacob, *Cell Calcium* **11** (1990) 241.
34. T. Takamatsu and W.G. Wier, *FASEB J.* **4** (1990) 1519.
35. G. Dupont and A. Goldbeter, *BioEssays* **14** (1992) 485.
36. T. Meyer and L. Stryer, *Proc. Natl. Acad. Sci. USA* **85** (1988) 5051.
37. A. Goldbeter, G. Dupont and M.J. Berridge, *Proc. Natl. Acad. Sci. USA* **87** (1990) 1461.
38. G. Dupont, M.J. Berridge and A. Goldbeter, *Cell Calcium* **12** (1991) 73.

39. E.A. Finch, T.J. Turner and S.M. Goldin, *Science* **252** (1991) 443.
40. G. Dupont and A. Goldbeter, submitted to *Cell Calcium* (1992).
41. S. Girard, A. Lückhoff, J. Lechleiter, J. Sneyd and D. Clapham, *Biophys. J.* **61** (1992) 509.
42. P. Nurse, *Nature* **344** (1990) 503.
43. A.W. Murray and M.W. Kirschner, *Nature* **339** (1989) 275.
44. G. Draetta, *Trends Biochem. Sci.* **15** (1990) 378.
47. C. Hyver and H. Le Guyader, *Biosystems* **24** (1990) 85.
45. T. Lorca, D. Fesquet, F. Zindy, F. Le Bouffant, M. Cerruti, C. Brechot, G. Devauchelle and M. Dorée, *Mol. Cell. Biol.* **11** (1991) 1171.
46. E. Karsenti, F. Verde and M.A. Félix, *Adv. Protein Phosphatases* **6** (1991) 453.
48. R. Norel and Z. Agur, *Science* **251** (1991) 1076.
49. J.J. Tyson, *Proc. Natl. Acad. Sci. USA* **88** (1991) 7328.
50. A. Goldbeter, *Proc. Natl. Acad. Sci. USA* **88** (1991) 9107.
51. M.A. Félix, J.C. Labbé, M. Dorée, T. Hunt and E. Karsenti, *Nature* **346** (1990) 379.
52. A. Goldbeter, *Comments Theor. Biol.* (1993) in press.
55. F. Moran and A. Goldbeter, *Biophys. Chem.* **20** (1984) 149.
53. M. Alamgir and I.R. Epstein, *J. Am. Chem. Soc.* **105** (1983) 2500.
54. O. Decroly and A. Goldbeter, *Proc. Natl. Acad. Sci. USA* **79** (1982) 6917.
56. O. Decroly and A. Goldbeter, *J. Theor. Biol.* **124** (1987) 219.
57. P. Tracqui, A.M. Perault-Staub, G. Milhaud and J.F. Staub, *Bull. Math. Biol.* **49** (1987) 597.
58. A. Goldbeter, O. Decroly, Y.-X. Li, J.L. Martiel and F. Moran, *Biophys. Chem.* **29** (1988) 211.
59. F. Alcantara and M. Monk, *J. Gen. Microbiol.* **85** (1974) 321.
60. K.J. Tomchik and P.N. Devreotes, *Science* **212** (1981) 443.
61. G. Gerisch, *Annu. Rev. Biochem.* **56** (1987) 853.
62. P.N. Devreotes and J.A. Sherring, *J. Biol. Chem.* **260** (1985) 6378.
63. R.E. Gundersen, R. Johnson, P. Lilly, G. Pitt, M. Pupillo, T. Sun, R. Vaughan and P.N. Devreotes, in *Cell to Cell Signalling: From Experiments to Theoretical Models*, ed. A. Goldbeter (Academic Press, London, 1989), p. 477.
64. J.L. Martiel and A. Goldbeter, *Biophys. J.* **52** (1987) 807.
65. A. Goldbeter and J.L. Martiel, *FEBS Lett.* **191** (1985) 149.
66. J.L. Martiel and A. Goldbeter, *Nature* **313** (1985) 590.
67. Y.-X. Li, J. Halloy, J.L. Martiel and A. Goldbeter, *Chaos* **4** (1992) in press.
68. A.J. Durston, *Dev. Biol.* **38** (1974) 308.
69. M.B. Coukell and F.K. Chan, *FEBS Lett.* **110** (1980) 39.
70. A. Goldbeter and B. Wurster, *Experientia* **45** (1989) 363.
71. J. Halloy, Y.-X. Li, J.L. Martiel, B. Wurster and A. Goldbeter, *Phys. Lett. A* **151** (1990) 33, and **159** (1991) 442.
72. Y.-X. Li, J. Halloy, J.L. Martiel, B. Wurster and A. Goldbeter, *Experientia* **48** (1992) 604.
73. P.H. Richter and J. Ross, *Biophys. Chem.* **12** (1980) 285.
74. A. Goldbeter, *FEBS Lett.* **43** (1974) 327.
75. M. Darmon, P. Brachet and L.H. Pereira da Silva, *Proc. Natl. Acad. Sci. USA* **72** (1975) 3163.
76. G. Gerisch, H. Fromm, A. Huesgen and U. Wick, *Nature* **255** (1975) 547.
77. Y.-X. Li and A. Goldbeter, *J. Theor. Biol.* **146** (1990) 355.
78. Y.-X. Li and A. Goldbeter, *Biophys. J.* **55** (1989) 125.

79. A. Goldbeter and X.-Y. Li, in *Cell to Cell Signalling: From Experiments to Theoretical Models.*, ed. A. Goldbeter (Academic Press, London, 1989), p.415.
80. E. Knobil, *New England J. Med.* **305** (1981) 1582.
81. Y.-X. Li and A. Goldbeter, *Biophys. J.* **61** (1992) 161.
82. M. Markus and B. Hess, *Proc. Natl. Acad. Sci. USA* **81** (1984) 4394.
83. M. Markus, D. Kuschmitz and B. Hess, *Biophys. Chem.* **22** (1985) 95.
84. M. Markus, D. Kuschmitz and B. Hess, *FEBS Lett.* **172** (1984) 235.
85. B.C. Goodwin, *Adv. Enzyme Regul.* **3** (1965) 425.
86. C. Walter, *J. Theor. Biol.* **27** (1970) 259.

INDEX

www.ingramcontent.com/pod-product-compliance
Ingram Content Group UK Ltd.
Pitfield, Milton Keynes, MK11 3LW, UK
UKHW061932100926
8556IPUK00007B/886

9 789810 210243